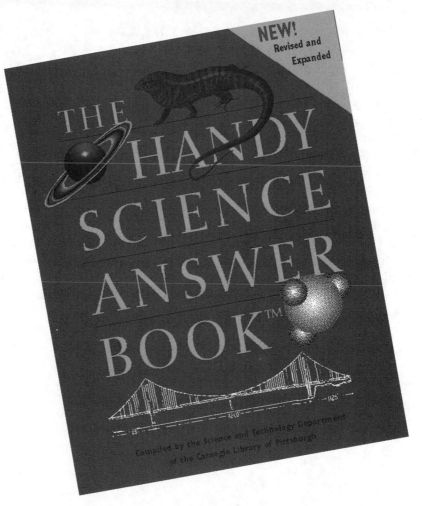

W9-BVF-031

About the Author

Intrigued by the weather since childhood, Dr. Walt Lyons received his Ph.D. in 1970 at the University of Chicago. He studied under Professor Ted Fujita, developer of the tornado classification system known as the Fujita scale. In his career he has been a university professor, broadcast meteorologist (WTMJ and WITI–TV in Milwaukee; KSTP and WCCO–TV in Minneapolis/St. Paul; and WLS/ABC in Chicago), and scientific researcher (working on programs with NASA, the U.S. Air Force, the Environmental Protection Agency, and the National Science Foundation, among others). He recently investigated "sprites," which are unusual glows high above thunderstorms caused by lightning discharges. Dr. Lyons is a frequent expert witness to the legal and insurance communities on meteorological matters and is currently president of Forensic Meteorology Associates, Inc., in Fort Collins, Colorado. He is a Certified Consulting Meteorologist and a Fellow of the American Meteorological Society. He also operates the Yucca Ridge organic garlic farm. He can be contacted at lyonsccm@csn.org.

THE
HANDY
WEATHER
ANSWER
BOOK

ALSO FROM
VISIBLE INK PRESS

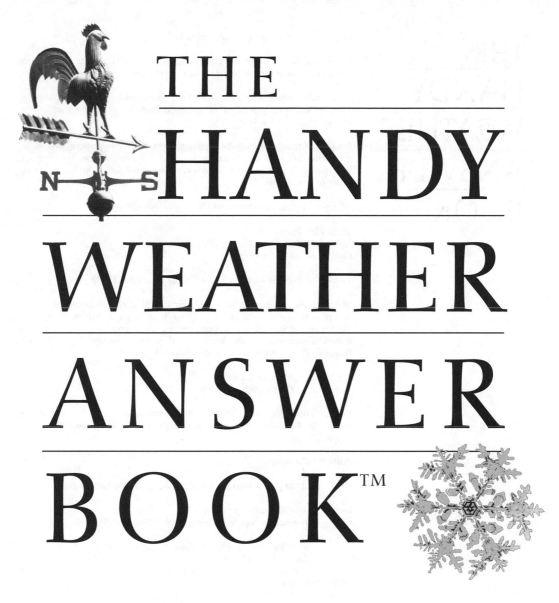

THE
HANDY
WEATHER
ANSWER
BOOK™

Walter A. Lyons, Ph.D.

VISIBLE
INK

THE HANDY WEATHER ANSWER BOOK™

Visible Ink Press™
42015 Ford Rd. #208
Canton, MI 48187-3669

Visible Ink Press is a trademark of Visible Ink Press LLC.

Most Visible Ink Press books are available at special quantity discounts when purchased in bulk by corporations, organizations, or groups. Customized printings, special imprints, messages, and excerpts can be produced to meet your needs. For more information, contact Special Markets Director, Visible Ink Press, at www.visibleink.com.

Art Director: Mary Krzewinski
Typesetting: The Graphix Group

Cover photos: Hurricane Diana, NASA; snowflake, W. A. Bentley; funnel cloud, NOAA; halo around the sun, lightning, and rainbow, Walter A. Lyons.

ISBN 0-7876-1034-8

Library of Congress Cataloging-in-Publication Data

Lyons, Walter A. (Walter Andrew), 1943–
        The handy weather answer book / Walter A. Lyons.
                p.   cm.
        Includes bibliographical references and index.
        ISBN 0-7876-1034-8
        1. Weather–miscellanea.   I. Title.
QC999.L96 1996
551.6–dc20

                                                96-30555
                                                CIP

Printed in the United States of America
All rights reserved

10 9 8

LIGHTNING AND THUNDER...139

OPTICAL PHENOMENA...161

TORNADOES...175

COLD AND WINTER STORMS...201

HEAT AND HUMIDITY...229

EARTHQUAKES AND VOLCANOES...243

AIR POLLUTION AND THE ENVIRONMENT...267

WEATHER AND THE HUMAN BODY...291

WEATHER FORECASTING...315

Farmer's Almanac . . . Internet weather sites . . . National Weather Service . . . television weather maps . . . long range forecasting . . . "nowcasting" . . . satellites

CLIMATE CHANGE...337

greenhouse effect . . . global warming . . . ice ages . . . El Niño . . . ozone hole . . . chlorofluorocarbons . . . nuclear winter

CAREERS IN METEOROLOGY...361

qualifications . . . training . . . employment opportunities . . . salary scale . . . radio and television . . . variety of forecasting . . . forensic meteorology . . . consulting

Introduction

We are all like fish swimming at the bottom of a giant ocean of air, an atmosphere that both sustains life and sometimes takes it away. The atmosphere is an endless source of curious, beautiful, mysterious, and sometimes deadly phenomena. *The Handy Weather Answer Book* is meant to present answers to many frequently asked questions about weather and the environment, helping the reader become better informed about the sky above, whether to enjoy its beauty all the more, or in some cases to survive its wrathful moments.

Do you ever just take time to look at the sky and admire its beauty and grandeur? It was the poet Ralph Waldo Emerson who said that "the sky is the daily bread of the eyes." Putting aesthetics aside, virtually all humans are affected by the weather, either as a factor in everyday comfort and enjoyment (finding a drizzly day at the beach is a bummer) or in far more dramatic ways that affect our livelihood, the nation's economy, and sometimes even our lives. Knowing about the weather—how to read the signs in the sky, and where to get reliable weather information—can save a vacation from disaster, allow us to protect our business assets or crops, or perhaps even prevent us from becoming one more fatality as the latest hurricane slams ashore.

I would like to express my concern that the reportage in this book of the more unseemly moods of Mother Nature that have caused much human misery and loss of life might appear to trivialize the sorrowful nature of these events. The recent spate of tornado and disaster videos being offered for sale to the public has been described by *Time* magazine, perhaps with some justification, as "weather porn." But it is the spectacular, the bizarre, the frightening, and the truly weird that tends to captivate us. And some of these facts, recalled in a fearful moment when the friendly skies turn on you with a vengeance, could be of value. So this book is also in part a survival guide to the tornadoes, microburst winds, smog attacks, lightning bolts, flash floods, three-inch hailstones, and hurricanes that do happen—and not always to someone else.

It might seem that all there is to be discovered in science has been discovered. That claim has frequently been made during the last century by some of the more arrogant members of the scientific establishment. Atmospheric scientists are generally more humble—we know for a fact that there are many things we simply don't understand about the complex interactions of our atmosphere with the Earth's surface and the not-so-empty space surrounding our planet. While many significant advances have been made, there is a huge world of discovery awaiting the curious. It was in fact a book somewhat similar in concept to this one, William R. Corliss's 1977 *Handbook of Unusual Natural Phenomena*—one of several of his wonderful collections of truly strange, odd, and hard to believe scientific tidbits—that led me into a major scientific endeavor. The Corliss book recounted that, for over a century, people had been reporting strange "lightning" shooting upward from clouds into the stratosphere and beyond. These reports were never taken seriously by mainstream scientists, just as early-nineteenth-century French scientists dismissed ignorant peasant tales of hot rocks falling from the sky (we now call them meteorites). By using the appropriate instruments (low-light television) and looking in the right places (above large thunderstorm systems) scientists over the past five years have documented a veritable zoo of strange lightning-related optical phenomena (now called sprites, blue jets, and elves) in the stratosphere and mesosphere above thunderstorms. Curiosity is clearly essential to the advance of science.

Atmospheric scientists deal with myriad problems ranging from forecasting tomorrow's weather, predicting air pollution impacts of new power plants, assisting in the structural design of buildings, helping prospect for new wind energy sources, basic research on cloud physics, and understanding the dynamics of climate and how human activities may, or may not, be significantly changing our environment. Meteorology is part of atmospheric science, which is in turn part of geophysics. Thus, interest in weather naturally leads one into many related fields such as oceanography, volcanology, astronomy, seismology, and space physics. In the physical world everything is ultimately related to everything. So in the study of weather, we often digress into seemingly unrelated areas. It was a German meteorologist named Alfred Wegener who noticed that the coasts of South America and Africa seemed to fit together like puzzle pieces. But he stuck his neck out and suggested that the continents might actually have moved about, and was roundly condemned as a total fool. Of course, Wegener's notion of how things work on a geophysical scale eventually proved correct. Aside from establishing the basis of plate tectonics (a keystone of modern seismology and volcanology), he also helped us understand ancient climates. Just how could coal—the product of decayed tropical vegetation—be present beneath what is now the south polar cap? Simple: The Antarctic used to be somewhere else.

The Handy Weather Answer Book is meant to be fun, to show that science can be entertaining, amusing, and fascinating. Learning and discovering new things about the natural world makes people see the world in a different way. If nothing else, I hope those readers who are a bit afraid of science will lose some of their apprehension. Sure, science has to be quantitative (the measurements and numbers are the key to everything), but understanding the foundations of science in their broader context is something that everyone must do—not just scientists. When you vote for a political candidate based on his or her environmental philosophy, are you sure that you really understand the issues at hand, or that the candidate really understands them? Raising the gasoline tax, funding basic research, property tax breaks for non-polluting industries, stricter air pollution controls on new power plants—these and many other important public policy issues can only be made wisely with a clear understanding of how the physical world works. And weather is a big part of that world.

This is not a text book, though the aim is to teach. The goal of *Handy Weather* is to pique your interest in weather and related phenomena in an entertaining way. The style is intentionally light, meant to be easy to read and sometimes even funny. I have made every effort to assure the accuracy and reliability of the contents. Of course, not every story in the weather record books, especially some of the truly weird ones that have been passed down over decades or more, can be verified. But new knowledge is created daily and old ideas are being discarded at the same rate. Weather records are being broken as this book is being written. If you discover an error, come across a fact that requires updating, or would like to contribute something to be included in a later edition, please pass it on to me care of Visible Ink Press.

While even the strangest things presented herein are believed to be accurate, being a little skeptical is always a good idea. In fact, that is how science works. Scientific facts have to fit the ever-expanding matrix of human knowledge about the physical world. They can't violate the laws of physics as we know them. If perhaps some unusual fact perks your interest in a topic—follow it up. I have included a Further Reading appendix, which supplies a long list of useful books to aid you in your weather exploration. But there are libraries that contain many more. And with the Internet, just rev up your search engine and you'll be amazed at what you can find. As a starting point, a few key weather sites are listed with their URLs in the chapter on Weather Forecasting. Be curious. Be skeptical. Be open to new ideas.

Mark Twain said that everyone talks about the weather . . . but do they really know what they are talking about? *Handy Weather* contains a collection of weather facts and trivia that could really spice up your weather conversations, and more important, feed your curiosity, knowledge, and understanding of the weather—one of the few things that all humans living on this planet have in common. Everybody talks about the weather . . . and now you'll have something to say that most people probably haven't heard before.

Acknowledgments

Writing a book is a lot of work. And you need a lot of help from many good people to get the job done. I am happy to report that a number of individuals did indeed make substantial contributions. Thomas Nelson and Liv Nordem Lyons of FMA Research poured over manuscripts, checking facts and searching for typos. Any mistakes still remaining in the text are, however, all mine.

I received input from a number of sources, including messages sent over the Net. Todd Glickman of the American Meteorological Society provided timely input on careers in meteorology. Suggested questions and answers came from Jack Bushong, Lawrence Estep, Larry Lowe, Karen Miller, Erica Page, and Tim Vasquez. Noel Risnychok of the National Climatic Data Center assisted by compiling a large amount of background data, weather information, and records. Professors Roger A. Pielke and Cecil S. Keen have been more than gracious in supplying many useful materials and ideas used throughout this book. Thomas Nelson also contributed up-to-date information on Internet weather sites. The staff of the National Weather Service Forecast Office in Cheyenne, Wyoming, were most helpful in obtaining copies of the latest National Weather Service public safety brochures and related documents.

Special thanks to Ken Fleck of Accord Publishing, Denver, Colorado, and Judy Galens of Visible Ink Press, Detroit, Michigan, who helped make this book happen. Thanks are also due to the many people at Visible Ink Press and Gale Research who assisted in numerous ways: James Craddock, Dean Dauphinais, Beth A. Fhaner, Brad Morgan, and Leslie A. Norback provided editorial skills, while Jeff Daniels, Pam Hayes, Randy Bassett, and Robert Duncan handled photographs. Thanks also to Anne Janette Johnson for lending her considerable skills to the proofreading and indexing of this book.

And I would like particularly to thank you, the reader, for picking up this book, spending some time with it, and thinking about some of the topics, which I hope you found interesting and informative. I hope I have been able to broaden your horizons about the natural world just a bit.

Credits

Photographs and illustrations in *The Handy Weather Answer Book* were received from the following sources:

National Aeronautics and Space Administration (NASA): **pp. 6, 71, 77, 81, 82, 86, 143, 318, 334**; Walter A. Lyons: **pp. 14, 18, 28, 31, 38, 40, 44, 49, 59, 74, 75, 76, 78, 79, 109, 112, 114, 117, 148, 162, 169, 171, 177, 181, 192, 198, 203, 209, 212, 217, 269, 271, 283, 324, 339, 363, 366, 369, 371, 374, 375**; National Oceanic and Atmospheric Administration (NOAA): **pp. 32, 33, 92, 99, 134, 140, 156, 187, 353**; AP/Wide World Photos: **pp. 120, 128, 129, 293, 299**; U.S. Coast Guard: **p. 123**; J. Pease: **p. 125**; © Gordon Garrado/Science Photo Library, National Audobon Society/Photo Researchers, Inc.: **p. 152**; W. A. Bentley: **p. 202**; Robert J. Huffman/Field Mark Publications: **p. 231**; U.S. Geological Survey Photographic Library: **pp. 243, 257, 263**; Robert A. Eplett/OES: **pp. 246, 252**.

THE
HANDY
WEATHER
ANSWER
BOOK

WEATHER FUNDAMENTALS

What is the difference between **climate and weather**?

The two are closely related, yet quite distinct. Weather refers to the state of the atmosphere at any given time, including things such as temperature, precipitation, air pressure, and cloud cover. Climate refers more to the totality of weather over a long period of time at one place or over a region. Climate comes from the Greek word *klima,* which refers to the elevation of the sun above the horizon. A "weatherperson" will try to provide you with a forecast for tomorrow's picnic, whereas a climatologist will get back to you next month with his or her estimate of what changes in atmospheric carbon dioxide might do to temperature trends in Finland over the next 300 years. Climate results from the accumulated impact of weather day after day.

The two fields often use very different tools. The weather expert uses technology to monitor the global state of the atmosphere at a given moment. The climatologist is more interested in annual or seasonal averages spanning hundreds or thousands of years. Climatologists often go to extremes to fathom what atmospheric conditions had been 2,000 years ago. Studies of trees rings, ocean sediment cores, air bubbles trapped in ancient glacier ice, and pollen preserved in amber from plants that withered millennia ago are just a few of the climatologist's tools. Both disciplines use supercomputers that simulate the behavior of the entire global atmospheric system, but on vastly different time scales (days versus centuries).

Why is the **study of weather** called meteorology?

In other words, what do clouds have to do with "shooting stars" (i.e., meteors)? At face value, not much. The origin of the term, however, goes back to ancient Greece. A meteor was any object that might appear in the sky, including clouds, rain, rainbows, etc. Today we use the term hydrometeors as a fancy word for raindrops. Aristotle even wrote a book called *Meteorologica,* which purported to explain all that there was to know about things in the sky. This first text book on atmospheric science had some interesting notions, such as "hurricanes are caused by evil winds falling upon good winds with a resulting moral conflict." Eventually the study of hot rocks falling from the sky (meteorites) became part of astronomy and the study of weather continued to use the term derived from ancient Greek.

What is **air**?

Air is a gas composed of several different chemical species, including about 78 percent nitrogen, 21 percent oxygen, 0.9 percent argon, and 0.03 percent carbon dioxide, with water vapor being highly variable but typically making up about one percent of the total. In the typical unpolluted atmosphere, there are also traces of neon, helium, krypton (not enough to be a problem for superman), xenon, hydrogen, methane, and nitrous oxide. The measurable properties of air include its temperature, pressure, and density. Polluted air can have numerous chemical species, but even in the most foul urban atmosphere, concentrations of contaminants are generally expressed as parts per million, billion, or even trillion. While the basic atmospheric chemical mixture has been stable over historical times, there is evidence that in early eons the amount of oxygen and carbon dioxide were substantially different than found today.

How heavy is something that is **lighter than air**?

For the record, a cubic yard of air at sea level weighs over two pounds. The air in a large room can weigh several hundred pounds. And how much does the entire atmosphere weigh? More than you can lift: 5.1 million billion tons. The volume of the troposphere, which contains most of the atmosphere's mass, as well as its weather, is about 5 trillion billion cubic yards. Since the atmosphere takes up a lot of space, its density (mass divided by volume) is relatively low, but a ton of feathers still weighs the same as a ton of lead.

The atmosphere exerts a pressure on all surfaces. At sea level this pressure is about 14.7 pounds per square inch. The planet Earth weighs a bit more than its atmosphere—some 588 quintillion tons.

How "thin" is air?

"Thin as air" is a relatively common expression. Compared to something like water, air may be considered "thin": Near sea level, the density of air is about one thousand times less than that of water. In other words, a thousand cubic feet of air weighs the same as one cubic foot of water. But in and of itself, air is actually quite dense. If you were one of the gas molecules that make up the atmosphere, as you moved about you might wish to have a little more elbow room. Typically, near sea level, air is so dense that a gas molecule only has to move about a millionth of an inch before meeting its nearest neighbor.

Speaking of thin air, the Colorado Rockies have a home field advantage, one that they have to share with visiting baseball teams. At more than 5000 feet, the air in the stadium is 15 percent less dense than at sea level ballparks, resulting in fly balls traveling almost 10 percent further. A 375-foot fly ball to left center in New York's Shea Stadium can be a 410-foot home run in Denver.

How does the **air pressure** decrease with **altitude**?

The air pressure always decreases with altitude. Near the surface the rate is about 0.01 inches of mercury on your barometer for each 10 feet of altitude. Take an elevator ride in a 50-story building and the barometer will fall 0.50 inches. The air pressure in Denver is only 85 percent that of sea level cities. By the time you reach 18,000 feet the pressure is half that of sea level; thus half of the atmosphere's mass lies below this level. At 40 miles, the air pressure is one ten thousandth of the surface. The Earth's atmosphere doesn't end, it just sort of peters out. At an altitude of 65 miles, the pressure is only about one millionth of what it is at sea level. Incidentally, the average air pressure over the planet at sea level is 29.92 inches.

What is the **dry adiabatic lapse rate**?

It turns out that when dry air (meaning no clouds) changes pressure the temperature also changes. You've had the experience of filling a bicycle tire. The pump is compressing the air and you have probably noted how warm the valve stem becomes. Similarly, when you let air out of the tire, notice how cool the rapidly expanding air feels. Since changes in altitude equate to changes in air pressure, for every thousand feet an air parcel moves upwards, its temperature drops 5.4°F (assuming no condensation has taken place). Similarly, if air flows downward, it warms 5.4°F for every thousand feet it descends. This is called the dry adiabatic lapse rate. On a sunny day, the temperature of the air at low and high elevations typically corresponds closely to this lapse rate. So on the top of an isolated mountain peak with an elevation of 5,000 feet, the air temperature could typically be about 27°F cooler than at the base.

How is **air pressure** used in **building design**?

A little air pressure can do a lot. Large domed stadiums, such as in Minneapolis or Pontiac, Michigan, are pressurized by giant fans. If you brought a barometer inside, you would find the pressure increased by only a few hundredths of an inch, yet that provides enough force to hold up fabric roofs weighing 35 tons. When patrons leave at the end of sporting events, they have to be careful going through the exits, because the building's air pressure sends 30–40 mph blasts of air with them out the door.

Which is denser, **dry or humid air**?

Strange as it may seem, at the same temperature dry air is "heavier" than moist air. This fact actually results in a type of front, called the dry line, which is a semi-permanent feature during spring and summer on the weather maps of the south central United States. It's a major factor in the weather of the southern high plains. On one side, there is sweltering humidity, while only a few miles to the west, the air can rival the Sahara in dryness. And since dry air is heavier than moist, the dry line often acts like a cold front, triggering severe thunderstorms and tornadoes.

Which way does air flow around **low and high pressure systems**?

What about those Hs and Ls that prance across the nightly television weather map? Just keep in mind that in the Northern Hemisphere, air spirals inward in a counterclockwise manner towards the center of low pressure (the L) and outward in a clockwise sense from high pressure (the H). Remember this and the whole map will make a lot more sense.

What are the various **layers of the atmosphere**?

The atmosphere with respect to the Earth is proportionally thinner than the skin of an apple, and it is layered like an onion. There are several classifications of atmospheric layers, but the most common is based upon whether the temperature is decreasing or increasing with height. The bottom layer, which typically extends to between 4 and 10 miles altitude, is called the troposphere. It is the layer in which most clouds and weather phenomena occur and in which the temperature usually decreases with altitude. The lowest portion of the troposphere, which varies from a few tens to thousands of feet in depth, is called the boundary layer. The next major layer is the stratosphere, a layer of increasing temperature which extends to a height of 30–35 miles. Above that we find the mesosphere, in which the temperature falls again, often reaching the coldest readings in the entire atmosphere (around -100°C) at about 60 miles. The outermost layer extending to several hundred miles up is called the thermosphere, in which the temperature (if you can call it that in the extremely thin air up there) increases to many hundreds or thousands of degrees Celsius. The boundary between the troposphere and the stratosphere is called the tropopause, with the stratopause and mesopause separating the upper layers.

If the atmosphere had a constant density rather than one that decreased steadily with height, it would be less than 8,000 meters thick. The carbon dioxide layer would only be 2.5 meters thick, and all of the ozone could be compressed into a layer only 3 millimeters deep. The water vapor thickness would vary from under a centimeter near the poles to perhaps 5 centimeters in the tropics.

What is an **inversion**?

The temperature within the troposphere generally decreases with height. But on occasions one will find layers in which the temperature actually gets warmer with height—an inversion of the normal situation. Inversions often form above the cold ground during the night, or when air flows over colder surfaces such as ice, snow, or cold water. Frontal surfaces aloft are usually marked by inversions. Inversions are important because they tend to have important effects on turbulence (reducing the mixing of pollutants) and often affect the propagation of light, sound, and radar waves. Inversions act very much like "lids" on the atmosphere because it is very difficult for air to move vertically through them. Thus they can trap pollutants near the surface. The Los Angeles smog, for instance, is in part the result of the persistent marine inversion from the nearby Pacific; this inversion constrains pollutants within the lower one or two thousand feet of the atmosphere.

What are the **scales of atmospheric motion**?

Just as the ocean has tides, swells, waves, breakers, and ripples, the air moves in a bewildering variety of scales of interacting motion, ranging from the molecular to the planetary. Processes at the level of individual molecules taking fractions of seconds are important in determining how clouds and precipitation form. By contrast, the global circulation refers to the large-scale movements of air over the space of thousands of miles and weeks and months of time.

Meteorologists talk about three major scales of motion: the macroscale, the mesoscale, and the microscale. The macroscale, sometimes called the synoptic scale, refers to processes you see on television weather maps, such as fronts, and high- and low-pressure systems. The microscale refers to processes that occur over tens to hundreds of feet, such as the formation of dust devils, or shallow fog that forms overnight in a valley. The mesoscale (meso means "middle") is the scale associated with much day-to-day weather phenomena, such as sea breeze fronts, thunderstorms, and cloud systems.

The different scales of atmospheric motion must be measured in different ways. The synoptic scale features are often defined by analyses of data from networks of weather stations often spaced hundreds of miles apart. The microscale can be studied at a single point, such as by equipping a tower with sensors at several levels above the ground. The mesoscale weather systems, with dimensions on the order of tens of

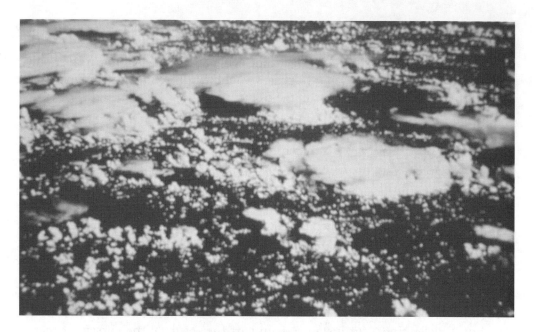

Many of the more common weather phenomena, like these thunderstorms photographed from orbit, take place in the middle scale of motion known as the mesoscale.

miles, are often too big to be perceived by looking out the window, but small enough to slip between the synoptic network of stations. Thus, remote sensing tools such as radars and satellites have played a key role in measuring and understanding these weather-making systems. Mesoscale weather systems, such as thunderstorms, sea breezes, and clouds, are those that we most often perceive as "weather" in our daily lives.

Is the **greenhouse effect** influencing our climate?

Yes, and it has been for billions of years. In the popular press, the greenhouse effect is generally referring to the possible increase in global temperatures as a result of humans dumping huge amounts of carbon dioxide into the atmosphere. CO_2 acts very much like glass in a greenhouse. It allows the sun's energy to warm the surface of the planet, but prevents some of the heat energy from radiating back into space. Thus the air temperature rises. But the greenhouse effect has been ongoing for billions of years and in fact is essential to maintaining our climate, and indeed life itself. Naturally occurring CO_2, water vapor, methane, and other "greenhouse gases" carefully regulate our planet's temperature balance. Our planet has just the right mixture of gases in its atmosphere, and is just the right distance from the sun to make life as we know it possible. We often hear of the "greenhouse effect" as if it were something bad. In fact, it is necessary for our survival. If gases such as water vapor and carbon dioxide did not have this property, our planet would be a frozen wasteland, with a mean temperature some 60°F colder than it is.

The planet Mars is a very cold place, but not just because it is further from the sun than the Earth. Its extremely thin atmosphere is inadequate to produce a significantly beneficial greenhouse effect. The thin air of Mars results in a planetary average temperature of -76°F, whereas the dense clouds of Venus, the second planet from the sun, result in temperatures as high as +900°F on its surface. The Earth's average temperature is in the 50–55°F range. Concern about global climate change stems from the unknown effects on the Earth's temperature of adding large amounts of CO_2 to the atmosphere as a result of combustion of fossil fuels and burning of forests.

What is the **jet stream**?

The jet stream is a high speed river of air found in the middle and upper portions of the troposphere. It essentially marks the boundary between major global air masses. The polar jet stream, the primary such air river affecting the United States, represents the boundary between polar air and warmer mid-latitude air masses. The jet stream can reach speeds well in excess of 200 mph near its core. The river snakes and weaves around the entire hemisphere in a sometimes discontinuous wavy pattern of troughs and ridges. Troughs are associated with surface low pressure systems and ridges with surface high pressure systems. As the jet stream goes, so do the surface weather features. Low pressure systems are steered by the jet stream aloft. If a large trough develops or moves into the western United States a cold air outbreak from the Arctic will usually follow behind it.

The strongest jet stream winds in the atmosphere, on the average, are most likely found over Japan, at a height of 20,000 to 40,000 feet. Here, wind speeds can sometimes reach 250 mph. American B-29 bomber pilots flying over Japan during World War II discovered the location of the planet's strongest winds. Westward flying bombers sometimes found themselves appearing to hover almost motionless over a target—and its defenders.

What is the **low-level jet stream**?

In the central United States, especially during summer, a nocturnal low-level jet stream is frequently found. Warm humid air streams northwards out of the Gulf of Mexico into the interior of the United States, transporting vast amounts of moisture. At the surface the winds may be nearly calm, but at the top of 1500-foot tall television towers in Oklahoma, anemometers may be showing winds screaming along at 50 mph. The low-level jet is often one of the key ingredients in severe thunderstorm and tornado outbreaks. The low-level jet also occurs during the daytime, but it is generally weaker and more diffuse.

7

How well mixed is the **atmosphere**?

There are literally trillions of gas molecules in the atmosphere. Wind and turbulence so thoroughly churns them up that the various gases that make up dry air (oxygen, nitrogen, argon) exist in the same ratios virtually every place in the atmosphere below 50 or 60 miles.

What is the **Intertropical Convergence Zone**?

The ITCZ, as it is called by those in the know, marks the clash of air masses moving equatorward from both hemispheres. The zone swings back and forth across the equator to mark the seasons. During much of the year and during summer over the Atlantic ocean, clouds associated with the Intertropical Convergence Zone girdle the Earth like a string of pearls in satellite pictures. The ITCZ is the breeding ground of tropical storms.

Is there **weather on other planets**?

There are storms in the atmospheres of other planets. On Jupiter, the Great Red Spot, which has been nearly stationary for 300 years, is the visible indicator of a giant atmospheric vortex. A large blue spot on Neptune was another example, but it was short lived and replaced several years later by a similar disturbance in the opposite hemisphere. The spot may be a hole in the methane cloud cover, allowing astronomers to see down to the surface or to the ammonium hydrosulfide cloud deck. Not to be outdone, Saturn's atmosphere has the Great White Spot (sometimes elongating into a streak) that reappears roughly every 30 years. Saturn's axis is tilted about 27°, so it too has seasons of sorts. However, it takes the ringed wonder almost 30 years for one orbit of the sun. Thus each season, like its year, is about 30 times that of Earth.

Giant dust storms kicked up by howling global winds sometimes obscure much of the face of Mars. The Hubbell space telescope has shown that the giant planet-wide dust storm found by the Viking, which landed in the 1970s, has now subsided. And the clearing has allowed the planet's atmosphere to cool down by up to 20°C from when Viking first planted its pod on the red soil.

A dense cover of clouds shrouds Venus in mystery and accounts for its brightness in the night sky. Lightning has been observed in the atmosphere of several planets, including Jupiter.

The Earth's moon, by contrast, is weatherless, given that its atmosphere is a close relative to a nearly perfect vacuum. While our moon has no atmosphere, not all of the solar system's moons are barren. Oxygen has been discovered in the thin atmosphere of Europa, one of Jupiter's moons. Io, another Jupiter satellite, is shrouded with sulfur dioxide from volcanic activity. Saturn's moon, Titan, has a dense atmosphere of methane and nitrogen, while Triton, a Neptune moon, has a very thin atmosphere of methane and nitrogen.

Is the Earth round?

No. The Earth is not flat, but it is not round, either. The planet's shape is that of an oblate spheroid. It is widest around at the equator. The equatorial diameter is estimated at 7,925.77 miles, whereas the pole-to-pole diameter is about 7,899.09 miles.

How old are the **Earth** and the **universe**?

Calculating the number of candles on the cosmic cake has been complicated by results from the Hubbell Space Telescope, which suggests the universe may "only" be 8–12 billion years old, billions younger than some earlier estimates. Earth is thus a relative newcomer, having been around for four to five billion years. The atmosphere has been developing for much of that period. Where have the gasses in the atmosphere come from? Conventional theories have assumed that volcanic eruptions and biological activity have provided most of what we breathe. However, some recent ideas propose that cometary bombardments may have supplied at least part of our atmosphere, including much of the water found on the planet.

What is the angle of the Earth's **axis of rotation**?

The third rock from the sun (the Earth) spins on its axis once per day like a top. But the axis is not straight up. "Up" in space is a hard place to define, but in this context it means the plane of rotation about the sun. The Earth's axis points toward the north star (Polaris) at all times, which means it is about 23.5° off the vertical. This apparently shoddy celestial workmanship actually is responsible for one of the prime drivers of the climate system—the seasons.

What is the **equinox**?

The autumnal and vernal equinoxes mark the astronomical beginning of fall and spring, respectively. At the spring and autumn equinoxes, the length of day and night are equal everywhere on Earth, each twelve hours long. If you're the type who appreciates consistency, live on the equator. Every day and every night on the equator is twelve hours. The sunrise and sunset times do not change with the calendar at 0° latitude as they do everywhere else on the planet. Also if you are not sure of the orientation of your property and don't have a compass, there is one way to tell: on the first days of spring and autumn, the sun rises exactly due east and sets directly due west.

What is a **solstice**?

In simple terms, the solstice is the date at which the sun appears directly overhead at noon the furthest north (the tropic of Cancer) and the furthest south (the tropic of Capricorn) during the year. The summer and winter solstices mark the beginnings of those astronomical seasons. Between these two latitudes are found the Tropics. Summer officially starts with the solstice on about the 21st of June, when the sun is furthest north. Yet the warmest part of the year over most of the United States doesn't occur until mid-July. Similarly the warmest part of the day is usually several hours after noon, when the sun is highest in the sky. The lag is due to the time required for ground and water to heat up. The longest day of the year occurs at the summer solstice in the Northern Hemisphere. On this day, north of the Arctic circle in Alaska there is continuous 24-hour sunshine. Along the U.S.-Canadian border the sun appears for 16.25 hours, and in southern Texas and Florida, 13.75 hours of sunshine are all that's available. The winter solstice is about 22 December. It is the shortest day of the year in the Northern Hemisphere. It is also the first day of the Southern Hemisphere summer, and on this day the atmosphere above the South Pole receives more light from the sun than any other place in the world, yet the temperature averages only about -10°F.

Is the **Earth's orbit** around the sun circular?

No. Our planet, like most, travels around the sun in an elliptical orbit at about 18.49 miles per second, meaning we are sometimes a bit closer to the sun than at other times. The average distance from the sun is 93 million miles. Around January 2, the Earth is at its closest point to the sun (91,400,005 miles), a position called the perihelion. As a result the Northern Hemisphere gets about 3 percent more heat energy from the sun in winter than if the planet's orbit were circular, making it just a tiny bit warmer during the winter on the top side of the planet. Aphelion, when we are furthest from the sun, occurs around July 6; at that point the Earth is about 94,512,258 miles from the sun. However, you still need to wear a sunscreen when you go to the beach. Interestingly, the discovery of the non-circularity of the planet's orbit caused great distress among philosophers and theologians at the end of the Middle Ages. They had trouble explaining why the Creator made something less than perfect (that is, an orbit that was not circular).

Which way is **north**?

That question may be deceptively simple. True north can be determined by pointing towards Polaris, the North Star. When you use a compass, don't forget it points towards a magnetic north rather than true north. The two can be 15° or more apart. The Earth's magnetic north pole lies some 200 miles north of Boothia Peninsula in Canada while the south magnetic pole can be found in South Victoria Land in Antarctica. Of course when you are standing at the South Pole, all directions are north.

Santa Claus had better be sure he has not set up residence at the North *magnetic* Pole. If he has, his home base will be a mobile home. The magnetic poles of the Earth are constantly shifting. And on occasion the polarity of the Earth's magnetic field suddenly reverses, so Santa could suddenly find his compass taking him to Antarctica.

Is the Earth's **rate of rotation** constant?

As a first approximation, the Earth rotates on its axis once every sidereal day, which at 23 hours 56 minutes and 4.09054 seconds is slightly shorter than the calendar day. At the equator, a point on the Earth is moving 1036 mph. At the poles, there is no rotational motion of the Earth. But the rotation of our planet does actually speed up and slow down in response to winds, tides, and changes in the distribution of mass in the planet. The planet's rate of rotation, and therefore the length of the day, can vary by several thousandths of a second over a year's time.

On the longer term, the Earth is slowing down, rotating slightly more slowly with each passing day. Four hundred million years ago the days were shorter, in fact, there were 400 of them in a year. In June 1992, the world's official time keepers added another second to their clocks to account for the very slight slowdown in the Earth's rotation since 1990.

What provides the energy to drive the **Earth's weather system**?

The sun. Direct sunlight is considered to be the source of almost all the energy received by our planet. Other factors such as starlight, moonlight (reflected sunlight), tidal energy, and heat from radioactive decay in the Earth's interior do equate to about 0.02 percent of the energy derived from sunshine. The amount of the sun's energy intercepted by the Earth in a single day is equivalent to burning 700 billion tons of coal. The unequal distribution of solar energy on the planet causes temperature imbalance, which in turn causes wind and weather. On the average, the Earth radiates more heat into space than it gains from the sun at latitudes poleward of 40°N and 40°S, and gains heat in between these latitudes. The excess heat supplied to the tropical areas must be transported by winds and ocean currents to the polar regions to keep them from becoming ever colder.

Although only one part in two billionths of the sun's energy output strikes the top of our atmosphere, that amount of energy is amazing. Solar energy at the top of **11** the atmosphere averages 5 million horsepower per square mile. The amount of sunlight reaching the Earth's surface is 6,000 times the amount of energy used by all human beings worldwide. The total amount of fossil fuel used by humans since the start of civilization is equivalent to less than 30 days of sunshine. The energy coming from the sun is called insolation, not to be confused with insulation.

Is the Earth's **heat budget** exactly balanced with the sun's energy?

The Earth is essentially in perfect heat balance with the sun. It radiates back to space the same amount of energy it intercepts from the nearest star—the sun. The amount of energy received from distant stars is trivial. Even the slightest long-term imbalance between incoming and outgoing energy could cause the planet to start heating or cooling with dramatic results. Changes in the chemical nature of the atmosphere can affect the balance between incoming and outgoing radiant energy, providing the basis for concern about global climate change resulting from both human and natural processes. This nearly perfect energy balance is not found on all the planets: Neptune radiates 2.5 times the amount of energy it receives from the sun, indicating some unknown heat source within the planet itself.

Could the **sun** ever go out?

The sun, our nearest star, is the source of virtually all energy on Earth, and should stay that way for a long while. In spite of its importance to the Earth, the sun is actually a very average star. No more than 10 percent of the stars in the universe are more than 10 times bigger or 10 times smaller than the sun. Along with the rest of the solar system, the sun is rotating around the center of the Milky Way galaxy at some 200 miles per second. The sun is the source of all light, warmth, energy, and in fact, all life on Earth. The ultimate "climate change" on Earth would occur if the sun "went out." We need not worry very much about this. The sun is expected to keep on shining pretty much the way it is now for 5 billion years or more.

What part of the country receives the most **solar energy** on a yearly basis?

Not Florida (too many clouds). Almost twice as much potential solar energy could be collected over large areas of interior California, Utah, Nevada, and Arizona than in the "Sunshine State." The three sunniest places in the world, as determined by measured annual solar radiation reaching the surface, are the Sahara and Arabian deserts, the southwestern United States and northern Mexico, and the interior of southern Africa.

Can birds tell the seasons?

Somehow many birds can tell the approach of the next season, be it winter or summer, and some can do so with uncanny accuracy. Most likely they sense the seasonal changes from the elevation of the sun. Each spring, swallows return to Capistrano on the same day in March. Less well known are the buzzards that return to their roosting sites in Hinkley, Ohio, every March 15th.

Where is the **sunniest place** on Earth?

During the endless sunshine of the Antarctic summer, the South Pole receives more sunlight than any place on Earth. So why does it stay snow covered? It is precisely because of the snow, which reflects some 50–90 percent of the sun's energy right back into space. Temperatures struggle to make it up to 32°F even on a "warm" summer day.

What is **albedo**?

Albedo is the percentage of radiation (typically sunlight) incident upon a body that is reflected from it. The albedo of bare ground ranges from 10–20 percent, a green forest 3–10 percent, dry sand 18 percent, and fresh snow 80–90 percent. It is estimated that about 30 percent of the sun's energy that reaches the planet Earth is immediately reflected back into space by reflection off clouds, snow, ice, land, and water surfaces. However, this number is still uncertain. One of the central issues in untangling the global climate change puzzle is the role of clouds. Some clouds tend to cool the Earth, whereas some tend to warm it up. On the whole, it is now felt the net effect of the Earth's clouds is to cool the planet.

How much **energy** is in the **wind**?

The total wind power of the atmosphere has been estimated at about 3.6 million kilowatts, yet less than one percent of the world's energy is currently derived from renewable wind energy. Efficient and reliable wind turbines, each of which can produce 500,000 watts of electricity, are now economically viable.

What is **Buys Ballot's Law**?

It is not some new-fangled legislation passed by Congress. Buys Ballot's Law is a rule devised by a Dutch meteorologist in 1857 that says that, in the Northern Hemisphere, **13**

This tree, permanently bent so that it is almost parallel to the ground,
is evidence of persistent steady winds in one direction.

when you stand with your back to the wind, the low pressure is on your left and the high pressure is on your right. In the Southern Hemisphere, the relationship is reversed. This law is true, at least as a first approximation, when one is dealing with large, well-organized weather features such as hurricanes or low pressure systems. At smaller scales of motion, the relationship between wind and pressure becomes more complicated.

Where should you not hold a **kite flying contest**?

There are two major bands of very light winds that gird the Earth, "the doldrums" near the equator, and the "horse latitudes" at roughly 20–25° north and south.

What are the **horse latitudes**?

Semi-permanent centers of high pressure straddle the globe between 25° and 40° latitude. In land areas, these are associated with deserts such as the Sahara. Over the Atlantic, sailors refer to these areas of weak winds as the horse latitudes. Apparently ships would be becalmed for extended periods and the horses on board would be thrown over as food and water supplies gave out.

What are the **trade winds**?

This large-scale component of the Earth's general circulation occupies much of the tropics. They blow from the subtropical high-pressure systems into the equatorial low-pressure troughs. In the Northern Hemisphere the trade winds blow mostly from the northeast. South of the equator they blow from the southeast. The winds can blow steadily for days and are among the most consistent on Earth. Visitors to Hawaii and Puerto Rico can often experience the brisk trade winds.

What are **aeolian sounds**?

The wind does actually "whistle through the wires." Strong winds flowing past roughly circular objects, such as electrical wires and tree branches, produce eddies in the air flow. These in turn can often be heard as distinct musical or humming sounds, which are called aeolian sounds or aeolian tunes.

What is the **"snow eater"**?

It is not a furry Arctic animal with an appetite for snow drifts. Rather it is a wind that is warmly welcomed by residents of the High Plains during the dead of winter. When the warm, dry, downslope "Chinook" winds blow during winter along the eastern slopes of the Rocky Mountains, up to a foot of snow cover can disappear in a day.

How far can **dust storms** travel?

During April of most years, huge clouds of dust kicked up by wind storms over the deserts of central Asia and China are blown far to the east—often reaching Hawaii. Not to be outdone, Saharan dusts routinely are tracked by weather satellites crossing the Atlantic Ocean, reaching Florida and beyond. One recent study even suggested some of the haze in the Grand Canyon region at times might be of African origins. Soil scientists claim that up to six inches of Georgia's topsoil may have a Saharan source.

What are **monsoons**?

In popular language the word monsoon is often used to denote a violent tropical storm, though this is not particularly accurate. A monsoon, derived from the Arabic word *mausim*, a season, is a large-scale wind system that lasts for an entire season. The name for these winds was first applied to winds over the Arabian Sea, which blow for six months from the northeast and then for six months from the opposite direction. They are the result of dramatic temperature differences between continents and the surrounding oceans. The best known are the monsoons of Asia. These persistent,

15

large-scale winds flow onto the heated continent during summer and off the continent during the winter. The summer monsoon is wet: In Asia it represents a great stream of humid tropical air moving inland over the sun-warmed continent. In India, waiting for the onset of the summer monsoon rains is a major pastime, if only because it means a break in the oppressive heat, not to mention salvation to the parched crops. When the air hits steep barriers like the Himalaya Mountain system, torrential rains are common. The town of Cherrapunji, India averages over 400 inches of rain per year. Cochin, India averages around 25 inches of rain in June and July, but during the dry winter monsoon, in January and February, less than one inch of rain falls. There is a summer monsoon that affects the southwestern United States, bringing increased moisture and numerous thunderstorms in mid- to late summer and early autumn.

What does weather have to do with the **Balanced Budget Amendment**?

The federal budget deficit is now in the $5 trillion range. A significant chunk of that can be blamed on Mother Nature's rampages of the past two decades. Since 1977, Washington has paid out some $120 billion in natural disaster relief payments, including hurricanes, floods, tornadoes, and droughts. That works out to about $500 for every person in the United States today.

Are raindrops **spherical**?

No, raindrops are neither perfect spheres nor are they tear shaped as one sees in cartoons or on greeting cards. Instead, as they fall, they become flattened into a shape rather resembling red blood cells or English muffins. The larger the rain drops, the more distorted they become, eventually breaking up into several smaller and more spherical pieces. The smallest cloud droplets are much closer to perfect spheres.

How fast do **raindrops** and **snowflakes** fall?

It depends on their size. The very smallest almost float, sifting Earthwards at much less than one mile per hour. The very largest fall at up to 16 to 20 mph. Thus a large drop descending from the base of a thunderstorm at 5,000 feet above the ground takes approximately three minutes to splash down. How large can raindrops get? The very largest measured rarely exceed a quarter of an inch in diameter. If they were to be any bigger, the instability caused by the turbulent motion of air around the drop as it fell would cause the drop to shatter into many smaller drops.

Snow flakes generally fall from the sky at speeds between one and five miles per hour.

What are the **common forms of ice** formed by atmospheric processes?

Ice is found in the atmosphere in a number of forms. This includes individual ice crystals, clumps of crystals (snow flakes), graupel (sometimes called soft hail), ice pellets (also called sleet), hail, rime (cloud droplets freezing on an aircraft wing during icing conditions), glaze (rain freezing on contact with cold surfaces), and frost.

What phenomena caused the **worst weather-related aircraft accident**?

It was fog, not a thunderstorm or wind shear, that caused the worst aviation accident in history. Two Boeing 747s collided in a dense fog while on the runway in the Canary Islands in 1977, resulting in the death of 582 people.

Does all **precipitation** falling from clouds **reach the ground**?

No. Sometimes a considerable fraction evaporates in the region between the cloud base and the ground. Streaks of falling rain evaporating before they reach ground are called "virga." In the low-humidity air of the Southwest, "dry" thunderstorms are common. They produce lightning but little rain and are therefore a significant cause of forest fires. Evaporating rain also chills the air, making it negatively buoyant and resulting in strong downdrafts beneath these clouds. Some become so strong that they are classified as microbursts, and are a major hazard to aviation.

If it is raining or snowing, isn't the **relative humidity** 100 percent?

Not necessarily. The cloud in which the precipitation formed certainly was saturated, but rain and snow can fall to lower altitudes where the humidity can be well below 100 percent.

Can the relative humidity ever be **higher than 100 percent**?

Strangely enough, the answer is yes. A 100 percent relative humidity means the air is saturated, that is, holds as much water vapor as it can at a given temperature and pressure. Within clouds, however, there is a phenomenon called "supersaturation" in which the relative humidity may be 101 percent or so. This greatly helps clouds and

17

Frost is not frozen dew, but ice crystals that have formed, in this case, on icy cold fence wire.

raindrops to grow. On the other hand, a relative humidity of less than 0 percent is not possible.

What is **drizzle**?

Drizzle is distinguished from rain primarily by the size of the drops. Drizzle drops are typically about 0.02 inches across. Rain drops can be many times larger. If the temperature of the surface onto which the drops fall is below freezing, you have freezing drizzle.

Can it **snow** when the temperature is 67°?

Sort of. Even on the hottest summer day, it can "snow." Much (but not all) rain that falls to the ground begins as snow crystals high in the cold clouds overhead. Temperatures within the top part of a large thunderstorm can be lower than -50°F.

Does **dew** fall?

It is a common expression in English to say that the dew has fallen, but it is scientifically inaccurate. Dew doesn't fall from the sky. Rather it forms in place by the process of condensation on cold surfaces, very much like moisture on the outside of a cold beverage glass. This fact was first demonstrated in England in 1814 by Charles Wells. Also, frost is not frozen dew, but ice crystals actually forming on the surface of the car, grass, or fence.

What causes **dew** to form on a glass containing a cold beverage?

There is always invisible water vapor in the air. At higher temperatures, air can hold much greater amounts of water vapor than when it is cold. In fact, air at 80°F and 100 percent relative humidity holds about 22 times more water vapor than saturated air at 0°F. One measure of the amount of H_2O molecules is called the dew point temperature. It is simply the temperature to which the air must be cooled (at constant pressure) in order to achieve saturation (100 percent relative humidity). This basic fact was first demonstrated in the year 1751 by a French physician named Charles Le Roy. So on a summer day when you pour that lemonade over some ice, the temperature of the liquid drops to near 32°F as does the outside of the glass. If the dew point of the air is greater than that of the glass, the air immediately in contact with the glass is chilled. If it is cooled to the dew point, then dew forms on the glass. Dew forms on the grass when, during the night, it cools to below the air's dew point temperature. What if it is below freezing? Then probably it should be called the frost point, but that term is not commonly used.

Water vapor can also condense on colder water. The water in Lake Superior is so chilly that during the summer, instead of losing water to the atmosphere by evaporation, the lake actually gains water, to the tune of several inches a month, due to condensation of moisture from the air onto the water surface. A simple rule: if the temperature of a surface is cooler than the dew point of the surrounding air, you will usually have condensation. So if you take film out of storage in your refrigerator (a common practice with photographers) on a muggy summer day, let it warm up before inserting it in your camera.

This rule also explains why eyeglasses fog up after being brought inside on a cold day. The glasses have been chilled below the dew point temperature of the air inside the building. Thus dew forms on the glass surface.

What do you see when you "see your breath"?

Condensed water vapor. The moisture in your breath is cooled by mixing with the environment, and if cooled below its dew point, a small cloud forms. When you collect a lot of "breaths" together you can get a much bigger cloud. In fact, during winter in the cold calm valleys of interior Alaska, large herds of caribou huddled together to keep warm create huge patches of ice fog that can fill a valley from their exhalations.

Does **water freeze** at 32°Fahrenheit?

Of course water freezes at 32°F, right? Not always. Salt water freezes at lower temperatures than fresh water, by several degrees, depending on the amount of salinity. Pure bulk water, such as in puddles, freezes at 32°F. But the minute droplets that comprise clouds can stay in the unfrozen state to temperatures as cold as -40°F. So water does not always freeze at 32°F, but it is correct to say that 32°F is the *melting* point of pure water ice in all cases.

Does water become **denser** as it becomes colder?

It would seem so, but water is a very odd substance. It is most dense at 38°F. As it cools below 38°F, it becomes lighter. This is good, because otherwise lakes would freeze from the bottom up rather than the top down, and almost all fish and marine life would be frozen solid every winter—if they ever could have developed in the first place. Ice is thus less dense than cold water and floats on the surface. The only other substance that is less dense as a solid than a liquid is the element bismuth.

What is water's "triple point"?

Water has the peculiar property of being able to exist simultaneously in the gaseous, liquid, and solid phases at a single temperature. This triple point is 32°F or 0°C.

Do lakes have **tides**?

Ocean tides, often reaching many feet, result from the play of the gravity of the sun and moon on the vast volume of water in the world's seas. Lakes are generally too small to have a detectable tide, except for huge Lake Superior which has a three-inch tide. There are also tides in the atmosphere, detectable as a routine daily fluctuation in the air pressure.

What is the **thermocline**?

There are temperature layers in bodies of water just as in the atmosphere. In a lake or ocean, warm water usually is layered above the colder bottom waters. The sharp temperature gradient is called the thermocline. The warm water above is called the epilimnion, while the chilly water underneath is the hypolimnion. Sometimes in the Great Lakes the thermocline is found just off the beach. You can step into 65°F water, wade a few yards offshore, and suddenly your toes are immersed in water that is 38°F cold.

How is the **world's water** distributed?

The water involved in all the world's clouds, rain, and snow makes up less than three tenths of one percent of the total free water. Ice caps and glaciers hold about 2 percent of the total. Most of the remainder is accounted for in rivers, lakes, and oceans.

Does water flow **downhill**?

It tries to. In the hydrometeorological model of the Earth's water cycle, river water flows downhill to reach the sea. But that has been getting more difficult lately. In the last 40 years some 15,000 large dams have been constructed worldwide, intercepting about 15 percent of the global water runoff.

What is the **fastest ocean current** in the world?

One candidate is the Florida current, part of the Gulf Stream system. It swirls through the Straits of Florida, past Miami, and northward towards Cape Hatteras, sometimes reaching speeds up to 5 knots.

At what temperatures does **sea water freeze**?

Sea water freezes at about 28°F, though the exact temperature depends on the salt content of the water. Around the globe, about 87 percent of all ocean water has a water temperature of 40°F or less.

Is there **salt** in the air?

When you go to the seashore for a little "salt air," you get exactly that. Near the ocean it is not uncommon to find more than 100,000 minute specks of salt per cubic meter of air. Sea salt can be found in the air thousands of miles into the interior of continents. It is one of the more important types of particles that serve as cloud nuclei, which allow clouds and raindrops to form.

The thin haze often noted at the seashore is the result of sea salt in the air. When the relative humidity of the air increases above about 75 percent, the tiny salt particles begin absorbing water, forming a myriad of minute haze droplets.

How have athletic shoes been used in charting **oceanographic currents**?

By accident. Oceanographers are always looking for ways to better chart the currents of the seas. Unlike landlocked rivers, which have obvious channels, the giant rivers that exist in the oceans are less well marked. In 1990, a Korean container ship bound for the United States ran into some difficulties, resulting in 80,000 Nike shoes being swept overboard. Oceanographers have since been retrieving the shoes on beaches from Alaska to Oregon to Hawaii. The shoes turn out to make nice "tracers," which allow validation of computer models of ocean currents.

The atmosphere is greatly affected by the underlying oceans. Computer models of water and climate processes are now beginning to link both the sea and the air as one system.

How much water is in the **world's oceans**?

The oceans of the Earth contain about 350 billion billion gallons of water.

The Atlantic Ocean loses more water from evaporation than it receives from precipitation. The amount of water that evaporates from the Earth's oceans and land masses in the course of the year would, if condensed at one time, form a layer about 40 inches deep over the entire planet.

What is the **speed of sound**?

The speed of sound varies with the type of medium transmitting the sound, and with the temperature of the air. At air temperature of 32°F the speed of sound in air is 740 mph (331 meters per second). As a rough rule, sound travels one mile in 5 seconds near sea level. In water, at 55°F it is 3,223 mph (1,441 meters per second), and in steel it is about 11,000 mph (5,000 meters per second). This is why you can hear a train coming at long distances by putting your ear on the railroad track.

What is **Universal Coordinated Time**?

For many scientific and practical reasons, meteorology and many other fields use a single time reference called Universal Coordinated Time (the abbreviation UTC coming from the French translation of the term). It has in the past been called Greenwich Mean Time (GMT) because it is the time at the observatory in Greenwich, England. The abbreviation "Z" for Zulu Time is also frequently used.

UTC is a 24-hour clock with midnight (0000 UTC) corresponding to midnight local time in Greenwich. By international agreement meteorologists around the world take coordinated (also called synoptic) surface and upper air observations at 0000 UTC and 1200 UTC (noon). Surface synoptic observations are also taken at 0600 UTC (6:00 A.M.) and 1800 UTC (6:00 P.M.).

What is a **rain shadow**?

When moist air streams up the slopes of a mountain, rainfall or snowfall is increased. However, when the air, depleted of much of its moisture, begins to descend down the lee slopes, the temperatures warm, the clouds dissipate, and very little precipitation falls. This "rain shadow" occurs on a variety of scales. The western mountains in the United States wring out much of the Pacific moisture that flows onshore nearly all year long, producing the western deserts and the low precipitation regions of the High

Plains. The Olympic Peninsula in northwestern Washington has some amazing climatic variations in the space of several counties. On the western slopes is a true rain forest. But in the "rain shadow" on the downwind slopes, less than 15 inches of rain per year allows cactus to grow. The upwind slopes of the Hawaiian islands are some of the wettest places on Earth, whereas the other side of the volcano resembles the dry grass lands of the western United States, cactus and all.

What time is it at the North and South Poles?

Since all the time zones converge at both poles, the answer is not at all obvious. To solve the problem, it was internationally agreed that Universal Coordinated Time (the time in Greenwich, England) would be in use at both ends of the planet.

What nation has the **worst weather** on Earth?

If one considers overall nastiness (heat, cold, flood, drought, tornadoes, tropical storms, etc.) the United States manages to come up number one in just about all polls taken.

Which continents have the highest and lowest average elevations?

The continent with the highest average elevation is Antarctica, a good deal of which is thick glacier extending to heights well above a mile. Australia has the lowest average elevation of any continent. These two continents are also the only two that lie entirely in the Southern Hemisphere.

Has **Inauguration Day weather** influenced our political history?

It was sunny but cold for Bill Clinton's 1993 inauguration. He was lucky. Deep snow covered the ground for JFK and Taft. Reagan's second inauguration was partially indoors due to temperatures in the teens. Franklin Roosevelt was soaked with rain in 1937. In 1841, William Henry Harrison, then the oldest man ever elected president, refused to give into the elements on a raw, blustery, and bitterly cold Inauguration Day. And he paid the ultimate price. He refused to ride in an enclosed carriage, much less wear a coat or hat, during a two-hour procession along Pennsylvania Avenue. He caught a "chill" and just one month into his presidency, he died from pneumonia.

What are **fronts**?

There are four major types of fronts drawn on weather maps: cold, warm, stationary, and occluded. They represent the boundaries between air masses with different temperature and/or moisture characteristics. In a cold front, the colder air is advancing, undercutting the air ahead of it. With a warm front, the warmer air is advancing by flowing up and over the shallow cold air, eroding away the boundary and replacing cold surface air with warmer and more moist air. In a stationary front, the warm and cold air masses are more or less in balance, though some stationary fronts can wobble back and forth several hundred miles in a day. In an occluded front, a cold front overtakes a warm front and forces the air in between to be pushed aloft. On a weather map, the "barbs" on a front point in the direction the cold air is moving. The term "front" originated by analogy to the battle fronts of World War I, the time at which the basic research on meteorological fronts was being conducted.

What is the **diurnal temperature range**?

The dinural temperature range refers to the gap between the lowest and highest temperature on a given day. The difference between the average daily high and low temperatures ranges from a little more than 10°F in tropical, oceanic areas to over 40°F in some deserts and in mountainous terrain. Cities along the Front Range of the Rockies in summer routinely experience morning lows in the 40s with daytime highs of 90 or above.

Does **San Francisco** have much "weather"?

Actually, they have more than their share. San Franciscans often claim they "don't have much weather around here." On a summer day, it is not the least bit unusual to have the shoreline on the Pacific ocean side of the city shrouded in fog and drizzle, with temperatures in the upper 50s, while just across the Bay the mercury has soared above 100°F under cloud-free skies. This disparity is a prime example of what are called microclimates, sharp differences in temperature, cloud cover, and precipitation due to proximity to the ocean and complex, mountainous terrain. And we would note that the Bay Area, along with much of the West Coast, can get whacked with some pretty vicious winter storms rolling off the Pacific from time to time.

What is the **urban heat island**?

Meteorologists have long known that cities are warmer than the surrounding countryside. The "urban heat island" is especially pronounced on calm, clear nights when even towns of a few thousand inhabitants can be several degrees warmer. Urban meteorologists will have no shortage of urban heat islands to study: by the year 2000 it is expected there will be 400 cities on the planet with populations over one million. During winter it is not unusual for the center of a major metropolitan area to be 10° or 20° warmer during the dawn hours compared to low-lying areas in surrounding rural countryside.

What is a **cyclone**?

The word cyclone has several uses in meteorology. It was once a common term for a tornado. In parts of the world tropical storms are still called cyclones. Technically, a cyclone is any organized area of low pressure. A typical low-pressure system that swirls across your nightly television weather map is technically called an extratropical cyclone.

What were **Benjamin Franklin**'s contributions to meteorology?

Just about everyone knows about Ben's kite experiment, which demonstrated that lightning was electricity (and also that he was darn lucky—a number of scientists were later killed trying to duplicate the feat!). He also invented the lightning rod, which is still effectively used on many buildings today. But Franklin made a much more fundamental contribution. He was the first to suspect low-pressure systems were giant rotating whirls of air. His view of the eclipse of 21 October 1743, in Philadelphia was ruined by a storm with northeast winds. Upon finding that the eclipse was seen in Boston, and that a northeast storm struck a day later, he concluded that the storm system had counterclockwise winds circulating about its center that moved from southwest to northeast. This basic understanding of storms was critical to advances in atmospheric sciences in the next century.

What is **latent heat**?

The condensation of water vapor gives off heat, as does the freezing of water into ice. When a cloud forms, for each gram of water vapor that condenses, about 600 calories of heat energy are released. The freezing of that water releases another 80 calories for each gram. Latent heat is the basic fuel source for growing thunderstorms and hurricanes.

INSTRUMENTS AND OBSERVATIONS

Who invented the **barometer**?

The barometer, which measures the pressure exerted by the mass of the atmosphere, was first devised in 1644 by Evangelista Torricelli, a student of Galileo, who had noted that a column of water in a tube could not be made to stand to a height greater than 34 feet. Torricelli experimented with fluids of different densities, including sea water, honey, and finally mercury (a liquid metal 14 times heavier than water). The mercury tube, when plugged at one end, filled with mercury, and placed upright with the open end in a container of more mercury, would only stand to a height of about 30 inches, or about 1/14th that of the water column. The mercury column was being held up by the pressure exerted on the reservoir of mercury by the atmosphere. Thus, the barometer was perfected. This instrument allows the continuous measurement of atmospheric pressure, which rises and falls with the passage of traveling weather disturbances in which the weight of the air column overhead is alternately heavier or lighter. Torricelli put it more elegantly: "We live submerged at the bottom of an ocean of elementary air, which is known by incontestable experiments to have weight."

What is the **highest barometer reading** ever measured?

Siberia has justly earned its reputation for cold. But cold air is also dense, and therefore the atmosphere weighs more . . . and the barometer soars to the upper end of the scale. On New Year's Eve of 1968, in Agata, Siberia, the hammer really fell. During a deep cold wave, the barometer set an all-time record for the Eastern Hemisphere, peaking at 32.01 inches.

27

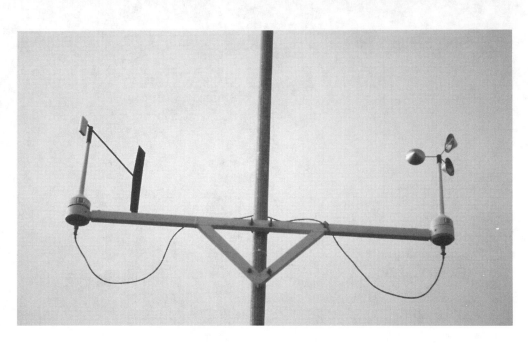

Wind direction is measured with a wind vane (left),
while wind speed can be measured with a three-cup anemometer (right).

What is the **lowest barometer reading** ever measured?

The lowest air pressure at sea level ever measured was 25.63 inches, in the eye of Typhoon Tip, as it roared across the Pacific east of the Philippines in October 1979. Maximum sustained winds were about 190 mph. This reading was obtained by an instrument called a dropsonde, which measures pressure, temperature, and humidity while it descends on a parachute after being ejected from a high-flying hurricane (or typhoon in this case) hunter aircraft.

Where in the 48 contiguous states does the **barometer change** the least and most?

The difference between the highest and lowest barometer readings ever recorded at a station can be called the barometric range. San Diego has the lowest range (1.02 inches) and Hartford, Connecticut has the highest range (3.02 inches) recorded in the 48 contiguous states (not including some hurricane landfall points).

What is the **average air pressure** at sea level?

The average sea-level pressure for the planet is 29.92 inches of mercury, or 1013.25

millibars. The millibar is one thousandth of a bar—a unit of pressure. On your home barometer, 0.03 of an inch on the pressure scale is equal to about one millibar.

What is wind?

It is the flowing of air, which is a fluid, albeit a thin one. This answer may seem rather obvious, but the question absorbed many ancient Greek philosophers. It was Anaximander who first proposed that wind was a natural flow of air rather than some mysterious phenomena related perhaps to the breath of the gods. (The theory that wind resulted from trees waving their leaves didn't get many converts).

One question that often causes confusion is, which way is the wind blowing from? If the National Weather Service reports a west wind, is the air coming from or going to the west? The arrow on a wind vane points toward the direction from which the wind is blowing. And when the Weather Service reports the wind direction is, say, westerly, that means the air is moving from west to east.

What is an **aneroid barometer**?

It is common practice even today to say "the mercury [meaning the barometric pressure] is standing at 30.07 inches. . . ." In fact, mercury barometers, while still the standard of accuracy, are rarely used in day-to-day measurements. The aneroid (meaning "without air") system is far more common. An aneroid system basically uses an evacuated metal chamber that expands and contracts as the outside air pressure fluctuates. This in turn drives a series of mechanical levers or other devices to produce a reading of pressure, with the display calibrated as if it were measuring the height of a mercury column. It would be more technically correct to give the air pressure in units such as pounds per square inch, but that convention is not often used. Air pressure is reported in a wide variety of units; aside from inches of mercury, you often hear millimeters of mercury, millibars, or hectoPascals.

How do you measure **wind speed**?

The basic wind speed measuring device is called an anemometer. The first known such instrument was devised in 1667 by the British physicist Robert Hooke. There are many variations on the theme but the classic rotating cup anemometer is still among the most widely used. Three or four cups are mounted to catch the wind and be rotated by it. Calibration of the anemometer in a wind tunnel allows the rotation rate to be converted to wind speed. Other techniques to measure wind speed include watching the drift of balloons (for upper air measurements), sonic anemometers, devices that are cooled in response to the speed of the wind, and plates that bend in response to the force of the wind. Many far more sophisticated systems are used to measure winds remotely, including radar, lasers, and sound pulses.

What **units** are used to **measure** wind speeds?

Most typically wind speed is measured in miles per hour, unless you are in the National Weather Service or Federal Aviation Administration or any other meteorological service dealing with navigation where knots (nautical miles per hour) are still used. One knot equals 1.15 mph. But scientists don't like either term, preferring to deal with meters or centimeters per second, or kilometers per hour, or feet per second. Vertical wind speeds can be expressed in centimeters per second or microbars per second (the rate of pressure change [altitude] over time).

What are some of the world's **highest recorded wind speeds**?

The highest remotely measured wind speed, using a small portable Doppler radar, was 287 mph, measured in a tornado.

The highest wind ever actually directly measured on the surface of the Earth using a standard anemometer was a 231 mph peak gust at the observatory atop Mount Washington in New Hampshire. This observatory is one of the windiest places on Earth, with 100 mph winds being almost commonplace.

Thule, Greenland, once reported a peak wind gust of 207 mph (we don't even want to think about what the wind chill was).

Sustained winds of 190 mph were believed present in Hurricane Camille as it made landfall on the U.S. Gulf Coast in 1969.

A gust of 186 mph was confirmed during the Hurricane of 1938 at Blue Hill Observatory, Massachusetts.

Port Martin, Antarctica once reported an average wind of 108 mph—for an entire day! It also logged a mean monthly wind speed of 85 mph.

If you are interested in some winds that are out of this world, consider Neptune, where space probes suggest winds may be howling at 1,250 mph.

Balloons have been used for over a century in weather research. Storing them has always been a bit of a problem.

How do you measure **wind direction**?

The basic instrument is called a wind vane, which is a free-swinging object mounted on its center of gravity on a vertical axis. The end of the vane that offers the greatest resistance to the motion of the air moves to the downwind position. In other words, it points in the direction the wind is moving. It may be in the form of a rooster, pig, or any number of fanciful shapes, but the principle is the same. When not on a barn roof for decorative purposes, most scientific wind vanes are far more austere in design. Wind direction can be measured using many other techniques ranging from analyses of drifting clouds, smoke, or balloons to state-of-the-art hot wire, sonic, and Doppler sodar (sound radar) and lidar (light radar) systems. Aircraft use precision devices based on inertial guidance (gyroscopes) or GPS satellites to estimate the wind direction and speed by noting the differences between their indicated air speed and the actual travel of the plane across the ground.

How is the exact time of **sunrise and sunset** measured?

There are several definitions of sunrise and sunset. In the United States, sunrise occurs when the upper limb of the sun's disk first appears above the sea-level horizon; sunset is when the lower limb *last* appears above the horizon. In Great Britain it is the center of the sun's disk that marks the sunrise or sunset time. But the visible sunrise

The world's first weather satellite, TIROS I (Television Infrared Observation Satellite), was launched by the United States on 1 April 1960.

or sunset is also a bit of an illusion. Since the sun's light is refracted (bent) by a long passage through the atmosphere when close to the horizon, the sun reaches the horizon about two minutes earlier than one would calculate from simple geometric calculations. Using sensitive light-measuring instruments, dawn can be detected in the eastern sky almost an hour and a half before sunrise.

What measurements do **grave diggers** provide?

Scientists will use any source of information they can get their hands on. And a key measurement for those dealing with soil problems and hydrology is the depth of the frost in the ground. Since there are no reliable measuring tools, they go to one group of people who would know. Thus, frost penetration is routinely reported by grave diggers. Incidentally, if one digs deep enough, rarely more than several tens of feet, the temperature of the soil becomes essentially constant. The constant temperature of the deep soil is very nearly equal to the annual average air temperature above.

How do you measure **snow fall**?

Weather instruments can be pretty fancy. Air pressure is measured with a barometer, temperature with a thermometer, humidity with a hygrometer, wind speed with an

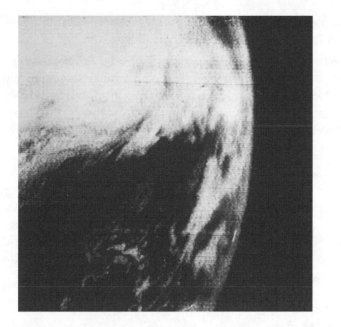

The first cloud picture taken by TIROS I in April 1960 is, by today's standards, fuzzy and difficult to interpret, but at the time it revolutionized weather forecasting.

anemometer, sunshine with a pyranometer, and snow depth—with a ruler, which proves that not everything has to be high tech. Snow remains one of the more difficult weather parameters to measure using automated systems. At the National Weather Service snow is typically measured using a ruler, taking readings at several sites, especially if there is drifting, and then computing the average depth. Sometimes, if a heated rain gauge is used, the water from the melted snow is used to estimate snow fall, assuming typically that ten inches of snow equals one inch of water conversion. Other devices have been tried to measure snow, including snow pillows, which record the weight of the snow as it accumulates. In mountainous areas, where even a yardstick is inadequate, tall poles with markers are placed so that observers can see just how far up the pole the snow has reached.

When was the **first weather satellite** launched?

On 1 April 1960, the TIROS I spacecraft was lofted from Cape Canaveral. The Television and Infrared Observation Satellite was crude by today's standards, but its TV images did reveal a high degree of organization to the Earth's weather and cloud systems that surprised many atmospheric scientists. These were not the first pictures of clouds from space, however. Cameras onboard World War II–era V2 rockets provided startling high altitude glimpses of the Earth's cloud patterns, which were an inspiration to many meteorologists and led to the design of the first dedicated weather observation system in space.

The first tropical cyclone to be detected by TIROS I was a typhoon, previously unreported, spotted on 10 April 1960, east of Australia. Since the launch of TIROS, no hurricane or typhoon has gone undetected. The improvement in the warning of severe tropical storms alone has probably paid for the investment in meteorological satellites many times over.

Have other nations launched **weather satellites**?

The Soviet Union was just behind the United States in launching a weather satellite. Geostationary satellites have been launched by Japan, the European Space Agency, and India. In September 1990 China launched its first experimental meteorological satellite (an earlier attempt failed after one week). The satellite is called *Feng Yun* which means "wind and cloud."

What is **GPS**?

GPS stands for global positioning system and refers to a series of Department of Defense satellites that were launched originally for the purpose of enabling military installations to precisely determine their location by monitoring the signals from several of the overflying GPS satellites. The system has proven so valuable that it now is used worldwide in hundreds of applications, with the potential to generate tens of billions of dollars in business revenues. GPS is being used, for instance, to track delivery trucks and trains. The extremely precise time provided by GPS is now the standard for scientific measurements that need precise timing. There are very slight distortions of the GPS signal as it passes through the atmosphere, and these distortions are actually being used by meteorologists as a probe of atmospheric structure. Hand-held GPS receivers can be purchased by hunters and boaters to prevent them from becoming lost in the wilderness, and similar receivers will become the basis of vehicle locating systems with dashboard mounted maps showing drivers their exact location.

What types of **weather satellites** are there?

There are two basic types of satellites—geostationary and polar orbiting, also known as GOES and POES. In 1946, Arthur C. Clarke (best known to the public as the author of the novel *Childhoods' End* and the short story upon which the film *2001: A Space Odyssey* was based) noted that an artificial satellite in orbit at 22,248 miles directly above the equator would orbit at the exact same velocity as the Earth rotates. Therefore, to an earthbound observer it would seem to hover in the same place in the sky. This notion eventually led to the communication satellites that now stud the Clarke Belt, as it is called. It also gave birth to the idea of the GOES satellite (Geostationary

Operational Environmental Satellite). From its point in space, a single geostationary satellite can view about 38 percent of the Earth's surface. About five GOES-type systems are currently operated by the United States, Japan, India, and the European Community. They continuously monitor the planet's weather at hourly or greater intervals. GOES satellites are the source of the cloud pictures on nightly television weather segments.

The second approach to monitoring weather via satellite is to launch a satellite into a polar orbit (or, if not passing over the poles, at least traveling at a great angle from the equator). The Polar Operational Environmental Satellites (POES) are represented by the successors to the original TIROS I: a series called Nimbus, many from Defense Meteorological Satellite Program (DMSP), NASA's Landsat series, and a host of other "birds" used to make environmental measurements.

What measurements are made by **weather satellites**?

From the beginning, satellites have had the task of recording cloud images. But there are many ways to do that. Even the first weather satellite took both visible light and infrared (heat) images. The infrared (IR) images, which can be taken in darkness, are necessary due to the simple fact that half of the planet experiences night at any given time, but weather occurs 24 hours a day. The IR sensors estimate the temperature of whatever surface is in the field of view. In the case of a cloud top, knowing the temperature profile in the underlying atmosphere allows scientists to estimate the height of the cloud.

Satellites can now perform many more tasks. Sea surface temperature is routinely mapped via satellite. Weather satellites can produce pictures with resolutions on the order of 1,000 meters (visible) and several times that in the infrared, while NASA's Landsat series can see objects only several tens of meters across. Satellites also make images in several wavelengths of the visible and infrared. This permits extremely detailed mapping and identification of what is actually on the surface—such as the type and health of a given crop or a forest. Radar sensors can determine the state of the sea, devices called sounders can approximately measure the vertical temperature of the atmosphere, and other devices are the key to mapping the amount of ozone in the stratosphere. And now, newer systems are being tested that will detect lightning from space.

Who invented **weather radar**?

No one actually invented radar for the purpose of tracking weather. Radar, an acronym standing for radio detection and ranging, was developed during World War II as a means of tracking airplanes and warships. During early experiments in the United States and Britain, strange radar echoes were noted that turned out to be unwanted reflections from rain storms. While this posed a problem for the military applications of radar, it was the basis of modern radar meteorology.

What is the Doppler effect?

Also called the Doppler shift, it was named in 1842 after Austrian physicist J. C. Doppler. The frequency of a sound, light, or microwave energy that reaches a receiver changes when the source of the energy and the receiver are in relative motion to one another. The familiar example is the whistle of an approaching train, which gets progressively higher in pitch (frequency) while it approaches and suddenly gets lower after it passes. Radars bounce microwaves off raindrops or ice crystals, and by measuring the difference between the transmitted and received frequencies, they can calculate the speed of the air in which the hydrometers (our fancy word for drops or flakes) are embedded. Conventional radars can only detect the intensities of rain or snow, while the Doppler can additionally measure the speed of the wind (though usually only where there are sufficient numbers of targets like rain drops, snow flakes, smoke, or insects). A single Doppler radar measures only the motion of the target either directly toward or away from the radar.

When was the first **operational radar system** installed in the United States?

Military surplus radars were occasionally pressed into service by what was then known as the U.S. Weather Bureau from the end of World War II. Yet it took a series of disastrous hurricanes along the U.S. East Coast in 1954 and 1955 before an operational radar designed for weather monitoring purposes was designed. The images of hurricane spiral bands swirling on experimental radar scopes as storms approached the coastlines prompted Congress to authorize a dedicated national weather radar grid. The WSR-57 radar (weather surveillance radar, 1957) was the backbone of the nation's weather radar warning system for almost 40 years. The last of the WSR-57s were finally being retired in the mid-1990s to be replaced by the more powerful NEXRAD system (Next Generation Radar). But the long bureaucratic process of approving the new system meant the '57s (as they were affectionately called), which still used ancient vacuum tube technology, had to be coddled and coaxed into working properly by dedicated technicians. Replacement parts became increasingly hard to find, and some even had to be handmade. But, in spite of problems in its later years, the WSR-57 was a wise investment that saved uncounted lives and minimized the effects of many weather disasters.

What is a **hook echo**?

Conventional radars, such as the WSR-57, could only see precipitation-sized particles. They could not "see" wind, and certainly not tornadoes. But it turns out the tornado creates a tell-tale radar echo pattern called the hook echo, so named because of its shape, which is also described as a "6" attached to a larger parent cell. The hook echo was first described by scientists from the Illinois State Water Survey when they tracked a tornadic thunderstorm near Champaign, Illinois. The swirling hook represented not the tornado itself but rather the larger low-pressure system in which it formed, the so-called tornado mesocyclone, a circulation about 10 to 100 times larger than the funnel itself. While not all tornadoes produce hook echoes and not all hook echoes produce tornadoes, it was a reliable enough signature to result in many life-saving warnings. Even in this day of modern, wind-measuring Doppler radars, the hook shaped appendage is a signature no radar operator ignores.

How has Doppler radar improved **tornado warnings**?

The NEXRAD systems, officially called the WSR-88D, which are currently being installed by the National Weather Service and the military, were designed in part to improve tornado warnings. The radars are sensitive enough to actually detect the tornado vortex itself, but only if the funnel is very large and quite close to the transmitter. Most of the time, NEXRAD is hunting for the tornado mesocyclone, the intense low pressure swirl surrounding the individual tornado that is buried within many severe thunderstorms, out of which many tornadoes are spawned. Thus, Doppler radar can greatly improve the lead times for tornado warnings and help save lives.

While even the powerful NEXRAD Dopplers can't detect every tornado, warning accuracies are typically jumping from around 30 percent to over 80 percent. The vortex can often be spotted inside the cloud before it descends to the surface, providing warning lead times of 20 minutes or more. In the ideal case, the circulation can be detected aloft many tens of minutes before the funnel actually reaches the surface. In years past the typical lead time of a tornado warning was on the order of one minute. NEXRAD is not perfect, however: many smaller tornado systems often form quickly and with little indication given by radar. But the "monster" tornadoes, with their higher wind speeds, wider paths, and longer path lengths, are detected with a very high probability. The chance of a major tornado striking a city without warning is far lower today than it was ten years ago.

Are **Doppler radars** cost effective?

The National Weather Service is well underway in completing its $4 billion modernization plan, which includes the new NEXRAD Doppler radars, hundreds of automated weather stations, and improved computers and communications. Initial estimates pre-

The powerful NEXRAD (Next Generation Radar) systems are now installed and providing Doppler radar coverage nationwide.

dict that the resultant benefits from improved local forecasts and warnings will save the U.S. economy $7 billion each year. Many more lives will be saved by improved warnings.

While billed as the front line in early tornado detection and warning, NEXRAD has many other uses. When heavy snow threatened to paralyze Washington, D.C. area traffic, the newly installed NEXRAD helped make a forecast upon which it was decided it would not be necessary to send federal workers home early. The estimated "cost avoidance" was $43 million in salaries.

The newly installed NEXRAD Doppler radar in Melbourne, Florida, some 125 miles north of Miami, was put to a long-range test on 24 August 1992. After 168 mph wind gusts destroyed the conventional radar on the roof of the National Hurricane Center (it landed in the parking lot below), the new NEXRAD was able to track Andrew's devastating path across south Florida.

What is the best **tornado detector**?

The human eye. As good as NEXRAD is, it does not detect all tornado-bearing circulations. Some twisters, especially the smaller ones, form very quickly from conditions that NEXRAD can't diagnose. Thus tornado spotters, trained by the NWS and emergency service organizations, still are the backbone of the nation's tornado warning system. When tornado watches or warnings are issued, these spotters, who are skilled at telling the difference between a real funnel and its many look-alikes, scan the skies and report immediately via radio to the local NWS office that has warning responsibility for an area. What should you, as an ordinary citizen, do if you spot a tornado? Head directly to a pre-arranged tornado shelter in your home or office. Then call 911 and describe calmly what you see, estimating the direction and distance to the twister as best you can. Use a cellular or cordless phone if possible (to avoid the lightning hazard). The police will then relay your report to the NEXRAD radar operator, who will use it to confirm their suspicions of a tornado. Do not try to call the National Weather Service itself.

Did a meteorologist sign the **Declaration of Independence**?

While the profession of meteorologist did not exist at the time, Thomas Jefferson was an avid weather watcher. He purchased his first thermometer on the 4th of July, 1776 (the stores were open because it wasn't a holiday yet). He also found time to sign the Declaration of Independence. He wrote in his diary that "it was a glorious morning, with the sun shining . . . the wind was from the southeast . . . the temperature rose to 72.5°F by the time Congress convened. . . . "

How do you measure the **amount of salt in sea water**?

Salinity is the measure of dissolved salts in water. An important parameter in many oceanographic and climate studies, the salinity of ocean water is variable. It has often been measured using electrical conductivity, or by determining the amount of chlorine or some other property. But a new technique has been developed that is far more sophisticated than dipping a sampling bucket in the water. An airborne scanning low-frequency radiometer (not a household item) has been flown on C-130 aircraft and can map ocean salinity at the rate of 100 square kilometers per hour.

Needles mechanically register the highest and lowest temperatures on this maximum-minimum thermometer.

What is the **wet bulb temperature**?

The wet bulb temperature is the lowest temperature to which air can be cooled by evaporating water into it. It is one of several measures of the moisture content of the air, in addition to dew point, relative humidity, specific humidity, mixing ratio, and water vapor density. The wet bulb is of practical significance because it can be relatively easily measured (see below). In desert areas, the wet bulb temperature is often many tens of degrees cooler than the regular temperature (also called the dry bulb temperature). This concept is the basis for many air conditioning systems used in the southwestern United States. Often called swamp coolers, these evaporative cooling systems simply pass air through a fabric-like material that is saturated, cooling the air stream down to nearly the wet bulb temperature. This type of air conditioning can cool a large building with minimal expenses: a few gallons of water a day and enough electricity to run a small motor.

What is a **psychrometer**?

A psychrometer is a device for determining the dry and wet bulb temperature of the air. Two thermometers are mounted near each other. One is a regular thermometer. The other has a wick-like covering that is saturated in distilled water. As the water evaporates, it cools the air in contact with the bulb to its wet bulb temperature. Knowing these two values, meteorologists can calculate the relative humidity of the air by either consulting tables or solving some equations.

The simplest form of this device is called a "sling psychrometer," which consists of the two thermometers mounted on a piece of metal that in turn swings rapidly in the air for several minutes to achieve the required degree of wet bulb cooling. In years past, the dexterity of meteorology students could often be measured by how few sling psychrometers they smashed into walls or tables while they made these measurements. Today automated systems exist, along with a multitude of other moisture sensors, including chemical-coated strips that change their electrical conductivity in proportion to the amount of moisture in the air. There are electronic systems that give a direct reading of the dew point—but you miss all the fun of whirling the psychrometer around in the air!

What is the Fahrenheit temperature scale?

It is antiquated, clumsy, and still in use in only one industrialized nation in the world today—the United States. The scale is set up so that a thermometer has the melting point of water at 32°F and the boiling point at 212°F. The weird benchmarks are rumored to have been selected in Paris by M. Fahrenheit, who was attempting to calibrate his new mercury thermometer. On a day when it was as cold as he thought it could ever get in Paris, he marked 0° on his mercury-filled tube. Then he took his own body temperature and called it 100°F.

What is the difference between the **Celsius and Centigrade** temperature scales?

The centigrade temperature scale was invented by the Swedish astronomer Anders Celsius in 1742. But the term "centigrade" is no longer in official use. In 1948, the Ninth General Conference on Weights and Measures decreed that the "degree centigrade" would thereafter be called the "degree Celsius" after the scientist who first proposed the 100-point scale (0°C is freezing, 100°C is the boiling point of water).

Thus, the centigrade scale and the Celsius scale are the same.

How do you convert **Celsius and Fahrenheit** temperature scales?

You are traveling in Europe one fine summer day, and you ask someone the temperature. They say it is 20°, and you want to know what that means in Fahrenheit. To convert: multiply the Celsius temperature by 1.8, or 9/5, and add 32. Twenty × 1.8 = 36, plus 32 = 68°F.

To go the other way (Fahrenheit to Celsius), subtract 32 from the temperature and multiply by 5/9. Thus 68°F is (68-32) × 5/9 = 20°C. By the way, the two scales do converge: if it's -40°, it doesn't matter which scale you are using. You're equally cold in either Celsius or Fahrenheit.

Still confused about Fahrenheit and Celsius? Remember this basic rule: Every change of 9°F is equal to 5°C. Here are a few benchmark temperatures:

-40°C	=	-40°F	Nanook stays in igloo
-20°C	=	-4°F	Chilly even for Minnesota
-10°C	=	+14°F	Need gloves and muffs
0°C	=	+32°F	Melting point of ice
+10°C	=	+50°F	Need light jacket
+20°C	=	+68°F	Room temperature, just right
+30°C	=	+86°F	Turn air conditioner on
+40°C	=	+104°F	Having a heat wave

Why can the **mercury** never drop below -40°F?

Mercury freezes solid at about -40°F. For colder temperatures, either alcohol-filled tubes or other devices, such as electrical resistance thermometers, must be used. So if the TV news says the mercury fell to -60°F in Saskatoon, you can give a knowledgeable little smile, knowing that isn't entirely accurate.

How did **amorous minks** nearly cause an international incident?

Meteorology has some pretty strange stories to tell when you dig behind the scenes. But oceanography has one of the most embarrassing ones, a revelation made by the Swedish government several years after the end of the Cold War. For decades they complained about detecting submarines from some "world power" cruising dangerously close to their coastlines. Turns out what Swedish sonar operators were tracking were the squeals and squeaks of mink—who love to frolic and mate in shallow coastal waters. No apology was offered to the Russians.

What is an **ombrometer**?

An ombrometer is the same as a micropluviometer. Those are the 25-cent words for a very sensitive rain gauge.

What is an **isobar**?

The standard joke is that it is where weatherpersons go for a cold one after they've busted another forecast. In actuality, "iso" is a prefix used in meteorology for a line on a weather chart connecting points that have the same value of a given parameter. So an isobar connects all points with the same air pressure. Drawing the isobars helps define high- and low-pressure systems. Some of the other "isos":

isotherm	=	equal air temperature
isotach	=	equal wind speed
isogon	=	equal wind direction
isallobar	=	equal change in air pressure
isohyet	=	equal rainfall
isodrosotherm	=	equal dew point temperature
isoneph	=	equal cloud cover
isopycnic	=	equal density
isoceraunic	=	equal number of thunder storm events
isobront	=	equal amount of thunder
isochasm	=	equal frequency of observing the aurora borealis
isochrone	=	equal time of some occurrence, such as a wind shift
isobathytherm	=	equal depth in water having same temperature
isopleth	=	line of equal anything

What is a **radiosonde**?

The balloon-borne radiosonde is the basic tool for measuring the winds, temperature, and moisture of the upper atmosphere. It was developed in Europe and Russia in the late 1920s and early 1930s and is still in widespread use. The small lightweight packages were designed to be powered by small batteries (often water activated).

Today's versions are called rawinsondes and are tracked by radar to provide wind speed and direction as well as the profile of temperature and moisture with height. These balloons can rise to above 99 percent of the atmosphere in some cases. Once the balloon bursts, the small white instrument packages descend back to Earth on small orange parachutes. While occasionally reported as UFOs or alien transmitters, most of the instrument packages are picked up and returned to the government so the device can be refurbished for another launch.

Rawinsonde balloons are launched at hundreds of sites around the world, even from some ships, twice daily at 0000 UTC (midnight) and 1200 UTC (noon). Their readings provide the basis of the upper air weather maps that are vital to weather forecasting.

Atmospheric scientists are forever designing new neasurement tools. This tethered balloon system was used to make temperature soundings in the lowest thousand feet of the atmosphere for air pollution studies.

How did meteorologists **probe the upper atmosphere** before radiosondes?

Long before meteorologists began using balloons with radio transmitters to broadcast reports of upper air weather conditions, kites were an important data gathering tool. The earliest such flight may have been in Scotland in the year 1749. How high can a kite be flown? Would you believe the world's record is 31,955 feet, set over Germany in 1919? But new experimental research kites might be able to go twice that high.

Manned and unmanned balloons, carrying sensors that had to be returned to Earth, were also widely used in the nineteenth century. Aside from measurements made on mountainsides, the first upper atmosphere temperature measurements were made in 1784 in France, using the newly invented hot air balloon.

Aircraft were quickly recognized as important platforms for making upper air weather observations. Today many jetliners are equipped with meteorological sensors that radio back important weather data while en route, as well as during ascent and descent.

What are some of the most unusual **oceanographic research tools** used lately?

Rubber duckies—and turtles, frogs, and beavers. Some 29,000 small plastic bathtub toys were swept overboard when a container ship crossing the Pacific met a fierce storm near the International Dateline in January 1992. As the toys started washing up on the North American coastline, they provided an excellent tool for oceanographers to test their models of ocean circulations and drift patterns.

How are **weather satellites** used in entomology?

A novel use of weather satellites is locust forecasting. By monitoring areas in the African continent that have regions of unusually lush, green vegetation, scientists can infer when conditions are favorable for the hatching of billions of locust eggs which may have remained dormant for up to 30 years waiting for the right conditions. Such locust plagues, which can reach biblical proportions, could result in famines. Early warnings help relief agencies to prepare. The locust swarms can sometimes even be seen from space. Radars are also used to track migrations of birds. Turns out migrating flocks of birds send down their own brand of precipitation that confuses the NEXRAD Doppler radar's computer: "Bird clutter" is now the new frontier of weather research. Weather radars have also been very helpful in bat research. It is not unusual in the southern United States to see expanding ring-shaped echoes emerge from the same point almost every night. These are bat swarms leaving their caves on their nightly mosquito-eating binges.

Can **environmental satellites** spot fires?

The fires during the Los Angeles riots following the Rodney King trial were large and hot—so much so that they were easily mapped by Earth resource satellites usually employed for mundane tasks such as mapping forests and crops. Actually such satellites have been mapping plumes of smoke from industrial sources for several decades. Forest fire hot spots can be tagged from space. Estimates of the size of areas burned in a conflagration can be obtained by comparing before and after pictures. Volcanic ash plumes are now routinely tracked by satellite. And the plumes of smoke from the burning of the tropical rain forests provide mute testimony to the despoiling of a vital ecosystem.

How can crickets be used to determine the temperature?

Count the number of cricket chirps in 14 seconds and then add 40. The number will be very close to the air temperature. An alternate approach is to note that they chirp 72 times per minute at 60°F. For every additional four chirps, add a degree to 60°F, and for every four chirps less, subtract a degree from 60°F.

What is **chaff**?

One of the more interesting "false echoes" one can see on a weather radar is from "chaff." Made from minute strips of aluminum-covered plastic cut to a certain length, chaff is routinely dropped in huge quantities from high-flying planes to give military radar operators some practice, and incidentally to confuse the weather radar observers since it often looks very much like precipitation.

What happens if a **weather satellite** fails?

Losing weather satellite coverage would severely hamper forecast operations, especially for marine and aviation interests, which often must operate in "data sparse" regions where satellite provides virtually the only weather information. In 1989, the GOES-6 failed, and then its replacement was destroyed in a launch vehicle accident. So, Europe lent the United States a geosynchronous weather satellite in 1992. The European METEOSAT-3 was moved from above the 0° longitude (the Greenwich, England meridian) to about 75°W (the U.S. East Coast).

How can you measure the distance to a **lightning strike**?

Count the number of seconds between the flash and the first clap of thunder. Divide the number of seconds by five, and you have the answer in miles. This calculation is based on the fact that light travels at the speed of light while sound moves at a tortoise-like 740 miles per hour, or roughly 5 seconds per mile. So if it's 10 seconds before you hear the thunder, the strike was two miles away.

If you hear the thunder, you know you can't be struck by that particular bolt of lightning. However, successive lightning strikes often jump around two or three miles from strike to strike. So if the last one was two miles away, you might be standing at ground zero for the next one.

How can you take **cloud pictures** from space at night?

One way is to use an infrared (heat sensitive) camera. And the conventional weather satellites do just that. But the Defense Meteorological Satellite, first launched in 1972, has gone one step further. It provides highly detailed visible cloud pictures over the Earth both day and night. The satellite's light detectors are so sensitive that they can spot a 100-watt light bulb from 500 miles up. Therefore, seeing clouds whenever there is moonlight is no problem.

What is the **coldest cloud top temperature** ever measured?

In 1990, a satellite's infrared heat sensor flying over Tropical Cyclone Hilda east of Australia measured the temperature of the storm's top at a record cold -152°F. The cloud top was estimated to be over 62,000 feet tall. At the base of the cloud the temperature was some 237°F warmer.

What is a **lidar**?

Laser beams are becoming an important part of meteorological measuring systems. By noting the reflections from a beam of light pointed into the atmosphere, scientists can measure the concentrations of various pollutants and, by noting the Doppler shift of the light, lidars (which are essentially light radars) can determine the velocity of the wind.

What is a **sodar**?

Sound waves can also be used in a radar-like application. Sound Detection and Ranging (sodar for short) is widely used to measure the vertical profile of wind speed and direction. Sodars emit chirps of sound (they can easily be mistaken for birds) and then scientists measure the Doppler shift of the sound waves that are reflected back by the turbulent eddies in the atmosphere, thereby computing wind speed and direction. Modern sodars can provide wind data for up to several thousand feet above the ground. They can also be used to detect inversions and other atmospheric structures.

What is the **Beaufort Scale**?

In the days before anemometers were widely used to measure wind speeds, observers would rely on quantitative rules such as the Beaufort Wind Scale. It was originally devised in 1806 by Admiral Sir F. Beaufort of the British Navy, and was based solely on human observation. For instance, if twigs and small branches were breaking off trees and it was difficult to walk into the wind, you would report a gale force wind (39–46 mph).

The following table shows the Beaufort numbers with equivalent wind speeds in miles per hour and knots, and descriptions of each level.

Beaufort number	MPH	Knots	International Description	Specifications
0	< 1	< 1	Calm	Calm; smoke rises vertically
1	1–3	1–3	Light Air	Direction of wind shown by smoke drift not by wind vanes
2	4–7	4–6	Light Breeze	Wind felt on face; leaves rustle; vanes moved by wind
3	8–12	7–10	Gentle Breeze	Leaves and small twigs in constant motion; wind extends light flag
4	13–18	11–16	Moderate	Raises dust, loose paper; small branches moved
5	19–24	17–21	Fresh	Small trees in leaf begin to sway; crested wavelets form on inland waters
6	25–31	22–27	Strong	Large branches in motion; whistling heard in telegraph wires; umbrellas used with difficulty
7	32–38	28–33	Near Gale	Whole trees in motion; inconvenience felt walking against the wind
8	39–46	34–40	Gale	Breaks twigs off trees; impedes progress
9	47–54	41–47	Strong Gale	Slight structural damage occurs
10	55–63	48–55	Storm	Trees uprooted; considerable damage occurs
11	64–72	56–63	Violent Storm	Widespread damage
12	73–82	64–71	Hurricane	Widespread damage

What is the **Fujita scale**?

Just as there is a Richter scale to gauge earthquake intensities, the "Fujita Tornado Scale," based upon damage, classifies twisters into six categories of wind speed (F0 through F5), ranging from 40 to 300+ mph estimated wind speed. The Fujita scale was devised by Professor T. Theodore Fujita of the University of Chicago. He is also called "Mr. Tornado" because he conducted much of the pioneering work on tornadoes that

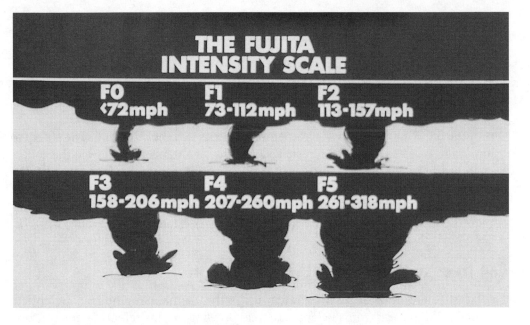

The Fujita Intensity Scale is the best tool available today for classifying tornadoes.

serves as the basis of today's rapidly improving understanding of severe storms. Since no one has ever been able to obtain wind measurements inside a tornado (except for some recent Doppler radar studies), Professor Fujita, along with Allen Pearson, the former director of the National Severe Storm Forecasting Center, devised a way to convert the degree and type of damage caused by a tornado into an estimate of the wind speeds inside the funnel. Correctly using the F-scale requires a lot of skill, but it is still the best tool available today. Almost all tornadoes for the last several decades have been classified according to the "F" rating.

Do the **colors on television radars** mean anything?

The colors on the television weathercaster's "color radar" do generally mean something. Typically there is a scale of six colors. The lightest precipitation rate, often shown as blue or light green, is less than 0.2 inches/hour. Yellow is medium heavy rain or snow. The heaviest, usually shown in bright red, can exceed 7 inches/hour of rainfall. Some of the newer Doppler radar displays use more than six colors.

What is a **trace of precipitation**?

A "trace" of rain is an amount too small to measure using a conventional rain gauge **49**

(< .01 inches) or ruler (< 0.1 inches for snow). Yet if a trace of rainfall were collected it could weigh almost 400 tons per square mile.

What **meteorological instrument** was invented by Samuel F. B. Morse?

One of the most important early weather instruments was the telegraph. The ability to communicate weather information via code nearly instantaneously allowed the collection of weather data in "real time" rather than waiting for weeks or months for the mail to arrive. By collecting and analyzing simultaneous weather reports from across a large area, advancing storm systems could be detected and monitored, and warnings issued. The first such telegraphic weather network was established by France in 1857.

Can **trees** be used as **weather instruments**?

The latest scientific weather instrument is . . . the Tasmanian pine tree. Scientists have been taking cores from ancient pine trees growing on this remote island south of Australia. The trees, which can live for well over a thousand years, provide a record of temperature conditions by examining the width of the annual growth rings. The results? Since the year 900, in Tasmania, the warmest period recorded was from 1965 to 1988. One possible explanation is that global warming is making its presence felt. In the desert southwestern United States, where very old trees can be found, cores have been used to show periods of extended droughts that severely impacted ancient Native American cultures.

Trees are also often used to make estimates of average wind speeds and directions at remote locations far from weather stations. Trees bent in one direction are excellent indicators of brisk, steady winds, which could mean a site would be excellent for wind power. On the island of Aruba in the Caribbean, the Divi Divi tree is famous for flagging the direction of the steady trade winds.

The massive earthquake that changed the course of the Mississippi River near New Madrid, Missouri, in 1812 also profoundly impacted the growth of bald cypress trees in western Tennessee as water levels rose and fell. The enormous earthquake left a clear signal in the tree ring growth patterns.

How did the end of the **Cold War** help atmospheric research?

The end of the Cold War made possible the deployment of new weapons in the fight against global environmental degradation. A stratospheric ozone mapping satellite (TOMS, the Total Ozone Mapping Spectrometer), built by the United States, was launched by a Soviet Cyclone rocket during the summer of 1991. The environmental monitoring system was blasted into orbit from the once top-secret Plesetsk Cosmodrome north of Moscow, hitching a ride with the Soviet's Meteor-3 weather satellite.

How would the **temperature of the entire planet** be measured?

Measuring the temperature of the Earth is no easy chore, but NASA's proposed Atmospheric Infrared Sounder, scheduled to fly on the first Earth Observation System Satellite and costing $145 million, will provide a global thermometer in space that monitors the temperature effects of increased greenhouse gases upon the planet's atmosphere.

What is **wind shear**?

Wind shear refers to sharp changes in wind speed and/or direction over short distances, either vertically or horizontally. Wind shear is a major hazard to aviation. It is usually associated with thunderstorms, however, dangerous wind shears sometimes occur in the vicinity of strong fronts and in mountain waves. Special Doppler radars are now being installed at major airports to help warn pilots of thunderstorm-induced wind shears. These radars will supplement the networks of anemometers that ring many larger airports.

What is a **barograph**?

A barograph is an instrument that records air pressure on a chart mounted on a rotating drum. A thermograph does the same for temperature. A hygrothermograph records both humidity and temperature. While these mechanical devices have been widely used for over a hundred years, the logging of weather data is rapidly being converted to all-digital technology, even at the level of the home weather enthusiast.

Which regions of the 48 contiguous states have the **driest and soggiest air**?

Based upon annual averages, the region of the United States with the lowest relative humidity is the interior of southern California and southern Nevada. The regions of the United States with the highest relative humidities are the coastal areas of Washington state and northern Oregon, as well as portions of western Maine.

Are weather stations being **automated**?

The United States will spend over $20 million to install up to 1700 automated weather stations across the nation. They will report weather typically every 20 minutes, rather than once per hour as scheduled at human operated stations. These stations will largely replace human airport observers, but not without some problems. No reliable way to automate snow depth measurements has been found. The optical weather type sensors sometimes get confused, for example reporting fog as volcanic ash. And operational meteorologists mourn the loss of the comments often appended to observations, such as "Tornado southwest, abandoning station!" or the remark once added by a Maine observer, "Moose on the runway."

How are **spider webs** used in atmospheric research?

Spider webs are excellent for collecting droplets of fog or cloud water. In fact, cloud physicists have used spider webs to collect samples as part of experiments in cloud water chemistry.

How are radars being used to **measure the winds**?

Meteorologists are always attempting to develop new instruments. A new technique uses radio waves transmitted vertically into the atmosphere. The device, called a Profiler, shows great promise for continuously monitoring winds to altitudes of many miles. Profilers could eventually supplant the rawinsonde weather balloon, now launched only twice daily.

How would you **measure the temperature** of an entire ocean?

Scientists monitoring possible global warming want to track the changes in ocean heat content. Taking temperature soundings over the vast Pacific is a practical impossibility. But, by measuring the speed of sound waves propagating through the ocean, which is a function of the mean temperature along the path between the sound's source and the microphone, a useful "ocean temperature" might be calculated. Proposed experiments to make periodic "booming" noises with giant underwater speakers have been put on hold, however, until the effects on marine life, especially whales, can be better understood.

THE UPPER ATMOSPHERE AND BEYOND

What are the **layers of the atmosphere**?

The bottom layer of the atmosphere, which typically extends to between 4 and 10 miles altitude, is called the troposphere. It is the layer in which most clouds and weather phenomena occur and in which the temperature usually decreases with altitude. The lowest portion of the troposphere, which varies from a few tens to thousands of feet in depth, is called the boundary layer. The next major layer is the stratosphere, a layer of increasing temperature that extends to a height of 30–35 miles. Above that we find the mesosphere, in which the temperature falls again, often reaching the coldest readings in the entire atmosphere (around -100°C) at about 60 miles. The outermost layer extending to several hundred miles up is called the thermosphere, in which the temperature (if you can call it that in the extremely thin air up there) increases to many hundreds or thousands of degrees Celsius. The boundary between the troposphere and the stratosphere is called the tropopause, with the stratopause and mesopause separating the upper layers.

How was it demonstrated that **air pressure** decreases with height?

Blaise Pascal, the famous seventeenth-century French scientific prodigy, was intrigued by the notion proposed by Evangelista Torricelli that we live at the bottom of an ocean of air that had a definite weight. So, just as water pressure increases with depth, the air pressure should decrease with height. Pascal reasoned that if one were to carry one of the newly invented barometers to the top of a mountain, the pressure should be higher **53** at the top than at the bottom. In 1648, Pascal was living in mountain-free Paris, but arranged for his brother-in-law to carry a barometer from the lowest point in the village of Clermont-Ferrand in the Auvergne region of southern France to the top of the local 4,888-foot volcanic peak, Puy de Dome. While in the village the pressure was 28.0 inches, at the mountain top it read 24.6 inches. This feat was witnessed by many local dignitaries, including some monks who were included to help convince the world that the experiment was legitimate. Today a permanent meteorological observatory is located atop Puy de Dome, and a large monument to Pascal can be found in downtown Clermont-Ferrand.

How fast does the **pressure decrease** with height?

By the time you fly to 18,000 feet, half of the atmosphere's mass is already below you. The pressure of the Earth's atmosphere drops to one percent of the average sea-level value at a height of about 31 kilometers (102,000 feet). At 370 miles, the atmosphere is a million million times thinner than at sea level.

The first manned balloon flight occurred in France in 1783 and soon became the rage, with even a horse going airborne by 1798. Yet lack of understanding of the fall in pressure with altitude proved fatal to several nineteenth-century European balloonists, then called aeronauts, who died from oxygen starvation as they floated to heights above 20,000 feet.

How was the **stratosphere** detected?

The stratosphere is a layer of the atmosphere in which temperatures increase with height. It begins typically about four to ten miles above the Earth's surface. Around the turn of this century, French meteorologist Teisserenc de Bort began launching balloons carrying temperature sensors, which showed that the troposphere's temperature did not continuously cool with height. There was, in fact, an apparent layer of warmer air at greater heights. Studies of the behavior of meteor trails in the 1920s led to the startling conclusion that some 30 miles above the Earth the temperature might sometimes be as high as 65°F. By 1930, the reason began to emerge: British scientists suggested that much of the sun's ultraviolet radiation was being absorbed in the stratosphere, in the process that generates ozone and heats the atmosphere dramatically.

What is the **thermosphere**?

The thermosphere is the fourth major layer of the atmosphere, above the troposphere, stratosphere, and mesosphere, and below the exosphere. It starts at around 50 miles altitude and is characterized by temperatures that get warmer with height.

Does it take longer to cook linguine in Denver than in Rome?

Yes, but this has nothing to do with the competence of Colorado chefs. Air pressure affects the boiling point of water, with boiling occurring at lower temperatures the higher one goes. The boiling point, which is 212°F at sea level, drops 1.8°F with each 1,000-foot increase in altitude. In fact, in high altitude cities such as Taos, New Mexico, Quito, Equador, and Mexico City, it takes much longer to cook pasta, potatoes, and other culinary fare: the water may boil faster, but because boiling occurs at a lower temperature, it takes longer to cook the food in question. Remember, if you want to use boiling water to calibrate a thermometer using the sea level boiling point of 212°F, you have to account for the pressure effect.

What is the **ionosphere**?

The ionosphere consists of layers of the upper atmosphere that have enough ions and free electrons to affect the transmission of radio waves. It begins at a level of 18 to 28 miles above the surface. These ions and electrons reflect radio waves back to Earth. If they didn't, commercial AM radio would be impossible for distances over about 60 miles because of the curvature of the Earth. The several layers of the ionosphere are called the D, E, F1, and F2 layers. Radio waves of high enough frequency will not bounce off the ionosphere. Thus, FM radio and television waves just go through the atmosphere and on into outer space.

How was the **ionosphere discovered**?

In the late 1800s scientists speculated that there might be electrical currents flowing somewhere in a conducting layer high in the upper atmosphere. In 1901, Italian radio pioneer Guglielmo Marconi succeeded in receiving in Newfoundland a radio signal transmitted from Cornwall, England—some 2,000 miles away. Two scientists named Kennelley and Heaviside resurrected the idea of a conducting layer through which radio waves of certain frequencies could not penetrate; they reasoned that the waves were thus bounced along between the Earth's surface and the ionosphere for considerable distances.

How **high is the sky**? Where does the atmosphere end?

There is really no exact answer to where one enters outer space. There are even minute traces of atmospheric gases at space shuttle orbital altitudes. By some estimates, interplanetary space begins where the Earth's magnetic field is no longer dominant, which is tens of thousands of miles from the planet. Near sea level there are about 20 billion billion molecules of air in a cubic centimeter of air. By the time one reaches above 350 miles, the number can still be counted in the millions. At sea level an air molecule can be expected to move only seven millionths of a centimeter before colliding with its neighbor. At 370 miles, the distance between molecules, called the "mean free path" has increased to about 10 kilometers (6 miles)! Another way of looking at this would be to note that an air molecule at sea level has about seven billion collisions with fellow air molecules per second, whereas at the fringe of space, a molecule can get downright lonely, only bumping into a neighbor about once a minute.

What is **atmospheric escape**?

At the very upper fringes of the detectable atmosphere, in the exosphere, air molecules are so far apart they can accelerate to very high speeds due to the lack of collisions with other molecules. Some actually can reach orbital or even escape velocity and drift off into space, just like a rocket to the moon. Only hydrogen and helium seem to be escaping in any significant amounts, and those are so small as to suggest our atmosphere will not be depleted anytime soon. However, it is believed that Mars, Mercury, and possibly our moon once had atmospheres that, because of the weaker gravity on these bodies, escaped out into space.

Where is the **coldest place in the troposphere**?

Over the equator. The troposphere is defined as a layer in which the temperature generally falls with height. The tropopause is deepest over equatorial regions, sometimes reaching as high as 10 or 11 miles, whereas it might be as shallow as four or five miles in polar regions. By the time one climbs from the steamy tropical jungles to the tropopause (the layer separating the troposphere and the stratosphere), readings can be as cold as -110°F. At higher levels, near the top of the mesosphere around 50 miles above the surface, temperatures as low as -225°F can be found.

Where is the **hottest place in the atmosphere**?

The very hottest temperatures in the Earth's atmosphere occur not over some desert sands, but high in the thermosphere, the layer above the mesosphere. At this altitude, about 180 miles above the surface, the extremely thin air reaches temperatures well above 3600°F. At that height, the concept of temperature becomes more of a scientific abstraction.

What is the ozone hole?

While most of the planet's ozone is found in the stratosphere, it is not evenly distributed there. It tends to be concentrated in the polar stratosphere. However, satellite monitoring since the late 1970s has noted an ever larger "hole" in the stratospheric polar ozone during certain seasons. It is believed that this hole stems from the certain class of chemicals known as chlorofluorocarbons (CFCs), which have been widely used in industrial and consumer products for decades. Ranging from deodorant propellants to air conditioning, the CFCs sent into the atmosphere have proven to endure for a very long time. As they drift into the upper atmosphere, they demonstrate a tremendous appetite for ozone, depleting it significantly. This process begins to change the entire chemical makeup of a very sensitive part of the atmosphere, causing more ultraviolet radiation to reach the Earth's surface. The decreased ozone levels, while most pronounced over the poles, are also showing up over the populated mid-latitudes. CFC use is rapidly being reduced by worldwide treaty agreements, and there are some signs that the rate of growth of the ozone hole has slowed down. Some estimates say it will take at least 50 years to restore the upper atmosphere to its original chemical state.

What is **ozone**?

Ozone is a form of oxygen. But while typical oxygen molecules have two atoms of oxygen, the ozone molecule has three. The gas is poisonous if inhaled by humans in significant quantities because of its extremely highly oxidizing properties. It can be formed by electrical sparks and also by complex photochemical reactions in the atmosphere. The ultraviolet rays of the sun acting upon oxygen in the presence of other molecules can lead to complex reactions that produce ozone.

What is the difference between **"bad"** ozone and **"good"** ozone?

From reading the daily press one might think that there are two types of ozone, **57** "good" and "bad." The bad ozone is found in photochemical air pollution—Los Angeles smog, for example. This ozone is formed by sunlight causing reactions with automobile and industrial pollutants. The federal air quality standard states that ozone values should not be greater than 120 parts per billion. Ozone can harm the respiratory system, damage materials such as rubber, and significantly reduce agricultural production. While normal amounts of ozone exist throughout the lower atmosphere, the amount found around large cities can be harmful.

The highest concentrations of ozone in the atmosphere are found about 12 to 20 miles above the Earth's surface. This is the "good" ozone, without which humans would all be dead, or more accurately, would never have evolved in the first place. Ozone is formed by atmospheric absorption of deadly ultraviolet (UV) radiation from the sun. Without the protection the ozone layer offers us from UV radiation, we would be blinded and sunburned into extinction. Even the very small amount of UV that does get through can cause a sunburn and, more seriously, the deadly skin cancer called melanoma.

The role of ultraviolet radiation in producing stratospheric ozone was first proposed by Professor Sydney Chapman, in 1930. As important as its role is, ozone only constitutes .00001 percent of the atmosphere. If it were compressed into a layer at sea-level pressure and room temperature, it would only be one-tenth of an inch thick.

How was the **ozone hole** discovered?

Sensors onboard the Nimbus 7 polar orbiting satellite, launched in 1979, showed the rapid yearly depletion of the ozone in the winter south polar stratosphere. Also, a worldwide network of about one hundred "Dobson spectrophotometers" peer upwards from the planet's surface measuring the total ozone in the atmosphere overhead.

What is the **solar wind**?

There is a lot going on in the "vast emptiness of space." It's not actually so empty; in fact, there is even "wind" in space. The vast outpouring of high-energy particles, mostly protons and electrons, along with ions of many elements, travel outward from the sun at speeds ranging from 200 to 600 miles per second. This "wind," which is really the outward expansion of the sun's atmosphere, streams past the Earth (and all planets) very much like a wind in the troposphere.

What is a **comet's tail** made from?

Comets are generally thought to be large chunks of rock and ice. These visitors swing by the Earth from time to time on huge elliptical orbits that take them way beyond the edge of the solar system. When nearing the sun, the heat begins vaporizing the comet's surface, producing a long tail. Just like smoke from a ship, the comet tail blows in the solar wind, always pointing away from the sun. When comet Hyakutake

The northern lights—aurora borealis—wave in the cold night sky high over Minnesota.

slipped past the Earth in March 1996, it produced a beautiful tail. It also surprised comet researchers, who detected both x-rays and the chemical species ethane and methane.

What are the **Van Allen belts**?

On 31 January 1958, the first U.S. satellite, Explorer I, weighing only 18 pounds, detected the radiation belts that are now known to surround the Earth. Two major zones have been discovered, with the Earth nestled in the hole of these giant bagel-shaped regions of high-energy particles. The first zone extends between 600 and 3,000 miles out, with the second from 6,000 to 40,000 miles. By trapping energetic radiation, these zones, or belts, shield the planet's inhabitants from being assaulted by high energy particles from the sun and beyond. When gusts of the solar wind do break through, not only do the auroras light up, but radio communications, magnetic compasses, and electric power grids are disrupted.

What is the **aurora**?

Most auroras (northern lights), taking place at heights between 50 and 100 miles above the Earth, occur in response to energetic particles from a solar storm, which cause the gases of the thin upper atmosphere to glow. The aurora can last anywhere between a few minutes and several hours. They can be so bright as to rival a full moon

59

and may cast shadows or allow one to read a newspaper. Aurora are most common in polar regions, with a doughnut-like ring of high occurrences centered over the geomagnetic poles. The various colors, of which green and red predominate, are the results of various light emissions from oxygen and nitrogen gases being energized by the solar particles. The aurora can take a wide variety of shapes including sheets, arcs, and waving curtains.

On a few extremely rare occasions there have been credible reports of the northern lights actually reaching to the Earth's surface. People were actually enveloped in the strange light and some heard crackling noises.

What is the best time to view the **northern lights**?

The best time for viewing the northern lights is not necessarily winter, as many believe. The longer nights and often clear skies of winter increase one's chances, but the key factor is the occurrence of a major solar "storm" a few days beforehand. Once a giant sun spot rotates into position to "spray" the Earth's upper atmosphere with high-energy particles, then the odds for seeing the northern (and southern) lights increase significantly. The ejected solar particles are channeled by the Earth's magnetic field into a ring that circles both the north and south geomagnetic poles. If you live below 30° latitude, you have little chance of seeing an aurora. In Boston, you might see some auroral activity on one out of twenty clear nights. Maximum activity in the Northern Hemisphere occurs from just north of Scandinavia to south of Greenland through Hudson Bay and on into Alaska.

Are there **southern lights**?

Popularly called the northern lights, the aurora borealis rings the Earth's north geomagnetic pole. But the same process occurs in the Southern Hemisphere, so there are southern lights, which are more properly called aurora australis.

Can you see the **aurora** during the daytime?

Only if you have ultraviolet or X-ray eyes. But the new Polar satellite has in fact been equipped with just such "eyes" and has been taking pictures of the entire auroral ring on both the day and night side of the Earth. The system is able to detect the auroral light by blocking out the light from the sun. The aurora is actually sometimes "brighter" on the day side than the night side.

Are there **dust storms** in space?

You might expect dust storms in the Sahara or the Gobi desert. But outer space? The

Galileo space probe, on its way to measure conditions around Jupiter in 1995, ran into the densest cloud of dust ever recorded in interplanetary space. Instead of being struck by one dust particle every three days—the average—the spacecraft plowed through 20,000 dust particles each day.

What are **cosmic rays**?

Cosmic rays are naturally occurring energetic particles, mainly ions, that bombard the Earth from outer space. Each second some one billion billion such particles with energies greater than one million electron volts hit the top of our atmosphere. Thereafter they produce showers of secondary cosmic ray particles below 30 miles.

Have **other planets** been discovered outside our own solar system?

Three down, several billion to go. Scientists have long theorized that planets should be orbiting stars other than our sun. With the discovery of at least three planets circling a pulsar with the catchy name of PSR B1257+12 some 1500 light years from Earth, we have identified some of them. But there may well be billions more waiting to be discovered.

Two Swiss astronomers reported evidence of planets having about the same mass as Jupiter orbiting a star only 42 light years away. If inhabited, the planet's residents are likely to wear asbestos clothing since the orbit around the parent star, 51 Pegasi, is only one-twentieth the Earth-sun distance.

Astronomers are fairly certain that there are at least two planets, each about three times the mass of Earth, orbiting a certain pulsar located in the constellation Virgo. They are too far away (1300 light years) to be seen, but their gravitation fields appear to be the cause of changes in the star's pulsations.

Is **outer space** empty?

Hardly. And some of the things we are finding out there are quite curious. For instance, researchers have found a stellar cloud of ordinary vinegar some 25,000 light years from Earth. Radio astronomers from the University of Illinois detected traces of acetic acid (vinegar) in a cloud of gas and dust named Sagittarius B2 North. Acetic acid may be one of the first steps in building the chemical needed for life to exist. If it mixes with ammonia you might get glycine, the simplest biologically important amino acid.

British astronomers have detected a huge cloud of alcohol in deep space, some ten thousand light years away. It's enough alcohol to fill four trillion trillion beer glasses.

What is the difference between a **meteor and a meteorite**?

They are essentially the same thing. A meteor burns up while traveling through the atmosphere. If enough of the mass survives its fiery traverse, it slams into the Earth and is henceforward called a meteorite. Incidentally, those who study meteorites are not meteorologists. Instead they are called meteoriticists.

Is there **salt in the stratosphere**?

Sodium is a component of table salt, and scientists were surprised to find quantities of it floating around some 85 km above the Earth's surface (actually near the top of the mesosphere) over Illinois. The likely source is meteors.

When did the **moon first visit the Earth**?

Humans first set foot on the moon in July 1969. Yet long before that, many people held pieces of the moon in their hands. In eons past, giant meteor impacts onto the moon have "splashed" debris upwards with such force that some of the fragments of lunar crust actually reached the Earth. These meteorite-like fragments, technically called tektites, can be found around the world, but are particularly easy to find in the layers of ice and snow of the Antarctic glaciers.

Have **Martians visited the Earth**?

Yes, but they were in the form of bacteria rather than little green men. Pieces of the red planet have arrived on Earth with some regularity. At least 12 meteorite fragments, which have been identified as originating on Mars, have been found on Earth. One of these was recently discovered to contain what may be fossilized bacteria that originated on Mars over three billion years ago. Scientists have theorized that it was perhaps ejected due to an asteroid or volcanic eruption. Such extraterritorial rocks are routinely found within the ice caps of Antarctica.

How many **moons** can a planet have?

Earth has one moon. But Saturn has at least 18, 2 of which were recently discovered by the Hubbell Space Telescope.

What else is in the **Earth's orbit** besides satellites?

Don't you just hate it when a rock kicked up from a truck driving in front of you dings your windshield? Could be worse, like when the space shuttle, traveling at 17,500 mph, hits a micrometeorite. In 1995, at least two space shuttle windows were pitted by close encounters of the debris kind. The U.S. Space Command estimates that aside from natural meteorites there are at least 7,800 pieces of orbiting "stuff" larger than baseballs sharing space with shuttle astronauts.

What is space junk?

The very word "space" connotes a vast, empty void. However, as the result of almost 40 years of missile firings, satellite launches, and orbital and interplanetary missions, the space around our planet is filled with junk. Debris from more than 120 satellite and rocket explosions is just part of the problem. The United States Air Force tracks the orbits of 8,000 pieces of space junk, but there are thousands more. Some are only the size of paint chips, but that is big enough to cause problems if they hit another vehicle going 22,400 mph.

Where do **meteors burn up** in the atmosphere?

The fastest meteorites slam into the Earth's atmosphere at over 40 miles per second. They often are completely burned up on their way down by the time they reach 75 to 35 miles altitude. While most meteors are quite small, many that are several meters in diameter continuously bombard the Earth's atmosphere. Almost all burn in spectacular fireball displays, releasing the energy equivalent to a small atomic bomb. Based upon reports from United States defense satellites, several hundred such events have occurred since 1975.

When did science finally accept the **existence of meteorites**?

Scientists by nature tend to be skeptical. But sometimes they tend to ignore some very good evidence, such as occasional reports from farmers about hot rocks falling from the sky. Ridiculous, said the French Academy, everyone knows that there are no rocks in the sky. By 1865, however, they finally came around to agree with the farmers about the existence of meteorites.

What are **meteor showers**?

Just as there are thundershowers and thunderstorms, there are also meteor showers and meteor storms. These occur rarely, only three times in the twentieth century, and result in not dozens but hundreds or even thousands of "shooting stars" per hour. The top three meteor storms in the United States were in October 1933, October 1946, and November 1966. In the last case, the Kit Peak Observatory in Arizona reported as many as 2,400 meteors per minute!

The best time to see a meteor? While one may flash through the sky at any time, the annual Perseid meteor shower in mid-August is the "Olympics of shooting stars." As many as 100 hunks of space debris slam into the atmosphere at up to 135,000 miles per hour, burning up as they go. They seem to emanate from the constellation Perseus (thus their name) but are actually debris from the passage of the comet Swift-Tuttle.

What are the Leonids?

Astronomers suspect that on 17 or 18 November of 1997, 1998, or 1999, a great meteor storm could erupt out of the constellation Leo during the early morning hours and be especially visible over the western United States. Usually the Leonid meteor shower is a pretty tame affair, but when the Earth passes close to the parent comet's tail, spectacular meteor storms can result. In 1833, the light from millions of "shooting stars" woke up and terrified Americans all across the nation. As many as 15,000 meteors per hour have been recorded in the past.

What was comet Shoemaker-Levy 9?

In 1994, Earth's astronomers were treated to the spectacle of the comet Shoemaker-Levy 9 striking the planet Jupiter. The several fragments of the comet made for enormous "dents" in the planet's dense cloud cover and ejected huge fountains of matter into space that were visible even from earthbound telescopes. One of the more intriguing results of the impact of the 20 cometary fragments that bombarded Jupiter was that somehow the planet's polar auroral displays were triggered.

Such an astronomical traffic accident was not without recent precedent. In 1979, a comet was photographed by U.S. satellites plunging into the sun. The results were unspectacular compared to the Jupiter bombardment.

What is the airglow layer?

The night sky is not pitch black. Take away the moon, the stars, and city lights, and there is still light. Where does it come from? It is called airglow, a faint, mostly green glow that comes from exciting atomic oxygen atoms some 60 or more miles above the Earth. Sometimes called the "permanent aurora," airglow is there most of the time. There are also other colors resulting from the excitation of other gaseous molecules and atoms. The amount of light is slight, equivalent to about one candle illuminating a football stadium.

On rare occasions one can actually detect structure in the airglow with the naked eye. If atmospheric waves are passing through the mesosphere, they can create alternately dimmer and brighter bands of green light, which are perceptible to the dark-adapted eye. Viewed with low light television from the space shuttle, the airglow layer looks like a thin bright band encircling the Earth.

Who discovered **Halley's comet**?

This is one of those trick questions. Sir Edmond Halley did not discover the comet now named after him. Chinese records describe the comet as far back as 1059 B.C.E. Pope Calixtus III called it "an agent of the Devil" in 1456. But Sir Edmond noted the repetitious nature of the big comet's appearances and predicted that the comet of 1682 would return in 1758. When it did, he became famous, albeit posthumously.

Has anyone ever been **killed by a meteorite**?

The answer is yes, and while once thought relatively rare, evidence is accumulating that suggests such encounters are more commonplace than previously believed. A University of Arizona professor has documented more than 130 meteorite falls causing death, injury, and damage, and it's no wonder: some meteorites hit the Earth traveling upwards of 160,000 miles per hour. Professor John Lewis recounts a tale of a monk killed in Italy by a meteorite in the 1600s. In 1490, in Shasni Province, China, "tens of thousands" of people were reported killed by a rain of "stones" the size of large eggs that pelted a town. Such blizzards of stones are real. In 1912, some 14,000 of them blasted Holbrook, Arizona. More than 100,000 were reported to have slammed into farm fields in Poland in January 1868. Fortunately, it was winter and no workers were present, otherwise a large number of people may have been killed.

The Draconids meteor shower, which occurs every year in October, was especially spectacular in 1992. Hundreds of shooting stars were seen, including some huge fireballs blazing across the sky. But a woman in upstate New York got a down-to-Earth visitation—a 30-pound meteorite that bounced off the hood of her car.

There's an old song, "Stars Fell on Alabama." But after an event in 1954, it should be revised to "Shooting Stars Fell on Alabama." In one of the few known cases in the United States of a person being struck by a meteorite, a woman was injured after a small meteorite crashed through her roof and ricocheted off a chair to strike her in the arm. There are unconfirmed reports from 1879 of an Ohio farmer actually being killed by a meteorite while he was out plowing.

What happened near the **Tunguska River** in Siberia in 1908?

Riders on the Trans-Siberian railroad were stunned on 30 June 1908, when a fireball as bright as the sun raced across the sky. After apparently exploding in the air above the ground near the Tunguska River, it flattened forests for over one thousand square miles. The fireball was seen in the sunlit sky for almost 1,000 miles, it was heard over 600 miles away, and people 30 miles away were injured by shock waves. Yet, a blast crater was never found. The best explanation of the phenomenon is that a small chunk of an asteroid or comet vaporized in the lower atmosphere.

What do **asteroids** look like?

Asteroids are too small and too far away to show up as anything more than bright specks in telescopes. NASA's Galileo spacecraft obtained the first close-up images ever of an asteroid, and these pictures of asteroid Gaspra caused gasps. Looking almost like a typical comic book fantasy, the asteroid was vaguely potato shaped, with dimensions of 12 x 7 x 7 miles. It was pockmarked with craters 300 meters across and larger. It had a smoother and more "sandblasted" looking surface than scientists expected. In February 1996, NASA launched the Near Earth Asteroid Rendezvous (NEAR) spacecraft that will spend nine months in 1999 studying the surface of a huge asteroid named 33 Eros. Astronomers have long guessed that some of these wandering pieces of space trash are made of metal and iron while others might be more like rocky chunks of boulders.

Have **asteroids** ever come close to **hitting the Earth**?

In June 1996, an asteroid with the prosaic name of 1996JA1 whizzed past Earth, missing by just 280,000 miles—or roughly the same distance away as the moon. This chunk of space rock was one third of a mile across, and may be the largest hunk ever observed to pass that close. But this wasn't just some freak occurrence. Just how close can an asteroid come to the Earth without astronomers knowing about it in advance? In June 1993, the answer was 90,000 miles. A 30-foot-wide chunk of space junk whizzed by several times closer than the moon and wasn't detected until it had already passed. On 7 December 1992, Asteroid "Toutatis" whizzed past the Earth. The two-mile-wide hunk missed our home planet only by about ten Earth-moon distances. And in 2004, it will swing by again, this time coming within four Earth-moon distances.

What are the chances of the Earth being **struck by an asteroid**?

There may be a lot more asteroids and comets out there than we think, and the chance of getting stoned on an astronomical scale is not small. Near Earth Objects (NEOs) have yet to be fully cataloged. There are at least 100 NEOs that are more than a half-mile wide, while some estimate that 20 times that number may be lurking in the void of space. And there are probably a few hundred thousand smaller chunks (down to boulder size) that have Earth-crossing orbits. The Tunguska asteroid was estimated to be about 150 feet; some scientists believe that a chunk this size hits the Earth every

300 years, with the likelihood of it hitting a populated area being about ten times that. Until there is a complete survey of all the NEOs out there, any risk assessment of the probability of the Earth taking a direct hit is still conjecture. But current thinking is that once in one thousand years a chunk more than 300 feet across will hit the planet, and NEOs that span two-thirds of a mile may impact once every 300,000. There is about one chance in 5,000 that in our lifetime an asteroid could slam into Earth that would be large enough to end human civilization.

In 1989, astronomers discovered a small 250-meter-wide asteroid *after* it whizzed passed the Earth, missing by only about 500,000 miles. If it had struck, going over 40,000 miles per hour, it would have made the largest H-bomb look like a firecracker.

What produces the sun's light?

The sun's light is produced by the same basic thermonuclear process that lies at the heart of the Hydrogen bomb—atomic fusion. The sun is composed of 73 percent hydrogen, 25 percent helium, and much smaller amounts of most other elements. The continuous fusion of hydrogen into helium going on in the center of the sun keeps it cooking at 27,000,000°F. The surface of the sun, also known as the photosphere, is comparatively cool, only 10,000°F. The sun's lower atmosphere, the chromosphere, is a few thousand miles thick, and at its base is only 7,800°F. But the temperature rises with altitude to the corona, the sun's outer atmospheric layer, which has a temperature of around 1.8 million°F. The sun has been cooking away for about 4.5 billion years. It is estimated that it has another 5 billion years to go. As it exhausts its fuel, it will cool, and expand into a red giant star. The outer surface will probably expand beyond the orbit of Earth.

How big is the sun?

Humans are probably a bit disappointed to find out that the sun is a very ordinary star, falling into the yellow dwarf category. Its vital statistics are still impressive, however. The sun is 868,000 miles across and weighs in at about 4,000,000,000,000,000, 000,000,000,000,000 pounds. The sun's mass is 334,000 times that of Earth and it is about 100 times the diameter of our planet. The sun rotates on an axis much like that of the Earth. The rotation rate is about once every 26 Earth days near its equator, but once every 34 days at its poles. This disparity suggests that the sun is definitely not a solid object.

What is the **solar spectrum**?

You see the solar spectrum, or at least part of it, whenever you see a rainbow. If you pass a sunbeam through a prism, it will generate all the colors that comprise the white light of the sun, from violet to red. If you then put the bulb of a thermometer in this beam, you will find the temperature goes up when moved from violet through red. However, it also goes up if you hold it beyond the red portion of the spectrum—in the infrared. The sun gives off vast amounts of invisible (to the human eye) infrared radiation. Only about 40 percent of the sun's radiation is emitted in the visible, with 50 percent at longer wavelengths, and the remainder below the violet: the ultraviolet.

What are **sunspots**?

Sunspots are sometimes described as storms on the sun. They are relatively dark splotches on the surface of the solar disk that usually occur in pairs. They were first noted by Chinese scientists as early as 28 B.C.E., though they thought the spots were giant birds flying in front of the sun. In 1610, Galileo noted sunspots with his telescope. He observed they often appeared on the left side of the sun, then moved to the middle, and finally disappeared off the right edge 13 or 14 days later. He realized this implied the sun was rotating.

Some sun spots can be more than 30,000 miles across. On rare occasions they can be so large as to be seen with the naked eye (although looking directly at the sun is something one should never do). Sunspots have lifetimes ranging from several days to several months. They are magnetic anomalies in the photosphere causing cooler regions that appear to be darker than the surrounding surface. The typical sunspot has temperatures around 6,700°F.

Do **sunspots** occur in a **routine cycle**?

Yes. There is a very well-known 11-year rise and fall in the number of sunspots. The cycle is rather irregular, however. In fact, between 1645 and 1715, in what is called the Maunder minimum, there were virtually no sunspots. During that period the weather in Europe was notably cooler. The debate as to whether and how sunspots can influence the Earth's weather still rages.

Do sunspots **influence weather**?

Yes, somehow. But no one is really sure how. There have been several studies trying to link cycles in weather to 11-year sunspot cycles. Some have been rather convincing, statistically, but without a physical mechanism to explain the processes, scientists are a bit reluctant to embrace the concept with open arms. Part of the problem is that in terms of pure energy, sunspots affect the sun's energy output by far less than one percent. This hardly seems enough to cause any major influences on weather, but part of the answer may have something to do with which wavelengths are affected by sunspots. By changing the heating rates of the upper atmosphere, some linkage mechanism yet to be uncovered may influence weather patterns at lower altitudes.

How did the sun make the **lights go out in Canada**?

We think of the sun as a source of light. Yet it was the sun, or more precisely, intense storms in the sun's atmosphere, that caused a nearly unprecedented geomagnetic storm in March 1989 that blacked out electricity to over six million Canadians. The intense magnetic fields induced current overloads in the transmission system, blowing fuses and breakers.

Is there **ice on Mercury**?

In one of astronomy's stranger recent findings, what may be water ice has been detected in the constantly shadowed polar regions of Mercury. Mercury is the planet closest to the sun, where equatorial temperatures can reach 800°F. So it seems hard to believe that ice could be found there, but the data have been found to be reliable.

What is the **mass of the universe?**

Scientists have been unable to determine the mass of the 90 percent of the universe that is unilluminated and therefore can be detected only by its gravitational effects. The "dark matter" is almost certainly out there. Attempts have been made to use the bending of light from visible stars as the light passes near the high gravity field of huge chunks of dark matter to locate at least some of the missing 90 percent.

What is **Earth shine?**

The moon's light is mostly directly reflected sunlight, but not entirely. Sometimes a faint illumination of the dark part of the moon's disk can be noted. This illumination is Earth shine, so called because it is produced by sunlight striking the "dark" side of the moon after first being reflected from the Earth's surface, clouds, and atmosphere.

Can you actually hear sounds from **northern lights and meteors?**

There have been enough reports of people "hearing" auroras and meteors on some occasions that it could be possible. But how this might happen is wide open to debate. The atmosphere is so thin at aurora and meteor altitudes that sound waves could not be formed. Possible candidates are ultra low frequency sound waves or radio waves that somehow can be sensed by the brain.

Are the **constellations** the same everywhere in the universe?

No. The bright stars constituting the various constellations are not neighbors in space. While appearing to be part of some pattern, they are located at vastly different distances from the Earth. Thus the constellations would not look the same from different parts of the universe.

Which planet is **furthest from the sun**?

Due to its highly elliptical orbit, Pluto, which typically occupies the position furthest from the sun, has temporarily swung inside the more circular orbit of the 8th planet, leaving Neptune the furthest away, at least through 1998.

Pluto is still a long way from the sun; in fact, it takes 25 hours to collect the same amount of the sun's energy that the Earth receives in one minute.

What are **nacreous clouds**?

The stratosphere is supposed to be cloud free, but there are rare appearances of clouds in the layer 12 to 20 miles above the surface. The clouds, which in form sometimes resemble cirrus or altocumulus lenticularis, show very strong irisation, that is, they have almost all the colors of the spectrum. Thus they are sometimes called "mother-of-pearl clouds." They appear only in the two hours after sunset or before sunrise. They are generally seen during winter in Scotland, Scandinavia, and Alaska, and only rarely at lower latitudes.

What are **noctilucent clouds**?

The highest clouds found in the atmosphere are called noctilucent clouds, and are found at heights around 47 to 56 miles, where temperatures hover around -200°F. They resemble cirrus clouds but have a bluish or silvery color, sometimes sporting a red-orange coloration. They are very rare, and generally can only be seen at twilight, especially when the sun is 5 to 13 degrees below the horizon. They are only observed in the summer months in both hemispheres and only between 50–75 degrees north and 40–60 degrees south. These clouds often move quite rapidly, sometimes well over 100 miles per hour.

How much "space dust" reaches the surface of Earth?

A lot more than you might think. The Earth has been slowly gaining weight: by some estimates over six tons of microscopic meteoric dust and debris sinter down into the lower atmosphere every day.

Do the oceans come from outer space?

The question is not quite as weird as it sounds. Most scientists believe that much of the water in the Earth's atmosphere came from outgassing from the crust as the planet cooled after its formation. According to one theory, however, much of our water is the result of massive bombardments by small icy comets several billion years ago.

Comet West.

CLOUDS

How did **clouds** get their **names**?

Cloud classifications were devised early in the nineteenth century by a London apothecary named Luke Howard and have been in use ever since. During the winter of 1802–1803, Howard addressed a learned group in London called the Askesian Society with his proposal to classify the myriad cloud forms into a rational system with suitable names to facilitate their observation and understanding. French naturalist Jean Lamarck had introduced a cloud typing scheme in 1802, but it was not widely adopted.

The names selected by Howard were derived from Latin. He classified clouds into three main categories. Cirrus, from the Latin for "lock of hair," were the wispy, high-level formations. The lumpy cloud masses found closer to the surface were named cumulus, after the word for "heap" or "pile." Clouds of large horizontal extent were called stratus, from the word meaning "layers." By combining other terms, cloud combinations could be described. The Latin word for "shower" was nimbus. So the cumulonimbus became the technical name for the thunderstorm. Extensive rainy cloud layers became nimbostratus. The nomenclature has been slightly modified, and all sorts of minor additions have been made by meteorologists ever since, but the basic system introduced by Luke Howard has stood the test of time and is used worldwide.

What are the **ten cloud types**?

Based on Luke Howard's scheme, the ten basic types of clouds were formally recognized by the World Meteorological Organization and published in the International Cloud Atlas (1956). They are often classified by their typical altitude of occurrence:

The marine stratus layer that blankets much of the West Coast
during spring and summer nestles up against the mountainous coastline north of Los Angeles.

High clouds
Cirrus
Cirrocumulus
Cirrostratus

Middle clouds
Altocumulus
Altostratus

Low clouds
Cumulus
Stratocumulus
Nimbostratus
Stratus

Clouds of great vertical development
Cumulonimbus

Clouds may look very "solid" from the outside. But the amount of condensed water is actually quite small, typically on the order of several grams per cubic yard. The various shapes the clouds acquire are often indicative of the internal and external mechanisms that formed them.

Low-level cumulus clouds and high icy cirrus clouds drift over Chicago on a warm summer day.

What are **the high clouds**?

The highest clouds in the atmosphere are cirrus clouds—icy clouds typically located at altitudes between 16,500 and 50,000 feet. They are always composed of ice crystals. The three "cirriform" cloud types are cirrus, cirrocumulus, and cirrostratus.

Cirrus are composed of detached elements in the form of white, delicate filaments, patches, or narrow bands. They often appear fibrous or silky. When cirrus-level clouds evolve into two small roundish elements, they can be termed cirrocumulus. Large patches or layers of relatively uniform cirrus-level clouds become cirrostratus. Since they are composed of ice crystals, cirrostratus often produce solar or lunar haloes, "rings" around the sun or moon. This feature distinguishes cirrostratus from its lower cousin, the altostratus, which is composed of water droplets. While the ice crystals in cirriform clouds often grow to precipitation size and stream downwards in long wispy tails, they always evaporate before reaching the ground. Cirriform clouds never produce precipitation at the surface, but they often herald advancing low-pressure systems with their rain and snow systems.

As shown above, there are ten main families or genera of clouds. Beyond that meteorologists have developed species (peculiarities of shape and internal differences), varieties (special characteristics of arrangement and transparency of clouds), and supplementary features and accessory clouds, including mother-clouds, the source of other cloud forms. So if you *really* know your cirrus, you would know its many

Cumulus clouds fill the sky over Milwaukee, Wisconsin.

species, varieties, and supplemental features including: cirrus, cirrus densus, cirrus duplicatus, cirrus filosus, cirrus firbratus, cirrus floccus, cirrus uncinus, cirrus vertebratus, cirrus intortus, cirrus nothus, cirrus radiatus, cirrus spissatus, cirrus cirrocumulogenitus, and cirrus cumulonimbogenitus.

What are the **middle level clouds**?

The middle level clouds, generally with bases more than 6,500 feet above ground, are altostratus and altocumulus.

Altostratus are grayish (never white) rather uniform layers, though sometimes striated or fibrous in appearance. They are sometimes thick enough to obscure the sun, and when it does shine through it resembles a light behind frosted or ground glass. Since there are few if any ice crystals, there are no halos. Precipitation rarely reaches the ground from altostratus, and if it does, it is usually very light.

Altocumulus, composed of water droplets, come in a bewildering variety of shapes, including layers, rolls, and rounded masses. Their outline may be sharp or soft and their thickness ranges from dense to nearly transparent. They sometimes produce a color phenomena called irisation in which the cloud takes on mother-of-pearl-like

A sea of marine stratocumulus clouds as seen from the perspective of an orbiting astronaut.

red and green tints. While streams of virga can sometimes be seen descending below cloud base, rarely will altocumulus generate precipitation that reaches the ground.

And just as with other cloud types, there are a plethora of nearly unpronounceable polysyllabic varieties known to the truly dedicated altocumulus watchers: altocumulus castellanus, altocumulus castellatus, altocumulus duplicatus, altocumulus floccus, altocumulus lacunaris, altocumulus lacunosus, altocumulus lenticularis, altocumulus opaucs, altocumulus perlucidus, altocumulus stratiformis, altocumulus translucidus, and altocumulus undulatus.

What are the **low clouds**?

The low clouds are stratus, nimbostratus, stratocumulus, and cumulus. Cloud bases for these are generally less than 6,500 feet above ground.

Stratus are grey clouds with generally rather uniform bases. Fog is simply a stratus cloud reaching to or forming on the ground. Stratus are usually composed of water droplets, but can, during very cold conditions, be made of ice crystals. They usually do not produce much precipitation, although drizzle and light snow can occur. Sometimes ragged elements of cloud are seen below base. These elements are called fractostratus or more commonly, scud. Stratus clouds are very common in marine areas. Often the low clouds that shroud coastal Los Angeles, San Francisco, and Seattle are stratus.

These towering cumulus clouds (cumulus congestus)
rapidly developed over the Minnesota prairie, becoming a thunderstorm in less than an hour.

Nimbostratus, rainy layers, are often vast, dank, dark, amorphous layers of clouds with very low bases, producing long periods of rain and/or snow. They are the clouds that produce those all-day rains that ruin outdoor plans. You generally can't see the sun through nimbostratus.

Stratocumulus, lumpy layers if you will, are like altocumulus in that they come in a wide variety of shapes and textures. They sometimes form from the spreading out and merging of cumulus clouds. They are usually composed of water droplets, and only rarely produce any significant precipitation. They are a common cloud during winter in the interior of the United States.

Cumulus clouds often look like heaps of mashed potatoes or heads of cauliflower, and often dot the sky on warm humid days in spring and summer. Their bases are usually lower than 6,500 feet above ground. They do not usually produce precipitation, but can grow into thunderstorms (cumulonimbus). Meteorologists often just refer to them as "Cu" (pronounced "Q"). They most often form from the heating of the ground by the morning sun. So on a clear day, you can often expect the first cumulus to pop up several hours after sunrise. Depending on the temperature and wind structure of the atmosphere, they can remain rather flat (cumulus humilis), start developing some vertical structure (cumulus mediocris), or obtain an even higher degree of vertical development (cumulus congestus). Weather people call these latter formations tower cumulus, or TCU for short.

This cumulonimbus cloud has the classic anvil-shaped top.

What is a **cumulonimbus cloud**?

The cumulonimbus is the Godzilla of clouds: the thunderhead. These are the clouds that meteorologists all go outside to look at. Usually abbreviated to "Cb" by meteorologists, these are the clouds needed if you want to have lightning, hail, or tornadoes. It is the only cloud that has groupies—every spring and summer hoards of severe weather junkies armed with camcorders drive hundreds of thousands of miles throughout the central United States chasing down Cbs likely to spawn photogenic twisters or windshield-bashing hail stones.

The cumulonimbus may have its base almost on the ground. Its top can range from about 20,000 feet to an as yet undetermined height. Radar operators have occasionally reported echo tops from 65,000 to 70,000 feet in the central United States. The tallest thunderstorms, featuring the very coldest cloud tops in the troposphere, may be found in the area between northern Australia and Indonesia/Papua New Guinea during the months of January and February. The cumulonimbus, seen at a distance, can sometimes be recognized by its distinct anvil-shaped cloud, which is essentially ice crystals from the top of the storm being sheared off and carried away by strong jet stream winds aloft. Sometimes intense updrafts will penetrate through the anvil like a giant cauliflower and punch several thousand feet into the stratosphere. Such an "overshooting top" is the sign of very intense updrafts. In order to distinguish a cauliflower-like cumulus congestus (TCU) from a genuine cumulonimbus (Cb),

79

watch the top, which in a cumulonimbus has to become soft and fuzzy, meaning it is becoming glaciated. In other words the water droplets in the rapidly rising cloud have started to convert into ice crystals. If you see lightning from a tall cumuliform cloud, it is then officially a cumulonimbus—a thunderstorm.

The cumulonimbus is often called the cloud factory because it gives birth to many other clouds. The anvils can shear off into layers of cirrus and cirrostratus that extend hundreds of miles downwind. Scud and fractostratus mark its rainy underside. The dissipating storm can generate vast flocks of stratocumulus and altocumulus that can persist for many hours. The rain-cooled air left behind often forms fog and low stratus clouds.

Cumulonimbus clouds can last for just a few minutes, or, when congregating into huge storm systems hundreds of miles across (called mesoscale convective complexes), can last for twelve hours or more.

What happens when an airplane flies through a cloud?

Usually nothing, but sometimes the cloud layer is composed of supercooled water droplets (at temperatures below freezing but still in the liquid form). Since these will freeze with the slightest disturbance—such as the passage of an aircraft—a region of snow flakes forms rapidly, and since they become quite large, they fall out leaving clear air behind. Sometimes a long streak resembling a boat wake, or even a perfect circular hole, is left behind.

What is a **mammatus cloud**?

The ominous looking, downward-bulging clouds sometimes seen clustered on the underside of thunderstorm anvils are cumulonimbus mammatus. In themselves they are harmless, but they are often associated with cloud systems that spawn severe weather.

What is a **cloud droplet**?

Clouds are made of trillions of minute droplets of water (or when cold enough, ice crystals). When air cools to the point of being saturated (100 percent relative humidity), water vapor condenses, though it can't do so on thin air. It needs a nucleus, a small speck of something upon which to condense. Cloud nuclei can range from sea

This view from Earth orbit shows the path of the Amazon river far below; the cooler waters of the Amazon inhibit daytime cumulus cloud formation.

salt to volcanic dust to grains of pollen and flakes of minerals. These nuclei are so minute they cannot be seen with the naked eye and there can be many millions in a cubic yard of air, but without them clouds and precipitation would not form. The droplets that comprise clouds are extremely small, often only a few micrometers across (a human hair is 100 micrometers or more in width). It takes about one million cloud droplets to provide enough water for one rain drop.

What are **supercooled droplets**?

The small cloud droplets that stay in a liquid state to temperatures as low as -40°F are called supercooled droplets. Aircraft flying through clouds of supercooled droplets can experience severe icing, as the water freezes instantaneously on contact with the wings and propellers. Winter fogs often contain supercooled droplets that freeze on impact as they drift through trees, past wire fences, and onto your car windshield. When the fog clears after such an event, the world can look like a crystal-covered winter wonderland.

What is **cloud seeding**?

In the late 1940s, Nobel Laureate Irving Langmuir and atmospheric scientist Vincent Schaeffer conducted a classic experiment that demonstrated that intentional human **81**

Cloud whirls form as maritime stratocumulus clouds flow around the rocky Galapagos Islands.

modification of the weather was possible. It was discovered, in experiments with clouds produced in home freezers, that supercooled droplets could be turned into ice crystals simply by dropping specks of dry ice (frozen carbon dioxide) into the cloud chamber. The passage of -109°F dry ice granules through the supercooled cloud chilled the droplets to -40°F, causing instant conversion to ice. The ice crystals then grew rapidly, reaching a large enough size to fall out of the cloud. Thus, by further cooling small supercooled droplets, precipitation could be generated. The theory was tested in the real world by dropping dry ice into a large solid deck of winter stratocumulus clouds. Within minutes a large dent and then a hole appeared in the clouds where snow had fallen out—the first human-made snow flurry. Other seeding techniques have been used, such as with the chemical silver iodide, which has a crystal structure very similar to water ice and fools the supercooled water droplets into freezing when they make contact. Other agents such as sea salt and large water droplets have also been used to induce precipitation in clouds.

Cloud seeding works well in winter clouds when the goal is to enhance snowfall from clouds that are already producing some precipitation. It is particularly effective in increasing snow packs over mountainous areas. During the warm season, growing cumulus clouds can sometimes be induced to provide more rain, but it is a more difficult and uncertain process. Suitable conditions for warm season rain enhancement are not sufficiently common to make precipitation enhancement by this method a totally reliable technique.

What is a contrail?

The plume of condensed water vapor in the exhaust trailing behind high-flying jet aircraft is called a contrail—short for "condensation trail." B29 pilots during World War II didn't appreciate contrails because they gave away their position. Contrails were intensively studied during the Cold War since they were sometimes a means for identifying high-flying bombers or reconnaissance aircraft. And the contrails, in addition to the other exhaust products generated by jet aircraft, are of some concern with regard to global climate change issues. There are so many contrails along some major jet flyways that the climatology of cirrus clouds is showing an increase. Also the various exhaust chemicals introduced into the upper troposphere and lower stratosphere could conceivably have some impact on atmospheric chemical and energy processes.

What other weather conditions can be affected by human intervention?

Supercooled fogs around airports have been routinely cleared for years by dry ice seeding. Unfortunately, 95 percent of the fogs in the United States are "warm," with temperatures above freezing.

In France, attempts at dissipating fog at airports by heating the atmosphere with jet engines worked, but proved very expensive (and noisy).

Dense but very shallow "ground fogs" can severely disrupt road travel. It has been found that the rotor downwash from low-flying helicopters can sometimes mix enough dry air into the surface layer to dissipate the fog bank.

How long does a **cumulus cloud** last?

The typical lifetime of a small cumulus is around 10 to 15 minutes.

Why do **clouds float**?

Clouds do actually fall, just very slowly. Clouds are composed of a myriad of minute water droplets and/or ice crystals. These are so small that their downward floating rate is negligible. The fall speed of a typical cloud droplet is so slow it could easily take an hour to fall only 30 feet. Updrafts also may carry them to higher altitudes. By comparison, raindrops and snow flakes grow large enough to fall with sufficient speed to reach the surface before they evaporate.

When is a **table cloth** a cloud?

In Cape Town, South Africa, a thin layer of clouds sometimes covers the flat, massive Table Mountain behind the city. As the air flows off the mountain top in all directions, it often looks like a huge piece of linen draped over the edge of the table-flat mountain, thus giving rise to the name Table Cloth.

What is a **mare's tail**?

There are many colloquial and local names for clouds. Long, wispy cirrus clouds that look like the tail of a horse are called "mare's tails".

Mackerel sky is so named because it resembles the pattern seen on a mackerel's back. It can result from either a large patch of cirrocumulus or small-element altocumulus patches.

HURRICANES AND TROPICAL STORMS

What is the difference between **a hurricane and a typhoon?**

Nothing except geography. Tropical storms occur in several of the world's oceans, and except for their names, are essentially the same type of storm. "Hurricane" and "typhoon" are among the regionally specific names given to strong tropical cyclones.

In the Atlantic and eastern Pacific (east of the international dateline), storms attaining sustained 74+ mph winds are called hurricanes. There are several theories about the origin of the word hurricane, but it appears to have originated either with the Caribbean Indians, who named their storm god "Huracan" or from the ancient Mexican word "hurrikan." Probably the first human record of hurricanes can be found in ancient Mayan hieroglyphics. In the western Pacific (west of the international dateline), the same storm is called a typhoon, perhaps after the Chinese word *ty-fung,* for big wind. In the Indian Ocean and Bay of Bengal, the storm is called a cyclone. In Australia, hurricanes and dust devils are referred to as Willy Willys.

What are the stages of **tropical storm development?**

Tropical cyclones come in a variety of strengths and sizes, and are classified by peak wind speed. An easterly (tropical) wave is an agglomeration of thunderstorms, sometimes called a hurricane seedling, out of which stronger storms sometimes emerge. A tropical depression is an organized cluster of thunderstorms with sustained winds below 38 mph. A tropical storm is more organized, with sustained winds in the 39 to

This computer-enhanced photo of Hurricane Diana, 11 September 1984,
gives a visual representation of a category three hurricane, with winds reaching 130 mph.

73 mph range. A hurricane has winds of 74 mph and up. Between 60 to 100 hurricane seedlings drift westward from Africa each summer, but only a small percentage (10 percent) actually develop into tropical storms.

What is the difference between a **hurricane watch** and a **hurricane warning**?

A hurricane *watch* means that hurricane conditions are *possible* in the specified area, usually within 36 hours. A hurricane *warning* means hurricane conditions are *expected,* usually within 24 hours. Hurricane watches and warnings are issued by the Tropical Prediction Center (formerly the National Hurricane Center), a component of the National Weather Service.

How accurate are **hurricane landfall forecasts**?

They are getting better all the time, but they are still far from perfect. The average error in predicting landfall some 24 hours in advance has now fallen to less than 100 miles. Thus, if a given site is suggested as the point where a storm may reach land, it could easily hit 100 miles either way up or down the beach. That is why landfall forecasts are often given in probabilistic terms. Prediction of intensity changes have proven even more difficult. There are many factors that influence a storm's intensity and some of these are very difficult to measure or predict.

What is the **Saffir-Simpson hurricane scale**?

The Saffir-Simpson Hurricane Damage-Potential scale was proposed in 1971 by engineer Herbert Saffir and hurricane expert Dr. Robert Simpson. The five-point scale is a way to relate measured storm characteristics (peak winds, lowest pressure storm surge) to its damage potential. Only two storms have hit the United States this century at category five strength, the 1935 Florida Keys Labor Day storm and 1969's Hurricane Camille. The categories are as follows:

ONE: Winds 74–95 mph. No real damage to building structures. Damage primarily to unanchored mobile homes, shrubbery, and trees. Aslo, some coastal road flooding and minor pier damage.

TWO: Winds 96–110 mph. Some roofing material, door, and window damage to buildings. Considerable damage to vegetation, mobile homes, and piers. Coastal and low-lying escape routes flood 2–4 hours before arrival of center. Small craft in unprotected anchorages break moorings.

THREE: Winds 111–130 mph. Some structural damage to small residences and utility buildings with a minor amount of curtainwall failures. Mobile homes are destroyed. Flooding near the coast destroys smaller structures, with larger structures damaged by floating debris. Terrain continuously lower than five feet above sea level may be flooded inland as far as six miles.

FOUR: Winds 131–155 mph. More extensive curtainwall failures with some complete roof structure failure on small residences. Major erosion of beach areas. Major damage to lower floors of structures near the shore. Terrain continuously lower than ten feet above sea level may be flooded requiring massive evacuation of residential areas inland as far as six miles.

FIVE: Winds greater than 155 mph. Complete roof failure on many residences and industrial buildings. Some complete building failures with small utility buildings blown over or away. Major damage to lower floors of all structures located less than fifteen feet above sea level and within 500 yards of the shoreline. Massive evacuation of residential areas on low ground within ten miles of the shoreline may be required.

What are **spiral bands**?

Hurricanes are more or less circular storms in which the air spirals inwards toward a nearly calm center. The clouds are often arranged in a banded pattern that swirls inward toward the center, sometimes resembling a spiral galaxy. Most of the heavy rain squalls and higher winds are found within the spiral bands. The outermost bands can be several hundred miles from the hurricane center. Tropical storms and hurricanes often begin with a hard, brief rain and wind squall, followed by a period of partial clearing, until the next band sweeps in and the pattern repeats.

What are the **eye wall and the eye**?

In the center of the most violent hurricane or typhoon is a nearly calm region called the eye. Typically 5 to 15 miles across, the eye may have sunny skies and minimal winds. The tall, nearly circular ring of thunderstorm-like cloud towers surrounding the often-clear center is called the eye wall, a giant "amphitheater of clouds" some 10 miles high. Within the storm's core, though winds are barely stirring, towering waves are crashing together from myriad directions. It is not unusual to find hundreds if not thousands of bewildered sea birds flying about, trapped within a giant cage of wind. Within the surrounding eye wall just a few miles away, devastating winds may be roaring at 150 mph or more. In some storms, the eye can be more than 40 miles across. The more violent the storm, however, the smaller the eye. The eye of Hurricane Camille (1969), one of the most intense Atlantic storms, was as small as 4 miles across at times.

If you are in the direct path of a hurricane, don't be fooled by the eye of the storm. The calm central portion of the storm may take minutes or an hour to pass by. Winds drop to near zero, and the sky often partially clears. People, lured outside by the calmness to inspect damage, are often trapped as the winds return to their former fury.

Why are **hurricanes** and other **tropical storms** given names?

The origin of naming tropical storms after people is uncertain. Among the first to do so may have been the residents of Puerto Rico, who would christen a storm after the name of the saint on the liturgical calendar the day the storm struck.

The U.S. practice may have started during World War II when military meteorologists informally named storms after girlfriends and wives. It became more widespread after the war as meteorologists found themselves tracking several storms at once. The names helped avoid confusion in communicating warnings.

Before 1951, many storms went unnamed. In 1951 and 1952, military alphabet names (Able, Baker, Charley, etc.) were used in the Atlantic basin. After 1953, Atlantic storms were named after women, with a new list of names prepared before each season. The practice started in the eastern Pacific in 1959 and in 1960 for the remainder of the North Pacific. Australian and South Pacific regions started using women's names in 1964. North Indian cyclones are not named.

In the late nineteenth century, tropical storms that struck Australia were given women's names by Australian meteorologist Clement Wragge. It was rumored he selected the names of wives of politicians and others he didn't like. During the 1940s, George Stewart wrote a book called *Storm* that involved a young meteorologist tracking a storm, which he called Maria, from its birth as an isobaric wrinkle in the North Pacific to its death over North America many days later. The book was later made into a movie by Walt Disney and may have further promoted the idea of naming storms.

Gender equality finally reached the tropical storm naming business in 1979. The National Weather Service, which had been using exclusively female names for hurricanes, began including male names. Recognizing the multicultural impact of Atlantic storms, they also began including some Spanish and French names.

A tropical weather system will not be given a name until its highest sustained wind speeds reach at least 39 mph.

What are some **tropical storm names** in other parts of the world?

Yasi, Atu, Drena, Hina, Epi, Guba, Ila, Kamo, Tako, and Upia. If you live in the Atlantic hurricane region, storms are given prosaic names like Bob and Betty. But around Fiji and Papua New Guinea, the names are far more colorful, like Tropical Storm Zaka, for instance.

What is the **average number** of tropical storms and hurricanes per year?

In a "normal" year in the Atlantic basin, there's an average of 9.7 named tropical storms, of which 5.4 reach hurricane strength. The most active basin is the northwestern Pacific ocean, averaging 26 tropical storms each year, of which 16 reach typhoon strength. Worldwide, in an average year, there are 84 tropical storms, of which 45 have peak sustained winds of 75 mph or more.

What was the most active **Atlantic tropical storm** season?

In 1933, 21 storms reached tropical storm strength with 9 becoming hurricanes. In 1995, there were not as many Atlantic tropical storms (19), but 11 of them reached full hurricane strength.

How is damage from hurricane winds related to **wind speed**?

The damage that can be caused by wind goes up with the square of the wind velocity. For instance, a 75 mph wind is 2.25 times more potentially destructive than a 50 mph wind. Based on damage statistics, the destruction from a category five hurricane (on the five-point Saffir-Simpson scale) striking the United States will average 250 times that from a nominal category one hurricane. While the winds in the strongest tempest may only be twice that of a minimal storm (75 mph), clearly the damage potential rises dramatically with each upward tick of the anemometer.

How fast can a **tropical storm** intensify?

While some take days to reach their peak strength, Typhoon Forrest (September 1983) in the Pacific ocean intensified from a nominal 75 mph to 173 mph sustained winds in just 24 hours. The central pressure dropped from 28.82 inches to 25.87 inches during the same time interval.

What were the largest and most compact **tropical cyclones** on record?

Typhoon Tip had gale force winds that extended outward for almost 700 miles on 12 October 1979; at the same time it was recording the world's lowest-ever barometer reading. But some tropical storms are far more compact. Cyclone Tracy, which devastated Darwin, Australia, killing 50 people on Christmas Eve, 1974, had gale force winds extending outward only about 30 miles.

What are the **largest rainfalls** ever to accompany tropical storms?

La Reunion Island, a small speck in the South Indian Ocean, holds a number of the all-time rainfall records for the planet. Some of the more notable deluges:

Duration	Amount of rain	Location	Date
12 hours	45.00 inches	Foc-Foc	7–8 January 1966 (Tropical Cyclone Denise)
24 hours	71.80 inches	Foc-Foc	7–8 January 1966
48 hours	91.10 inches	Aurere	8–10 April 1958
72 hours	127.60 inches	Grand-Ilet	24–27 January 1980 (Tropical Cyclone Hyacinthe)
10 days	223.50 inches	Commerson	18–27 January 1980

What constitutes the **"fuel"** of hurricanes and typhoons?

Water vapor. As water evaporates from the warm oceans, rises, cools, and condenses, it releases vast amounts of latent heat into the atmosphere, some 580 calories of heat for every gram of liquid. A hurricane can produce up to 20 billion tons of rain water per day. One of the reasons hurricanes rapidly weaken as they move inland is that they lose their source of nearly saturated warm air that they have while out at sea.

What is the **storm surge**?

The storm surge, the wind and pressure-caused rise in sea level associated with hurricanes, can sometimes reach 25 feet above mean sea level. Giant wind-whipped waves then form on top of the storm surge dome. The storm surge is sometimes mistakenly called a tidal wave.

As impressive as the winds are in a hurricane, it is the coastal storm surge that causes most of the damage and loss of life. The low pressure in the eye literally sucks the ocean surface upward, rather like liquid through a straw, some several feet. As the storm approaches land, the bulge can reach 15 feet or more due to the added effect of the wind pushing the water ahead of the storm on the right side of the track (in the Northern Hemisphere). When Hurricane Hugo hammered the South Carolina coast in 1989, it did so with 135 mph wind gusts and a storm surge that elevated the sea level some 20 feet above normal tide levels. Hurricane Camille in 1969 produced a 25-foot storm tide on the Mississippi coast.

The storm surge is the deadliest aspect of a tropical storm, with wind-driven waves piled atop a high mound of water inundating vast tracts of low-lying coastal terrain. The maximum probable storm surge along the U.S. coastline is on the order of 20–25 feet. In 1876, some 250,000 people drowned in the delta region of the Ganges River of India as a cyclone reportedly swept inland with a 36-foot storm surge. In the Bay of Bengal, a cyclone smashed onshore in 1737 with a storm surge estimated at 40 feet. Huge areas of coastline were inundated, and over 300,000 people drowned. A similar death toll has been estimated from the 1970 Bangladesh cyclone. A tropical cyclone that struck Bathurst Bay, Australia, in 1899 was reputed to have a storm surge of 42 feet.

During tropical storms, the coastline is the most dangerous place to be. But how far inland is far enough? In 1977, a severe cyclone swept the Andhra Pradesh region of India. The storm surge was over 18 feet high. Boatmen on a river some 20 miles inland reported the sea wave reached that far, destroying boats and buildings alike.

Has the hurricane **disaster potential** increased?

Hurricanes will haunt our future. Not only does it appear the recent 30-year lull in Atlantic tropical storm activity is about over, but there are now so many more targets for the storms to hit. More people today live between Miami and Ft. Lauderdale than lived on the entire coastline from Texas to Virginia in the mid-1930s. More than 50 million Americans now live in the immediate vicinity of hurricane-prone shorelines. The evacuation planning problem is made worse due to the fact that summertime vacationers swell the population of some coastal resort areas by 10 to 100 times their offseason numbers. The majority (perhaps as high as 85 percent) of coastal dwellers

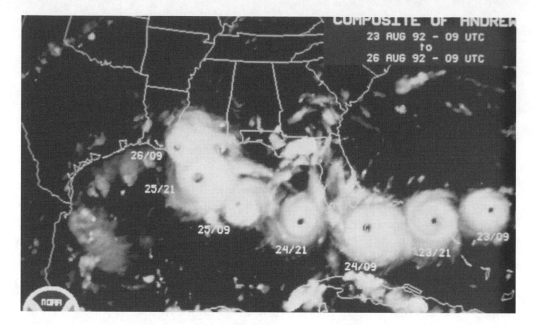

This composite image from the National Oceanic and
Atmospheric Administration shows Hurricane Andrew's path of destruction.

today have never experienced a major hurricane. Planners are concerned that many
will discount the risk and try to ride out storms, holding "hurricane parties" in their
beachfront condos. If there is any good that came from Hurricane Andrew, it is that
the entire nation had a glimpse of what a hurricane could do.

Do **hurricane evacuations** help save lives?

Hurricane Andrew, with winds estimated at up to 177 mph in south Florida, killed less
than three dozen people. This relatively low number attests to the effectiveness of
massive evacuation plans, which resulted in over two million people being moved
inland, out of harm's way.

It pays to know when to run away from danger. With the approach of Hurricane
Hugo in 1989, the governor of South Carolina ordered the evacuation of over 250,000
people from the dangerous coastal area. Of those who stayed behind, 10 were killed.
Yet this death toll is many, many times lower than if the population had tried to ride
out the storm.

Is there any continent that **does not have hurricanes**?

Tropical storms of hurricane strength can strike every continent except Antarctica. It is unusual for a tropical storm to strike the north coast of South America, but it does happen. There are no tropical storms in the south Atlantic, so Brazil and Argentina are not at risk. At least 15 percent of the world's population lives on shorelines that are vulnerable to such tempests.

The storms that strike southern and especially northern polar regions can be every bit as intense as hurricanes. In fact, polar lows, as they are often called, have been seen to develop eye-like structures, suggesting that they may have much in common with their tropical cousins.

Did **Christopher Columbus** encounter any hurricanes?

Columbus was lucky on his first few journeys, managing to avoid intercepting any tropical tempests, though he battled several on later voyages. By 1495, the small town of Isabella, founded by Columbus on Hispaniola, became the first European settlement destroyed by a hurricane.

What is the official **Atlantic hurricane season**?

The first of June marks the official start of the hurricane season, which extends until the end of November. In 1995, the National Hurricane Center moved into a new "storm proof" building near Miami. During 1992's Hurricane Andrew, their old home was damaged by the howling winds, and the storm-tracking radar was literally ripped off the roof by the winds. The most active month for North Atlantic hurricanes is September, with August ranked second and October third. Mid-September marks the absolute peak of the Atlantic hurricane season.

When is the **Pacific typhoon season**?

Typhoons can occur during almost any month of the year. The small island of Guam in the western Pacific was struck by no less than five typhoons throughout 1993.

Are there hurricanes in the **eastern Pacific**?

Though they rarely affect land, the eastern Pacific ocean west of Mexico is a real breeding ground for hurricanes. In a typical year, 16 or 17 tropical storms and 9 hurricanes form. As many as 14 full-fledged hurricanes have formed in the eastern Pacific. Hurricanes hardly ever happen in Hawaii, but they occasionally strike and can be quite severe. On 11 September 1992, Hurricane Iniki caused $1.8 billion in losses on the island of Kauai, Hawaii. The records show that 4 major hurricanes have struck the Hawaiian Islands during the twentieth century.

Are hurricanes a **major source of flooding**?

Hurricanes traveling inland pose a major flood threat. Rains from weakening Hurricane Diane (1955) killed 200 and caused $4.2 billion in losses. Hurricane Agnes (1972) caused $6.4 billion in damage and 122 deaths throughout the northeastern United States. For many storms, the inland flooding causes more economic disruption than the coastal winds and waves. This is especially true of weaker hurricanes, which, while they never have much wind, still carry a potent precipitation punch. Mud slides (or more properly ash slides) resulting from typhoon rains after the eruption of Mt. Pinatubo in the Philippines caused almost as much havoc as the volcano itself.

What are the **longest-lasting** Atlantic and Pacific tropical storms?

An Atlantic hurricane named Faith in 1966 roamed the ocean for 26 days. But the longest-lasting Atlantic tropical storm was Ginger, in 1971, which spun around the open ocean for 28 days. In the Pacific, a 1994 storm named John hung on for 31 days. It formed off the west coast of Mexico on 11 August, traveled over 4,000 miles, and reached peak winds of 170 mph while passing south of Hawaii. And since it crossed the dateline twice, it changed status from hurricane to typhoon and back to hurricane.

What do **snakes** have to do with hurricanes?

These tubular reptiles do not like salt water on the whole, and they make as much effort as humans to find high ground during hurricane storm surges. Often the two species spend some apprehensive times together in all-too-close proximity when stranded in trees or on floating roof tops.

Are **tropical storms** big rain makers?

Hurricanes and even weaker tropical storms are prodigious rainmakers. Tropical Storm Claudette (1979) dumped some 43 inches of rain near Alvin, Texas. Over half a billion dollars of damage resulted. Tropical Storm Jerry sloshed its way across southern and central Florida in August 1995. It could barely muster 40 mph winds, but dumped rains of up to a foot. So flooded were lowlands that wildlife fled, including the infamous "walking catfish" that slithered out onto the runways at West Palm Beach International Airport to disrupt takeoffs and landings.

What are the **lowest barometer readings** in Western and Eastern Hemisphere tropical cyclones?

The lowest barometric pressure recorded in the Western Hemisphere was 888 millibars (26.17 inches) during Hurricane Gilbert (1988).

On 12 October 1979, the most powerful tropical storm ever recorded was prowling the Pacific. Typhoon Tip had a central pressure of about 25.69 inches (870 millibars) and sustained winds of 190 mph.

Do hurricanes always **weaken** as they move inland?

While hurricane winds usually decrease rapidly after landfall, sometimes the gusts take quite a while to subside. Hurricane Hugo (1989) battered Charlotte, North Carolina, some 175 miles inland, with gusts near 100 mph. Someone forgot to tell Hurricane Hazel, which came ashore in North Carolina in 1954, that it was supposed to weaken after moving onshore. While moving north into Canada through Virginia, Pennsylvania, and western New York, it lashed Washington, D.C., with sustained 78 mph winds and gusts to 98 mph. Philadelphia saw gusts to 94 mph and the anemometer atop New York City's Empire State Building whirled at 113 mph. Severe damage was reported in the Toronto area with 75 mph winds.

What are the **ten most intense storms** to strike the United States?

The ten most intense storms to make landfall in the continental United States between 1900 and 1994 are:

Hurricane	Year	Category	Millibars	Inches
1. Florida (Keys)	1935	5	892	26.35
2. Camille	1969	5	909	26.84
3. Andrew	1992	4	922	27.23
4. Florida (Keys)/Texas	1919	4	927	27.37
5. Florida (southeast)	1928	4	929	27.43
6. Donna	1960	4	930	27.46
7. Texas (Galveston)	1900	4	931	27.49
Louisiana (Grand Isle)	1909	4	931	27.49
Louisiana (New Orleans)	1915	4	931	27.49
Carla	1961	4	931	27.49

The last four were tied in terms of air pressure. The lowest sea-level air pressure ever recorded in the United States was 26.35 inches within the eye of the September 1935 hurricane as it crossed Matacumbe Key, Florida. Hurricane Camille (1969) had a minimum pressure of 26.28 inches. Hurricane Andrew (1992) is in third place at 27.22 inches, recorded in South Dade County, Florida.

How busy were **hurricane forecasters** in 1995?

The 1995 Atlantic hurricane season was the second most active in 125 years of record-keeping. Of the 19 named storms, 11 reached hurricane status. At one time forecasters at the Tropical Prediction Center in Miami were busy tracking one dissipating tropical depression (once Tropical Storm Jerry), one tropical wave, and four named storms (Tropical Storms Karen and Luis and Hurricanes Humberto and Iris). The seven named storms during August tied a record for any month. By contrast, eastern Pacific tropical storms, which average 16 (9 of hurricane strength) were below normal in 1995, with only 10 named storms, 7 of them hurricanes. Total United States hurricane-related economic losses in 1995 were $5.2 billion.

Why will **Hurricane Andrew** never strike again?

Some hurricanes get their names "retired" like baseball players' numbers; truly memorable storms have their names removed from the rotating list. So Floridians don't have to worry about Hurricane Andrew striking again. The reason names of major hurricanes are "retired" from the rotating list of Atlantic tropical storms is that research on these storms continues for many years and it becomes confusing if several major storms with the same name are being studied at one time. Other "storms" we won't see again include: Agnes, Alicia, Allen, Anita, Audrey, Betsy, Beulah, Bob, Camille, Carla, Carmen, Carol, Celia, Cleo, Connie, David, Diana, Diane, Donna, Dora, Edna, Elena, Eloise, Fifi, Flora, Frederic, Gilbert, Gloria, Gracie, Hattie, Hazel, Hilda, Hugo, Inez, Ione, Janet, Joan, and Klaus.

What are some **hurricane safety tips**?

If a major hurricane is threatening your area, here are some tips. First, if there are evacuation orders, obey them. Otherwise, fill containers with clean water for three to five days' supply (five gallons per person). Have a supply of "baby wipes" to use for cleansing if bathing facilities are lost. Fill sinks and bathtubs with extra water for dish and clothes washing. Adjust thermostats on freezers and refrigerators to get as cold as possible before the loss of power. Eat perishable items. Store up on supplies of batteries, flashlights, and first-aid kits. Make sure you have essential medicines, canned food and a can opener, rainwear and sleeping bags, a battery-powered radio, and any special infant items that might be needed. Also make sure you know how to turn off your gas, water, and electricity if authorities order an evacuation.

Where is the **National Hurricane Center** located?

The National Hurricane Center, recently renamed the Tropical Prediction Center, recently moved into a new storm-proof building on the campus of Florida International University near Miami. Its 10-inch walls are designed to withstand 130 mph winds, and the structure is far enough inland to avoid the flooding from storm surges. It has windows with heavy roll-down shutters, and even a room for forecasters' families. This accommodation solves a problem that emerged during Hurricane Andrew: forecasters wanted to stay on duty to help out, but were worried about the fate of their families at home.

Can we **control hurricanes**?

Several ideas have been proposed over the years. Cloud seeding has been tried but results are inconclusive. Using nuclear weapons to tame the tempest has been suggested, though this is not likely to be an ideal solution. A mature hurricane releases energy equivalent to 400 twenty-megaton hydrogen bombs. In fact, one extra megaton might just add fuel to the fire, making the vortex radioactive to boot.

One idea does hold some promise. Since water vapor evaporating from the warm ocean is the fuel of hurricanes, it has been proposed that spreading a thin chemical film that retards evaporation in the path of an oncoming storm could rob it of much of its energy. The flaw is that no thin film is known that can withstand the huge waves found in even a modest storm.

What is a **hypercane**?

Hurricane Andrew would be a mere sneeze compared to the "hypercane" hypothesized by Massachusetts Institute of Technology (MIT) Professor Kerry Emmanual. It has been theorized by scientists at MIT that if a giant meteor struck the tropical oceans, aside from making monstrous waves, it would warm large regions to temperatures above 120°F. Since hurricanes generally become stronger as the underlying ocean temperature rises, theoretical calculations predict that "hypercane" winds could exceed 500 mph. The "good" news is that the intense vortex would likely be only 15 or so miles across. Computer models suggest that the hurricane-like vortex would eject clouds and moisture some 25 miles into the atmosphere. In another variation on this theme, it is conceivable that during long-ago geologic times, undersea volcanoes could have heated patches of ocean surface to over 122°F.

Has there ever been a year without an **Atlantic hurricane**?

Only once in this century have no full-fledged hurricanes formed anywhere in the Atlantic basin. There were only four tropical storms during the entire 1907 season, none of which reached hurricane strength. In terms of the U.S. coastline specifically, **97** there have only been 18 years this century without at least one hurricane striking. In the Gulf of Mexico, there have only been three hurricane seasons in the twentieth century without any tropical storms: 1927, 1962, and 1991.

Where do people go to church to **ward off hurricanes**?

Hurricanes haunt the history of St. Croix in the Virgin Islands. So every July 26th the islanders take the day off to attend church services on Hurricane Supplication Day, to pray for protection from hurricanes.

What are the **ten most damaging hurricanes** to strike the United States?

From 1900 through 1994, the top ten most damaging storms, in 1990 U.S. dollars (with the exception of Hurricane Andrew), were:

Hurricane	Year	Category	Damage (in billions)
1. Andrew	1992	4	$25.000
2. Hugo	1989	4	7.155
3. Betsy	1965	3	6.461
4. Agnes	1972	1	6.418
5. Camille	1969	5	5.242
6. Diane	1955	1	4.199
7. New England	1938	3	3.593
8. Frederic	1979	3	3.502
9. Alicia	1983	3	2.391
10. Carol	1954	3	2.370

Final figures are not in yet, but 1995's Hurricane Opal, which smashed up the Florida Panhandle, may ring up some $3.0 billion in losses, moving it into ninth place.

Who pays for **hurricane damage**?

At least half of all hurricane damage is covered by insurance. So it is the insurance company stock holder and, ultimately, the rate payers, who have to cover the loss. Hurricane Andrew was devastating in many ways, including to insurance companies. Who paid out the most? State Farm led the way with $3.5 billion, followed by AllState at $2.1 billion. The federal government picks up a large chunk of the rest.

Hurricane Hugo prepares to strike Charleston, South Carolina, in 1989.

How do hurricanes **rank in costliness** compared to other disasters?

Between 1988 and 1993, there were at least seven weather-related disasters in the United States causing losses of one billion dollars or more. Of these, four were hurricanes.

Drought of '88	$40 billion
Hurricane Andrew	$25 billion
1993 summer floods	$12 billion
Hurricane Hugo	$7.2 billion
Blizzard of '93	$6 billion
Hurricane Iniki	$1.8 billion
Hurricane Bob	$1.5 billion

Are hurricanes the biggest **weather-related killers**?

Hurricanes dominate the headlines, due to their extreme winds and great potential for damages. But with modern hurricane evacuation planning limiting much of the loss of life, winter storms can be the more deadly tempest. The Great Blizzard of 1993 in the East took 243 lives, three times the combined toll from Hurricanes Hugo and

99

Andrew. A great Nor'Easter in March 1962 swept from Florida to New England, killing 33 and eroding beaches, causing some barrier islands to disappear and leaving behind almost a quarter billion dollars in property losses. In terms of sheer *economic* damage potential, however, the hurricane ranks as the number one severe weather threat.

What was the **record year** for Pacific hurricanes?

The 1992 eastern Pacific tropical storm season was a record breaker with 27 tropical cyclones. In fact, the year's list of storm names was totally used up and a contingency list had to be employed—the Greek Alphabet. 1992 was also a banner year for western Pacific typhoons. The island of Guam was affected by 6 of them, including the devastating Typhoon Omar.

What was the **"Storm That Wouldn't Die?"**

Super Typhoon Gay, in late November 1992, traveled across thousands of miles of the Pacific, with winds up to 225 mph. Its remnants then formed into a massive storm off the Aleutian Islands. After reforming again in British Columbia and crossing California, it gave birth to another low in Texas that became a great Northeast storm along the U.S. East Coast with wind gusts reaching 90 mph on 11 December 1992.

What were the **deadliest years** for U.S. and Atlantic basin hurricanes?

The three deadliest years for U.S. hurricanes were 1900, 1928, and 1938. Looking at the Western Hemisphere as a whole, the following are the ten deadliest years:

Year	Number of deaths	Location
1780	22,000	Martinique and Caribbean
1900	6,000–12,000	Galveston, TX
1974	8,000–10,000	Honduras
1930	8,000	Dominican Republic
1963	8,000	Haiti, Cuba
1776	6,000	Martinique
1775	4,000	Newfoundland Banks
1899	3,433	Puerto Rico, Carolinas
1928	3,411	Florida, Puerto Rico, Caribbean
1932	3,107	Cuba, Jamaica

What was the **greatest hurricane disaster** in the northeastern United States?

Over 600 were killed on Long Island and in New England when the "Great Hurricane of 1938" swept ashore. In those days before radar and satellites, warnings were virtually non-existent. The noon forecasts called for scattered showers and warm temperatures with "small craft warnings." By nightfall, many of the small craft had been driven several miles inland.

What were the **most hurricanes present** at one time in the Atlantic?

Only once were four hurricanes present at the same time in the Atlantic Ocean—on 22 August 1893. On 11 September 1961 there were three hurricanes, with a fourth storm very close to hurricane strength.

Which area is the luckiest in terms of **avoiding hurricane strikes**?

Perhaps the luckiest area so far in dodging the hurricane bullet is Pensacola, Florida, which has gone over 60 years without major hurricane activity. The region around the Kennedy Space Center has also been spared for most of the twentieth century. The Miami area had been doing well since Donna in 1960, until Andrew came along in 1992. But not far away, two destructive hurricanes in just nine days struck the Apalachicola-Tallahassee, Florida, region.

What year had the **most hurricanes** in the Atlantic basin?

The most active year for full-blown hurricanes anywhere in the North Atlantic, Caribbean, or Gulf of Mexico was 1969. In that year, a dozen storms reached full hurricane strength. Only in 1907 and 1914 did any storm fail to reach hurricane force.

What is the **earliest** an Atlantic hurricane has formed?

The official start of the hurricane season is 1 June. For a long time, the earliest known hurricane of the year to form in the Atlantic ocean was on 7 March 1908, but the hurricane season got off to an early exceptionally early start in 1955, with the first hurricane occurring on 2 January.

How many **category five hurricanes** have struck the United States this century?

Only two category five storms—the most severe—have struck the U.S. coastline this century. To achieve this ranking, sustained winds must be 155 mph or higher and a storm surge of 18 feet or more must inundate coastal regions. There have, however, been over 20 storms that have reached this level of fury over the open ocean in the Atlantic and Gulf of Mexico.

What was the **deadliest U.S. hurricane?**

The Galveston, Texas, hurricane of September 1900 was the United States' worst weather disaster. No part of the city—built on a sand bar—was more than 9 feet above sea level. Not a single building on the island escaped damage and more than half were swept into the sea. At least 6,000 people perished, according to the official report. There is evidence that the actual death toll may have been as high as 8,000 to 12,000 people. This storm single-handedly made September the deadliest month on the calendar for hurricane-related deaths in the United States.

How many **lives have been lost** to hurricanes on the U.S. mainland?

The 35 deadliest U.S. hurricanes in the last century have taken at least 17,700 lives. The ten deadliest storms to strike the U.S. mainland between 1900 and 1994 are:

Hurricane	Year	Category	Number of deaths
1. Texas (Galveston)	1900	4	6,000–12,000
2. Florida (southeast)	1928	4	1,836
3. Florida (Keys)/Texas	1919	4	600
New England	1938	3	600
5. Florida (Keys)	1935	5	408
6. Audrey	1957	4	390
Northeast U.S.	1944	3	390
8. Louisiana (Grand Isle)	1909	4	350
9. Louisiana (New Orleans)	1915	4	275
Texas (Galveston)	1915	4	275

Where are the **deadliest tropical storms** in the world?

As tragic as the U.S. hurricane toll is, it is dwarfed by the death toll from single tropical storms that sweep out of the Bay of Bengal to strike the populous Indian subcontinent, sometimes killing 100,000 or more.

In May 1991, a tragic tropical storm, called a cyclone in Bangladesh, swept ashore, bringing 145 mph winds and 20-foot waves to the low-lying coastal plain and mud flats that were home to at least 10 million people. Many people simply had no place to go, no buildings or ground higher than the advancing sea. The death toll will never be known for sure, but certainly is in the hundreds of thousands.

It is also estimated that over 500,000 persons perished in the flooding and storm surge associated with a cyclone that struck East Pakistan (Bangladesh) in 1970.

Has a hurricane ever struck **southern California**?

Not yet. Though rare, tropical storms can strike the U.S. West Coast. In 1939, such a tropical storm, though not at hurricane force, took 45 lives in southern California due to winds and flood waters. On 10 September 1976, Tropical Storm Kathleen caused widespread flooding in California (11 inches of rain on Mt. Wilson) and produced 72 mph wind gusts at Yuma, Arizona.

Los Angeles and hurricanes usually don't come up in the same sentence, but they could someday appear together in a headline. Satellites have shown that frequent hurricanes do spin up just off the west coast of Mexico. Occasionally one drifts as far north as San Diego and even Los Angeles.

How much damage could a **severe hurricane** actually do?

If one assumes the most intense possible storm were to strike the most densely populated coastal region of the Gulf of Mexico or the Atlantic Ocean, a $100 billion storm is not out of the question. Hurricane Andrew's $25 billion damage toll was more than twice the combined losses from San Francisco's most devastating earthquakes, in 1906 and 1989 (in 1991 dollars). Moreover, had Andrew's track moved just 30 miles northward into the heart of metropolitan Miami, the price tag could well have been over $60 billion.

Between 1949 and 1990, there has been approximately $77 billion (in equivalent 1990 dollars) in damages from 126 U.S. hurricanes and tropical storms. Of the total, 27 percent was caused by just six category four and five landfalling storms. Between 1984 and 1993 in the United States, 40 percent of privately insured losses due to damages have resulted from hurricanes.

At the Richelieu Apartments in Mississippi,
25 people gathered for a "hurricane party," ignoring official pleas to leave before Hurricane Camille struck.

How many **direct hurricane hits** has the United States had?

Between 1900 and 1991, 152 hurricanes have made direct hits on the U.S. coastline (excluding Hawaii).

What was the decade with the **most U.S. hurricane hits**?

In this century, the decade with the most number of direct hurricane hits on the U.S. coastline was the 1940s (23). The most quiet decade was the 1970s, with only 12 storms crossing the shoreline. The most hurricanes to strike the U.S. coastline in one season was 6—this happened twice, in 1916 and 1985. The 1890s were also a decade of very high storm activity.

How deadly are **Florida hurricanes**?

Florida has suffered three of the five deadliest U.S. hurricanes of this century. In 1928, a hurricane crossing Lake Okeechobee took 1,836 lives. Six hundred were lost in a 1919 storm and 408 perished in the violent storm that swept across the Keys in 1936.

The Richelieu Apartments after Camille swept the coastline. Only two people survived the "party."

Did hurricanes **affect the colonization** of what is now the United States?

Hurricanes have played a significant role in our history. Sir Francis Drake, while attempting to settle the North Carolina coast, was buffeted by violent hurricanes in 1586, 1587, and 1591. In 1609, a fleet of ships carrying settlers from England to Virginia encountered a hurricane. Some of the ships were grounded at Bermuda and the passengers became the island's first inhabitants. Their stories helped inform Shakespeare's writing of *The Tempest*. The French lost their bid to control the Atlantic coast of North America when a 1565 hurricane dispersed their fleet, allowing the Spanish to capture France's Fort Caroline near present-day Jacksonville, Florida.

What are the most **hurricane-prone** parts of the U.S. coastline?

According to climatology, the areas of the nation most likely to feel the impacts of hurricanes are the southern tip of Florida, the outer banks of North Carolina, and the upper Texas Gulf coast into Louisiana.

Can recent history be used to **predict hurricane threats**?

The frequency with which hurricanes strike portions of the U.S. coastline goes through various bursts and lulls. North Carolina was walloped six times between 1953 and 1955, and yet has had periods as long as 21 years without being struck.

What factors influence the **number of hurricanes** that form in the Atlantic?

Researchers have uncovered a strong link between the rainfall patterns in Africa's Gulf of Guinea and western Sahel regions and the chances for damaging hurricanes striking the U.S. East Coast. The prolonged West African drought may well account for the dearth of East Coast storms in the 1970s and '80s. And as the Sahelian drought began to ease in the late 1980s, along came Gloria, Bob, Hugo, and Andrew. Other factors such as El Niño and high-level winds are also known to affect hurricane numbers.

What were the **peak winds** in Hurricane Andrew?

No one may ever know for sure how strong the winds were that lashed south Dade County during Hurricane Andrew. Virtually every anemometer failed. One automated station on the Fowey Rocks, some 10 miles offshore from Miami, recorded a five-second peak gust of 169 mph—before its 30-foot mast was bent over at a 90-degree angle. Some amateur meteorologists watched their home anemometers gust to 177 mph (values corrected by wind tunnel calibration) before their homes and their wind sensors were destroyed. Peak gusts may have briefly hit over 200 mph.

What impact did **Hurricane Hugo** have on forests?

In 1989, Hurricane Hugo took 49 lives and affected 300,000 families along its path. It also had a devastating effect on the forests of South and North Carolina. Almost 9,000 square miles of forest (seven times the area of Rhode Island) were flattened. Hurricane Hugo's 135 mph winds destroyed more board feet of lumber than the combined effects of the eruption of Mount St. Helens and the 1988 Yellowstone National Park fires. Total damages approached $8–9 billion.

How much **energy is released** in a hurricane?

Hurricanes are not to be trifled with. One reason is their enormous energy. According to one estimate, a hurricane's energy output can be rated on the order of 100 billion kilowatt hours each day. As a testament to the forces involved, when a savage hurricane swept across the Caribbean island of Barbados in 1831, it caused extreme damage—and either wind or wave moved a solid piece of lead, weighing some 400 pounds, over 500 yards.

Do hurricanes affect the **coastline**?

Yes, and quite literally. In 1853, a barrier island called Isle Derniere off the Louisiana coastline was split into four segments by a hurricane. Scientists studying these rapidly eroding islands thought that at least two of the Isles Dernieres would disappear between 2001 and 2016. Then came Hurricane Andrew, contributing mightily to the islands' erosion.

Several of the inlets and cuts in the Outer Banks of North Carolina were carved by passing hurricanes during the last 150 years. Sediment cores from Florida's west coast indicate huge freshwater floods during strong hurricanes more than a thousand years ago. Geologists have found layers of sediment in an Alabama lake that were probably carried inland by intense hurricanes some 3,000 years ago.

What can a **severe hurricane** do to a region's electrical grid?

Hurricane Andrew caused some major power shortages. Florida Power & Light reported losing 1,900 transmission towers, 8.5 million feet of distribution lines, 18,700 utility poles, and 16,800 switches. Up to 1.4 million customers lost power. At Turkey Point, near Homestead, nuclear generators, which took a direct hit, survived with no real damage. But a 100-foot tall smoke stack for the adjacent coal-fired generators was so severely damaged it had to be destroyed.

What were some other **unexpected impacts** of Hurricane Andrew?

Refugees. The wrath of Hurricane Andrew, the greatest single natural disaster in U.S. history, initially left more than 250,000 people homeless, creating a refugee population the size of the city of Orlando.

Hurricane Andrew set a record for damage from a hurricane. It also created a record amount of trash. It has been estimated that if all the debris created by Andrew in South Florida were placed in one pile, it would tower more than 300 stories high. But the huge piles of debris left behind by Hurricane Andrew also contained "treasure." More specifically, the tons of aluminum scrap (previously house sidings, lawn chairs, awnings) were scavenged by entrepreneurs and sold for salvage. Some individuals made up to $1,000 a day or more until the booty was finally gone.

How fast do hurricanes **move forward**?

Typically, hurricanes move forward with speeds in the 10–15 mph range. The "Great Hurricane of 1938," sometimes called the "Long Island Express" raced up the East Coast at over 60 mph. Sometimes storms can almost stall for a day or more, and sometimes they go nowhere fast, traveling in loops.

How have **tropical and extratropical storms** affected military history?

At the onset of the Spanish-American War, President William McKinley declared he feared a hurricane more than the Spanish Navy. He was probably justified in his concern. During World War II, in the Pacific, one of Admiral William Halsey's worst defeats came at the hands of a typhoon. Three destroyers were sunk, many ships damaged, almost 800 sailors drowned, and the fleet had to stand down from a major attack.

And if it weren't for a great storm, Americans today might all be speaking Spanish. In 1588, the Spanish Armada, 130 ships strong, sailed for an invasion of England. But raging winds and mountainous seas smashed many of the vessels into the Scottish coast.

In the thirteenth century, the Great Kublai Khan's armies swept out of Mongolia, bent on conquering the world. When they turned on Japan, a great fleet of 1,000 ships sailed for the island nation. The fearless samurai fought bravely in a battle that seemed destined to be lost . . . until a great typhoon smashed the invader's ships, drowning over 100,000 warriors.

When has a devastating hurricane had a positive impact?

In 1976, Typhoon Pam ravaged the island of Guam with 160 mph winds and more than 30 inches of rain. There was one fatality, and 3,000 homes were damaged or destroyed. Yet the island soon prospered as the result of insurance payments and massive federal construction aid. The high unemployment rate soon dropped by over 50 percent.

Do hurricanes have any other **beneficial aspects**?

Yes. Rain. While sometimes they bring floods, more often than not the rains of a tropical storm moving inland break a drought, often saving huge agricultural crops that were in danger of being lost.

Do hurricanes still pose a **hazard to shipping**?

Modern forecasting and communications have helped eliminate many of the tragic shipwrecks that once claimed the lives of so many seamen. But disasters in the open ocean

This aerial photo of an island of the Bahamas shows deep cuts in the island and shallows gouged by the raging winds and waves of Hurricane Andrew.

still can happen. In 1990, Hurricane Bertha stayed at sea, where 30-foot waves snapped a Greek freighter like a match stick, with six crewman lost while abandoning ship.

Do the **Philippines** have many tropical storms?

The Philippines are situated right within the main Pacific typhoon pathway. During 1990, 14 typhoons affected the island. The worst came on 12 November, as Typhoon Mike's 144 mph winds killed more than 400 people and damaged or ruined over 600,000 homes. The island of Luzon is affected by an average of 20 tropical storms each year.

Do hurricanes happen in **Great Britain**?

Hurricanes are very rare in Great Britain, but on 25 January 1990, hurricane force winds, gusting to 120 mph, caused widespread havoc throughout Britain. London's double decker buses had to be taken off their routes for fear they might topple over. Damage was in excess of $1 billion and at least 45 died. The remnants of some Atlantic tropical storms do indeed reach into northwestern Europe as powerful storms.

What were some of the consequences of the **Hurricane of 1938**?

Fishing boats were seen lying across the tracks of the New York, New Haven, and **109**

Hartford Railroad in several places, including Stoningham, Connecticut. Many equally bizarre occurences, most with tragic overtones, were recorded throughout Long Island and New England after the passage of the Great Hurricane of 21 September 1938.

In the downtown business district of Providence, Rhode Island, late on the afternoon of 21 September 1938, many cars weren't found by their owners, because hurricane tides and winds piled water eight to ten feet deep in the city streets and floated the cars away.

Do hurricanes produce **special kinds of waves**?

A hurricane might typically move forward at a speed of 250 miles a day. The giant waves created within its center radiate outward at up to 900 miles per day. Before weather satellites, the arrival of these distinctive swells was the only advance notice of a hurricane to sailors and residents of coastal areas.

What threats do hurricanes pose to **inland areas**?

Hurricanes are usually thought of as hazards to coastal areas. Yet, their dying remains can have major impacts far inland as well. Torrential rains from the decaying Hurricane Agnes in June 1972 caused over $3 billion of flood damages in Virginia, Pennsylvania, and other nearby states.

Are there hurricanes on the **equator**?

No. If you wish to avoid the risk of being hit by a tropical tempest, go to the equator. No tropical cyclone has been observed within about 5° latitude of the equator.

Do hurricanes roam the **South Atlantic**?

No. There is only one recorded case of a tropical storm, and a very weak one at that, in the South Atlantic. It was spotted off the coast of Congo in 1991. The reasons for this "hurricane-free zone" are several. Often thought to be the result of cold ocean water, it is more likely that upper-atmosphere wind conditions are unfavorable. Also the Intertropical Convergence Zone (ITCZ) almost never dips south of the equator, and thus the chief breeding grounds for tropical storms is simply not present over this area of water.

THUNDERSTORMS, HAIL, AND FLOODS

What are the **five horsemen** of thunderstorms?

Wind, flood, hail, lightning, and tornadoes. All are spawned by cumulonimbus, or thunderstorm clouds, sometimes all at the same time. This cloud is an amazingly efficient weather factory.

What is a **thunderstorm**?

Technically, according to the National Weather Service (NWS), a thunderstorm occurs when an observer hears thunder. Radar observers use the intensity of the radar echo to distinguish between rain showers and thunderstorms. Lightning detection networks now routinely track cloud-to-ground flashes, and therefore thunderstorms. Thunderstorms arise when clouds develop sufficient upward motion and are cold enough to provide the ingredients (ice and supercooled water) to generate and separate electrical charges within the cloud. The cumulonimbus cloud is the perfect lightning and thunder factory, earning its nickname, "thunderhead."

Thunderstorms are like nature's heat pumps. At the very top of giant thunderheads, air temperatures can sometimes drop below -100°F. Sometimes, on a hot summer day, this air originates near the ground at 100°F. Thunderstorms carry the sun's energy from the surface into the cooler upper reaches of the atmosphere. Without this "convective heat transport" it is estimated that the mean temperature of the planet would increase by over 20°F, making many areas uninhabitable.

111

What appears to be smoke is actually mammatus formations on the underside of a cumulonimbus anvil cloud.

What is a **severe thunderstorm**?

By definition, the National Weather Service classifies a thunderstorm as severe if it contains hail of three-quarter inches or larger, and/or wind gusts of 58 mph or higher, and/or a tornado. Severe thunderstorm watches, meaning conditions are suitable for severe storm development during the next several hours, are issued for areas several hundred miles on a side by the NWS Storm Prediction Center in Norman, Oklahoma. A severe thunderstorm warning is issued by the local NWS office, usually for several counties over an hour or so, based upon spotter reports or radar indications of conditions exceeding severe levels. If there is a distinct threat or actual observation of a tornado, a tornado warning is issued. Tornadic storms also produce hail, winds, downbursts, and lightning, and those hazards should likewise be considered.

What creates a **thunderstorm**?

Warm, moist air rising in sufficiently large volume with a high enough velocity results in a thunderstorm. The fuel for these storms is warm, moist air present near the surface of the Earth. If the atmosphere around the cloud is unstable, that is, the temperature of the air falls faster than that of the rising parcel of air within the storm, then the updraft becomes ever more warmer than the air outside, and therefore more buoyant. The release of latent heat when water vapor turns to liquid and then the liquid to ice further warms the rising parcel, stoking the "fires" of the updraft.

A trigger is often necessary to get the warm bubble of air rising in the first place. Sometimes it can be a warm air thermal rising from a large, heated field or a sunlit mountain top, or the upward motion produced by fronts pushing air together so it has no place to go but up.

Thunderstorms are complex mixtures of water, ice, and wind. As the electric field builds within a cloud before a lightning discharge, the ice crystals are aligned by the electric field. As soon as the discharge occurs, radar studies show the ice crystals become randomly oriented again until the next charge builds up.

The energy in even a modest thundercloud can be impressive. The first atomic bomb was detonated in the desert near Alamagordo, New Mexico, on 16 July 1945. Though the energy released was awesome, it was several times less than that generated by the almost daily thunderstorms that dot the New Mexico mountains on a typical summer day.

How **high** do thunderstorms go?

Almost all thunderstorm clouds grow to heights above 20,000 feet. The more intense ones continue upwards until they hit the top of the troposphere, called the tropopause. Since penetrating into the stratosphere takes a lot of energy, many cumulonimbus clouds flatten out on the tropopause into the shape of an anvil with the tip streaming off downwind. If the storm is unusually intense, the updraft may punch into the stratosphere in cauliflower-like turrets. These "trop busters" are usually severe storms, with internal updrafts perhaps exceeding 100 mph. At any given place and time the height of the tallest storms is thus controlled by the height of the troposphere. Over the United States the tops of the stronger storms range from 40,000 to 65,000 feet from spring through summer and from north to south, respectively. There are some radar reports of echoes exceeding 70,000 feet, but if these reports are correct, this would be a very rare event. In any case, most thunderstorms are high enough that commercial jet traffic does not fly over but rather circumnavigates them since there can be "surprises" inside thunderstorm tops, including extreme turbulence, hail, lightning, and wind shears. Pilots must be sure to fly far enough away from thunderstorms not only because of turbulence, but also because hail can be ejected five miles or more from the main storm cloud.

Can there be **thunder during snowstorms**?

While not very common, occasionally thunderstorms can be embedded in winter storm systems, producing lightning and thunder. Winter "thunder snows" often result in intense snowfall rates sometimes exceeding three inches per hour. On occasion

Towering cumulus clouds define a fast-moving cold front.

thunder and lightning will be present in the most intense squalls that drift off the Great Lakes in fall and winter.

How many **raindrops** fall in a year?

For a location having 30 inches of rain a year, each square yard receives about one billion drops.

What is an **acre-foot**?

Those responsible for measuring water for irrigation, rainfall runoff volume, and reservoir capacity use the odd-sounding unit of an acre-foot. One acre-foot is 43,560 gallons of water, enough to cover an acre of land with one foot of water.

Are there **different types of thunderstorms**?

Basically any cloud that generates lightning, and therefore thunder, is a thunderstorm. However, there are a wide range of sizes and shapes of such storms, indicative of the atmospheric processes that give them birth. There are dozens of informal classifications of storm types. The air mass thunderstorm is perhaps the most common. It is relatively small and short-lived and forms in semi-random patterns within large, moist, high-pressure systems. Sea breeze thunderstorms are named for their triggering mechanism, as are cold frontal storms. Squall lines are long thin chains of storms, which on occasion have been known to extend for more than 1,000 miles along or ahead of cold fronts. Larger, non-frontal thunderstorms are often called mesoscale convective systems or, the biggest of them all, the mesoscale convective complex (MCC). These monsters can be the size of several small states and live for twelve hours or more. Supercells, which often rotate as a whole, are usually relatively small in size but long-lived, often producing tornadoes and major hailstorms. Sometimes these cells split into two pairs, one moving to the right and the other to the left. Trying to decipher which "mode" of convection the atmosphere will take on a given day is one of the challenges of contemporary forecasting.

What is the moist tongue?

In meteorology, it refers to the strong surge of moisture-laden tropical air that streams poleward, often ahead of an advancing low-pressure system. Over the United States in spring and summer, the moist tongue is centered about 5,000 feet over the central plains and is a key ingredient in thunderstorm development. It is also associated with a low-level jet stream. The persistence of such a moist tongue during the summer of 1993 contributed to the great floods of that year.

What are **downbursts** and **microbursts**?

Inside a thunderstorm there are powerful updrafts and, as the storm matures, downdrafts (what goes up does come down). The updrafts can reach many tens of miles per hour. Turns out storm downdrafts can be equally intense. The downdrafts are caused by factors such as the drag from heavy masses of rain and hail, and especially the fact that falling precipitation evaporates and cools the air, making it heavier than its environment. Most thunderstorms generate downdrafts, the cooling, outward-rushing air that often breaks the heat of any oppressively hot summer afternoon. The leading edge of the downdraft is called the gust front. It is sometimes marked by spectacular cloud features called shelf or roll clouds. On some occasions, downdrafts can become very intense, slamming into the surface with wind gusts well in excess of hurricane force. That is a downburst. The smallest of these are called microbursts, some of which may be only several hundred yards wide. Recent research has shown that much storm damage once ascribed to tornadoes is actually the result of microbursts. Their winds can equal that of small tornadoes and, to the untrained eye, the damage looks as if a **115** tornado went through the area. They can also be accompanied by very loud roaring noises.

The second greatest cause of aircraft accidents, after pilot error, are the extreme wind shears associated with thunderstorm downburst winds. Since 1964, at least 29 major airline accidents have been caused by downbursts.

How strong can **thunderstorm winds** be without a tornado?

Wind speeds above 120 mph in downbursts are not that uncommon. In 1995, downburst winds were clocked at 136 mph at Grissom Air Force Base, Indiana, with some estimates as high as 140 mph in Miami County, Indiana. Just five minutes after President Reagan landed in Air Force One at Andrews Air Force Base near Washington, D.C., the highest recorded thunderstorm microburst wind ever clocked—gusts of 149.5 mph—struck there, causing considerable damage.

What is a **"heat burst?"**

A heat burst is a rare phenomenon during or after a thunderstorm in which the temperature rapidly but briefly jumps to extreme values. One case, in which the temperature was estimated at 140°F, occurred near Kopperl, Texas, in 1966. The cause is thought to be related to extreme downdrafts associated with the storm. Though starting out cold, in a downdraft the air temperature increases about 5.4 degrees for every thousand feet of descent. If it descends far enough it can become very warm.

How many raindrops fall in a **typical thunderstorm**?

In a typical heavy thunderstorm (one inch of rain), about three million raindrops fall on each square yard.

How much does a **rainstorm weigh**?

One inch of rainfall over an acre of land weighs 226,000 pounds. If that same rainstorm occurs uniformly over a ten by ten mile area, a relatively modest size for a shower, we are dealing with 7.2 million tons of rain water.

When does a **flood have teeth**?

Heavy rains from tropical storms are not unusual in Thailand. However, in October 1995, torrential downpours had one unexpected result: a crocodile farm was flooded, allowing hundreds of the toothy reptiles to escape into the surrounding neighborhoods.

A supercell, the most severe of the severe thunderstorms, advances across the American prairie.

What is the U.S. **record for rain** in an 18-hour period?

In an 18-hour period, Thrall, Texas, received 36.40 inches of rain on 9 September 1921.

What is a **derecho**?

A derecho (der-ray-cho) is a type of severe storm. It is often very long lasting, covering many hundreds of miles, while generating a continuous series of very strong downburst winds. In July 1995, an extremely severe derecho storm swept through upstate New York. Wind gusts up to 106 mph were recorded. Striking mostly in rural areas, the storm devastated over one million acres of trees, felling tens of millions, piling trees 10 to 20 feet high in places. Worse, five people camping in the area were killed by the falling timbers.

What is an **"air pocket"**?

Airline passengers often describe hitting "air pockets" while flying as if there were potholes in the flight lanes. The turbulent ups and downs in airplanes come from hitting air with rapidly varying up- and downdrafts. These vertical motions can be formed in numerous ways, with thunderstorms being perhaps the prime cause. Airline passen-   **117** gers should keep seat belts buckled at all times, even when the buckle sign goes off. Sudden turbulence can strike without warning and cause some nasty surprises. In 1995, 20 people were injured when an airliner at 35,000 feet above Green Bay, Wisconsin, dropped suddenly in an "air pocket" and some "unsecured" passengers were bounced off the ceiling. Six of the injured were taken to hospitals after an emergency landing at O'Hare. In 1996, Air Force One was flying near thunderstorms over New Mexico with President Clinton on board when it encountered severe turbulence that bounced a number of staffers around the aircraft.

How many **thunderstorms** occur each year?

Nearly 2,000 thunderstorm cells are estimated to be present over the planet at any given time. It is estimated that globally there are 16 million thunderstorms each year. In the United States, central Florida has almost 100 thunderstorm days annually. Other areas with large numbers of thunderstorms include much of the Gulf Coast region, the Rocky Mountains, and adjacent High Plains. Kampala, Uganda, may hold the world record for thunderstorms, averaging 242 rumbly days each year, though portions of Indonesia may have them beat. Between 1916 and 1919, the city of Bogur averaged 322 thunderstorms per year.

Should you **stand under trees** during thunderstorms?

Most people seem to know that they shouldn't stand near trees in a thunderstorm due to the lightning hazard and the fact that trees blow down. In a recent 15-year period in Ohio alone, over 40 persons were injured and at least nine were killed by trees toppled during thunderstorms. Many of the victims were in vehicles. If you pull your car over during a downpour, be sure you are not in the potential path of a tree that could fall as the next gust of wind strikes.

In June 1993, a severe thunderstorm with several microbursts downed more than 13,000 trees in the northern part of Cincinnati and its suburbs. Over 60,000 utility customers lost power. A woman was killed when a tree crushed her car, which she had parked because the rainfall was too heavy for driving.

In June 1992, thunderstorms sweeping across Gebhard Woods State Park, Illinois, knocked down numerous trees. But one was special—the largest tree in the state, a 138-foot-tall cottonwood thought to be 200 years old.

How can thunderstorms cause **record low temperatures**?

Rain-cooled air masses in a thunderstorm's outflow are often the source of the coolest summertime readings in tropical areas. Temperatures only have to be cooled into the low 70s or upper 60s in order for a new daily minimum to be set.

Where can you go to **avoid thunderstorms**?

Some people enjoy thunderstorms, and some don't. For the latter, St. Paul Island in the Bering Sea off Alaska might be an appropriate destination. On 8 November 1992, they reported a thunderstorm—the first one in 40 years! The coastal deserts of Chile and Antarctica are also pretty free of thunderstorms.

Is the **Sunshine State** a good name for Florida?

Florida may be the Sunshine State, but that sun also triggers many thunderstorms, especially during the rainy season. April through October sees three-quarters of all the state's precipitation fall. Florida has more thunderstorms and lightning than any other state.

How long does the **average thunderstorm** last?

It varies considerably, from a single clap of thunder to an all-day affair. On the average, most storms last only one to three hours, with individual cells within the storms lasting some 10 to 30 minutes. Thunderstorms, while they may bring very inclement weather, at least have the virtue of being brief. In New York City, 80 percent of all thunderstorms last three hours or less.

How much rain does the **Amazon rain forest** get?

In much of the Amazon basin of Brazil rainfall averages over 80 inches per year. Surprisingly, portions of coastal Alaska average upwards of 100 inches of precipitation per year.

What is the **smallest rainstorm** in history?

In 1925, an English meteorologist reported that a rain shower fell from a small patch of altocumulus clouds at 12,000 feet altitude, moistening the sidewalk for a distance of only 20 yards.

Are there ever thunderstorms in **sunny southern California**?

Thunderstorms are rare during the winter in most parts of the United States, and they are rare in any season in coastal southern California. Yet, in Los Angeles, the "peak" of their meager thunderstorm season occurs during winter, as large Pacific storms sweep ashore. On 10 February 1992, along with several inches of rain and driving winds,

After several days of heavy rains and severe storms in February 1992, Ventura Highway, in southern California, closed due to flooding.

lightning knocked out power to over 100,000 customers in the L.A. basin. Over the years more than a few tornadoes have also touched down in the L.A. basin during the winter months. During the summer, thunderstorms are rather common further inland in southern California over mountainous areas.

What are some **record heavy rains** over short time periods?

It rained 1.23 inches on 4 July 1956, in Unionville, Maryland. Nothing unusual about that, you might say. Well, it fell in exactly 60 seconds—still the nation's one-minute rainfall record. Barst, Guadaloupe, West Indies was reputed to have received 1.50 inches in one minute on 26 November 1970.

Many parts of the western United States routinely receive less than 12 inches of precipitation in an entire year. But that amount of rain fell in 42 minutes on Holt, Missouri, on 22 June 1947. And 15.78 inches deluged Muduocaidang, Inner Mongolia, in just 60 minutes on 1 August 1977. Alvin, Texas, has the distinction of the heaviest 24-hour rainfall total in the United States, a whopping 43 inches. The Canadian 24-hour record is 19 inches, on Vancouver Island in British Columbia.

Of course, record rainfalls are all relative. In Los Angeles on 8 July 1991, a "deluge" of 0.13 inches fell, the most in a July "storm" since the quarter-inch "inundation" of 1886. But while the amount was puny, the event was noteworthy since it has only rained 13 times in July in downtown L.A. since 1877.

What are the **threats from landslides**?

When asked to name natural hazards most people will include tornadoes, hurricanes, floods, and lightning. Landslides don't usually make the list, but they should. Approximately 25 lives are lost and there are $2.5 billion in damages each year caused by catastrophic earth movements, often triggered by heavy rains or earthquakes. On 5 July 1992, motorists on I-65 through downtown Indianapolis experienced a new wrinkle—not road construction but road destruction as torrential rains caused an embankment to fail and a mud slide covered the road, closing it for three hours.

Some so-called natural disasters really have man-made origins. In 1966, torrential rain caused a massive mudslide in Aberfan, Wales. Over half a million tons of hillside slid into town, killing 144 people, including many children buried alive in their school house. Aside from the heavy rains, the major contributing cause was due to the dumping of coal mine waste into a giant heap immediately in back of the town for over 50 years.

On the average, where is the **wettest place on Earth**?

If you like the sound of raindrops dancing on the roof while you lie in bed on a lazy morning, perhaps you should move to Tutunendo, Colombia. It may be the rainiest place on Earth, averaging 463.4 inches per year.

The domestic site with the greatest annual average rainfall appears to be Waialeale, Hawaii. It averages approximately 460 inches per year. The rainfall is caused by persistent warm, humid tropical winds being lifted up volcanic mountain slopes.

Does a **rising barometer** always portend fair weather?

Generally it does, but there is a big exception. The pressure usually jumps dramati- **121** cally as big thunderstorms approach and pass overhead. This jump results from the increased weight of the atmosphere caused by the rain-cooled air in the downdraft.

What city has recorded the most rain in one year?

Cherrapunji, India, in the Himalayas, gets soaked by the annual monsoon rains. In 1942, the region experienced 1,042 inches of rainfall. Of that total, 364 inches fell in one month. In a more "normal" year, Cherrapunji averages 450 inches of rain, the most in Asia. The wettest year in the United States occurred from December 1981 to December 1982, when Kukui, Maui, Hawaii, received 739 inches of rain.

Which is the **least rainy city**: New York City, Miami, or Seattle?

Seattle. In terms of total annual rainfall, Seattle has only about 39 inches, whereas the Big Apple gets about 42 inches and Miami's heat is cooled by almost 60 inches annually.

Do you need a thunderstorm or hurricane to cause **widespread forest damage**?

No, some extratropical cyclones (large low-pressure systems) often cause considerable forest damage, especially in the western United States. An intense Pacific storm pounded the Washington-Oregon coastline on 12 October 1962. Sometimes referred to as the Columbus Day "Big Blow," this storm, with winds gusting up to 120 mph, persisted for five hours. Approximately 11 billion board feet of timber were blown down.

What are some of the **world's driest spots**?

The Sahara desert may sound like a good candidate for the driest place on Earth, but it's downright soggy compared to the deserts in northern Chile. Years often pass between rain "storms," and some places have never seen a drop. The driest place in the United States is Death Valley, California, where the annual precipitation is 1.63 inches. In terms of drought, Bagdad, California, holds the record for the longest dry period in U.S. record, at 767 days.

Often, flooding results from days of heavy rains or melting snows, with rivers gradually rising and going over their banks.

What is the **driest continent**?

Antarctica. Even though largely snow covered, the cold air holds so little moisture that annual precipitation totals only a few inches in most places. Melting and evaporation are so meager that the south polar ice sheets are up to thousands of feet thick.

How many **kinds of floods** are there?

There are several kinds of floods. The "traditional" flood results from days of heavy rains and/or melting snows, with rivers gradually rising and going over their banks. These can usually be predicted with considerable skill, providing adequate warnings that result in saving lives and reducing loss of property.

Coastal flooding can occur during hurricanes or even during large-scale storms such as Nor'Easters, which can cause elevated tides along the beaches for days at a time. Rapid rises in the sea level can result from tsunamis, which are caused by undersea earthquakes or landslides.

Flash floods are a different matter. They usually result from rapidly changing weather situations, such as the sudden development of an intense local storm over the drainage basin of a small stream or river. Rivers can rise way above flood stage in a matter of hours if not minutes. But not all flash floods are caused directly by heavy rain. Ice and log jams can suddenly let loose huge torrents of water. Natural or constructed dams can collapse due to earthquakes or landslides.

123

Which took more lives in 1992, **Hurricane Andrew** or **floods**?

While 1992 will be long remembered as the "Year of Hurricane Andrew," that violent storm took far fewer lives than flooding did. At least 87 flood-related deaths were reported in the United States in 1992. Over the last three decades, our nation's annual death toll from flooding has been 138 compared to only 27 for tropical storms.

What should you do if you encounter a **flash flood** while hiking in a canyon?

If you are hiking along a stream in a canyon and distant thunderstorm rains send a wall of water rushing downstream towards you—the dreaded flash flood—don't try to outrun the water. Rather, climb up the hillside. It may sound obvious, but over a hundred people could have done that during the 1976 Big Thompson, Colorado, flash flood. Some tried to run, but perished.

Why do so many people die in **flash floods**?

Aside from the factor of surprise (many people are caught sleeping), people just don't appreciate the power of moving water. Even six inches of fast-moving flood water can knock you off your feet. Most automobiles will float and can be swept away in only *two feet* of water. Never try to walk, swim, or drive through the swift currents of a flash flood. Nearly half of all U.S. flash flood fatalities are auto related. Never attempt to drive over a flooded road. The depth of water is not always obvious. Also, the road bed may have been washed out under the water. Dry creek beds in the southwestern United States, called arroyos, can go from dusty bone dry to a ten-foot-deep torrent of water within a minute as the thunderstorm rains drain down from surrounding higher terrain.

Is flooding in **"urban canyons"** a common occurrence?

Flash floods are very common in the canyon lands of the western United States, but urban areas are also flood prone. The huge, paved areas in urban settings increase rainfall runoff by up to six times that of rural regions. In 1987, many drivers in Minneapolis lost their cars in freeway underpasses as they tried to navigate the ever-deepening waters. Hot Springs, Arkansas, might have considered changing its name to Deep Springs after 13 inches of rain flooded large portions of the town, washing cars through the downtown streets 19–20 May 1990. In September 1977, a creek flowing through one of the most fashionable parts of Kansas City became a maelstrom, trapping people in cars and buildings, killing at least 23. Torrential rains (4 to 7 inches)

The aftermath of a desert flash flood. Dry stream beds can become raging torrents in a matter of minutes after thunderstorms dump torrential rains.

soaked the southern suburbs of Cleveland, Ohio, on 13 August 1994. One insurance company alone expected up to $2.5 million of damage per inch of rain.

What was the **Black Hills flash flood**?

On 9 June 1972, what seemed to be just an ordinary afternoon thunderstorm occured in the Black Hills of South Dakota. This storm, however, dropped some 15 inches of rain in five hours as it stalled over the relatively small (100 square mile) drainage basin of Rapid Creek. The debris carried downstream towards Rapid City clogged a spillway, allowing the waters to pool in a reservoir. When the dam broke, a mass of water and debris surged into the residential and downtown sections of Rapid City. The **125** resultant flash flood killed 238 people and caused $164 million in damages. Many of the losses occurred in areas where homes were built in known flood plains. There were rumors that cloud seeding experiments caused the disaster, but they have been proven unfounded.

What were the **Johnstown Floods**?

Note the plural. Johnstown, Pennsylvania, has twice suffered severe flooding. The first was perhaps the worst flash flood ever in the United States. On 31 May 1889, a wall of water some 40 feet wide swept through Johnstown killing over 2,200 people. In July 1977, another flash flood devastated the area. This time, improved warnings helped limit the death toll, but 77 people still lost their lives.

Which country has the **worst floods**?

Floods are a longtime problem in China. Large populations crowded along river valleys are a recipe for a flood disaster. In 1887, over 900,000 Chinese perished in major floods that devastated the nation. Even in modern times, tragic floods have struck. In 1954 alone, swollen rivers claimed more than 30,000 lives.

Coastal flooding associated with cyclones has also killed hundreds of thousands of people this century in Pakistan, India, and Bangladesh.

How much flooding resulted from **Hurricane Diane** in 1955?

In August 1955, Hurricane Diane swept through New England with winds and rains. Estimates are that in Connecticut and Massachusetts alone, 2.5 trillion gallons of rain water fell, enough to furnish Niagara Falls with a continuous flow for over two weeks. At least 186 lives were lost due to these floods.

Has **El Niño** caused flooding?

The worldwide impacts of El Niño events in the Pacific ocean ranges from extreme droughts to record-setting floods. The damage from the series of California floods that resulted from Pacific storms, just one of many such catastrophes caused by El Niño during the winter of 1994–95, was well in excess of $2 billion.

Has **a baseball game** in the Astrodome ever been rained out?

A baseball game at Houston's Astrodome was indeed "rained out." No deluge fell inside the domed structure, but flash-flooded streets made travel to the sports complex

almost impossible.

What are some examples of **severe flash floods** in other nations?

In September 1992, over six inches of rain fell in less than 20 hours in Islamabad, Pakistan. The resulting floods killed at least 1,184 people and damaged some 2.8 million homes. Over four million were left homeless. By comparison, Hurricane Andrew destroyed or damaged some 136,000 Florida homes.

What happened in the **Big Thompson Canyon** on 31 July 1976?

The Big Thompson River has its headwaters on the Continental Divide above Loveland and Fort Collins, Colorado. On the evening of 31 July–1 August 1976, a tremendous thunderstorm formed over the headwaters of the river, dumping nearly 12 inches of rain in just several hours. The scenic Big Thompson Canyon was filled with 3,500 tourists and campers, many preparing to partake in Colorado's centennial celebrations the next day. The canyon is basically a rock-lined channel, and most of the water surged downhill, wiping out much of Highway 34 along its banks. A wall of water carrying boulders, cars, and houses swept away everything in its path. The water-level gage was washed away, but high water marks suggested the previous flood record was bested by over ten feet. At least 139 people drowned. Damage estimates came in at $36 million, including 316 homes, 45 mobile homes, and 52 businesses.

What was Colorado's worst **flash flood**?

Terrible as it was, the Big Thompson disaster was not Colorado's worst flash flood. In 1921, days of heavy rain caused the Arkansas River to break through the Schaeffer Reservoir containment, sending a massive torrent towards the then-booming city of Pueblo. A warning siren was blown, but some people considered it an invitation to come watch a spectacle, and many were swept away from the river's banks. The exact death toll is unknown, but over 250 people were lost. Nearly 1,000 railroad cars were toppled or washed away. The property loss of $19 million (a staggering sum at the time) was nearly two-thirds of the entire assessed valuation of the city, which at that time was Colorado's second largest.

How can floods on land affect **marine life**?

The natural world is filled with strange and unexpected connections. For instance, rainstorms in Iowa can affect fish in the Atlantic ocean. The torrents of water exiting the mouth of the Mississippi during the summer 1993 floods caused a 10-mile-wide "river" of fresh water to wind its way through the Gulf of Mexico, around the tip of

A man treks past flooded mail boxes in downtown West Des Moines, Iowa, during the Great Flood of 1993.

Florida and out into the Atlantic Ocean. Salt water fish were observed moving further out to sea to escape the fresh water.

During the summer of 1994, torrents of water drained into Apalachicola Bay as a result of floods in the southeastern United States, all but ruining the summer oyster harvest in the waters off the Florida Panhandle. This region provides up to 10 percent of the nation's fresh oysters.

The Great Flood of 1993 in the United States affected pearl production in Japan. High waters carried predatory zebra mussels from Lake Michigan to St. Louis via the Illinois river. The zebra mussels began attacking a native mussel whose fragmented shells are used in Japan as the "seeds" for its $800 million cultured pearl industry. Local hunters also found mallards were gorging themselves on the abundant zebra mussels. Although they grew plump, the ducks also gained a fishy, repulsive taste.

What were some of the consequences of the Great Flood of 1993?

The Great Flood of 1993 resulted in a sixth Great Lake. At least that is how it looked to sensors on one of the National Oceanic and Atmospheric Administration (NOAA) satellites designed to monitor global soil moisture. Much of Iowa was so wet it was hard to distinguish its appearance from Lake Michigan or Superior.

During the Great Flood of 1993, the Mississippi River flooded the streets of Davenport, Iowa (among many others), making passage by boat the only transportation possible.

The 1993 flood saw the Mississippi River swell to widths as great as seven miles. Over the period of the flood, an area twice the size of Massachusetts was under water. Surprisingly, even larger inundated areas resulted from the floods of 1926 and 1973.

The Great Flood of 1993 severely disrupted transportation: rail, barge, highway, and airplanes. The thunderstorms that caused the floods also disrupted flights at Chicago's O'Hare airport on at least three occasions in June 1993. At least 255 commercial flights were canceled and 41 more were diverted. Also, the downtown airport at St. Paul, Minnesota, was under several feet of water.

Some of the grim statistics from the Great Flood of 1993 in the Mississippi and Missouri River basins: 48 fatalities; $15–20 billion in property and crop losses; 56 **129** small river towns completely inundated; 85,000 people evacuated from homes; 8,000 homes destroyed and contents damaged in 20,000 more; 404 counties declared disaster areas; 2,000 loaded barges stranded.

The devastation of riverside cemeteries was another grim side effect of the 1993 Midwest floods. The Hardin Cemetery in Missouri had more than 700 graves opened. Some caskets were deposited 14 miles away, and some were never found.

Trees were among the many casualties of the Great Floods of 1993 in the central United States. In Iowa alone, substantial tree deaths and reduced growth were found on over 110,000 acres of woodlands that were repeatedly submerged during the flooding.

Des Moines, Iowa (29.67 inches), Madison, Wisconsin (21.49 inches), and Columbia, Missouri (25.81 inches), all reported their wettest summers ever during the year of the Great Flood of 1993. This flood was all the more extraordinary as it had no contribution from snow melt, the usual culprit in Mississippi and Missouri river flooding.

One town just got fed up with all the water from the 1993 Mississippi River floods and voted to leave. Valmeyer, Illinois, some 30 miles south of St. Louis, voted on a $16 million project to move the entire town to higher ground.

During the height of the Great Flood of 1993, when even the massive dikes in the St. Louis area were beginning to look vulnerable, the folks downstream in New Orleans were not too concerned. The lower Mississippi can handle three times the volume of water that was squeezing through the narrow part of the river in Illinois and Missouri.

What was the weather like for the rest of the United States in the **summer of 1993**?

The summer of 1993 was one of extremes in the United States. The period June through August (meteorological summer) was bone dry in parts of the West. Denver had only 3.14 inches of rain, and Rifle, Colorado, was moistened with less than an inch (0.87 inch). But it was the wettest summer ever in waterlogged Waterloo, Iowa, with 31.00 inches.

While cities in the southeastern United States sweltered during the hot, humid, and hazy summer of 1993, the West chilled out: it was the coolest summer on record in Boise, Idaho; Missoula, Montana; Salt Lake City, Utah; Spokane, Washington; and Williston, North Dakota. Unfortunately, the mean between the two extremes was the Mississippi River Valley—and the clashing air masses produced the Great Flood of 1993. Iowa had its wettest month ever, since 1895. During the same month, Ohio, less than a day's drive down the Interstate, had its driest August on record since 1895.

What is hail?

Hail is made up of spherical balls of ice that fall from spring and summer thunderstorms. Hail forms as the result of small frozen raindrops or graupel being continuously recycled through multiple up- and downdrafts. They continuously accumulate new layers of ice until they become so heavy that they can no longer be supported. Hail is not to be confused with sleet, which is frozen raindrops that fall during winter storms.

Are floods common in **Australia**?

More often than not, Australia can be described as a hot, dusty, dry desert. These associations would normally be fairly accurate, but in March and April 1990, as much as three feet of rain fell over portions of the state of Queensland, with large portions of the state being submerged under water for a time. Heavy rains from tropical cyclones can occur in the northern part of the country. Thunderstorms accompanied by hail, tornadoes, and flash floods are also common.

What is the impact of **hail** upon **crops**?

Worldwide, annual crop losses from hail represent about one percent of the total annual agricultural production.

Hail causes nearly $1 billion in damage to property and crops in the United States each year. Illinois farmers spend the most on crop hail insurance, some $60 million annually.

Are hailstorms a **threat to animals**?

The natural world can be a cruel place. On 15 July 1978, a hailstorm spewing stones as large as baseballs killed more than 200 sheep in Montana. A hailstorm in Alberta, Canada, on 14 July 1953, was found to have killed some 36,000 ducks and thousands of other birds. Another storm several days later killed another 30,000 ducks in the same area.

Are **humans at risk** from hailstorms?

The deadliest hailstorm on record was perhaps on 30 April 1888, in the Moradad and Bareilly districts of India, where 246 people perished. There are reports of 100 persons **131** killed and 9,000 injured in Sichuan Province, China, on 22 March 1986. On 14 April 1986, 92 persons perished in a hailstorm around Gopolganj, Bangladesh. Some of the individual stones were reported to weigh in at 2.25 pounds!

By contrast there have been only two known hail fatalities in the United States this century. One happened in Texas during the 1930s. The other occurred on 30 July 1979 in Fort Collins, Colorado, where a baby was struck by grapefruit-sized hail, to the despair of its mother, who was rushing to carry the child to safety. In addition, 25 others were injured, including an 84-year-old man whose arm was broken.

What part of the United States has **the most hail**?

About 4,800 hailstorms strike the nation every year. Of these perhaps 500 to 700 produce hailstones large enough to cause damage or injury. Hailstones form within thunderstorms. Yet the region with the greatest number of thunderstorms, Florida, has one of the lowest hail rates in the nation (less than one event per year at any one place). Cheyenne, Wyoming, is the U.S. city with the most hailstorms, averaging about ten hail days per year. Hail is most frequently found in "hail alley," which covers portions of eastern Colorado, Nebraska, and Wyoming. It is also rather common throughout the High Plains, the Midwest, and sometimes the Ohio Valley. The lowest frequency of hail is along the immediate Pacific shoreline. What hail falls on the West Coast is usually small, accompanying the winter thunderstorms that blow inland during winter storms.

Summer is baseball season, and Amarillo, Texas, was pelted with baseball-sized hailstones on 19 June 1955, damaging 5,000 homes and 1,500 cars in the time it takes to play three innings. Considering a baseball-sized hailstone hurtles out of the sky at the speed of a major league fastball, one might wonder, since most of us don't wear batting helmets, that there haven't been more hail-related tragedies.

Are **large hailstones** possible in Florida?

In spite of its many thunderstorms, Florida, as noted above, has relatively few hailstorms. But sometimes mid-latitude weather systems do occur there, especially during winter and early spring. Grapefruit is nothing unusual in Florida, unless it happens to be grapefruit-sized hail. The Orlando area was bombarded with hail on 25 March 1992, with some stones reaching the size of that large citrus. Drifts of hail reached several feet deep in some places. Damages totaled many tens of millions of dollars.

What regions of the world are **hail prone**?

Northern India appears to be the region where the greatest frequency of extremely large hail events occur. India also appears to have the dubious distinction of having

the most human fatalities from hail.

The world's "hail belts" are generally found at mid-latitudes, often downwind of large mountain ranges. The High Plains of the United States and Canada, central Europe eastward to Ukraine, the Himalayan region, southern China, and portions of Argentina, South Africa, and southeastern Australia all have to deal with outbursts of hail.

Keriche, Kenya, averages 132 days per year with hail. This may be by far the highest frequency anywhere in the world. One theory holds that large amounts of pulverized tea-leaf litter from the local tea plantations get stirred into the atmosphere and serve as excellent "ice nuclei" once in the rain clouds overhead.

Are all **hailstones** round?

More often than not, hailstones are basically spherical. But not always. They sometimes have strange protruding spikes, or they are star shaped or highly oblong. In 1979, in Norwich, England, a nearby thunderstorm generated flakes of ice about two by four inches in size, which fluttered out of the sky like falling leaves.

Has hail ever affected **world history**?

Probably in more ways than one. During the summer of 1788, severe hailstorms wiped out much of the crops in the region around Paris. That, along with other factors, contributed to a severe shortage of food, which further fueled civil unrest—leading to the French Revolution.

What do the **rings** mean in hailstones?

Alessandro Volta, in the year 1806, is believed to be one of the first scientists to study hailstones. Upon cutting some specimens open he noted the alternating bands of clear and cloudy ice, indicative of multiple trips through up- and downdrafts before the stone became too heavy to be supported by the storm's updrafts. He also correctly speculated that at the center of each stone is an "embryo," often composed of a snow flake or a frozen raindrop. As many as 25 concentric rings have been counted in larger specimens.

What is the **most damaging hailstorm** ever?

It would appear that Munich, Germany, holds that record. In 1984 a massive hailstorm caused at least $1 billion in damages.

The costliest hailstorm in U.S. history pounded the Colorado Front Range, including the City of Denver, on 11 July 1990. The resulting property losses were at least $625 million. Forty-seven people were injured at an amusement park, some seriously, when a power failure trapped them on a Ferris wheel and they were battered by softball-sized hail.

A three-inch hailstone.

How are **hailstone sizes** categorized?

There was a lot of hail in the central United States during the hot month of July 1991. Marble-sized hail reports were legion. Golfball-sized hail was reported in Illinois, baseball-sized hail struck in Iowa, and softball-sized hail fell in Colorado. There was even one unconfirmed (and highly suspect) report of *basketball*-sized hail near Manhattan, Illinois. Weather observers report hailstone sizes either in inches or in the following equivalents:

0.25 inches	pea
0.50	marble
0.75	penny/large marble/dime
0.88	mothball/nickel
1.00	quarter
1.25	half dollar
1.50	walnut
1.75	golf ball
2.00	hen egg
2.50	tennis ball
2.75	baseball
3.00	tea cup
4.00	grapefruit
4.50	softball

These classifications are used in the United States. Other countries probably use different terms.

How **large** can **hailstones** become?

No one knows for sure the absolute upper limit of hailstone size. In the spring of 1995, less-than-credible reports circulated on the Internet from the Guandong Province, China, of falling hailstones, some of which were reported to be the size of basketballs.

The largest *documented* hailstone ever found thudded to Earth on 3 September 1970, in Coffeyville, Kansas. It was 17.5 inches in diameter and weighed 1.67 pounds. The record Canadian hailstone fell in Cedoux, Saskatchewan, and weighed in at 10.23 ounces. The largest hailstone that has ever fallen? Reports of hailstones in India "the size of elephants" have been dismissed. But a 1925 report from Germany of a single stone weighing 2.04 kg (approximately four pounds) is believed to be possible. It could have been several large hailstones frozen together.

And then on 30 April 1985, a 13-year-old boy in Hartford, Connecticut, was shocked when a 1,500-pound block of ice fell from the sky into his back yard near where he was playing with a friend. The source of the six-foot-long and eight-inch-thick slab, which probably had an impact speed of 200 mph, was never identified, but it was not a natural hailstone.

What is graupel?

Graupel is another form of frozen precipitation that is sometimes called soft hail. The word graupel is derived from a German word for barley, which it resembles. Graupel is usually present inside thunderstorms, and occasionally reaches the surface. Typical times for graupel showers are on cool afternoons in spring, or near mountain tops when you are much closer to the base of the cloud, or perhaps even inside the cloud. Graupel rarely get larger than a quarter of an inch, and are often shaped like rough pyramids. At times heavy graupel showers can whiten the ground.

How deep can **accumulations of hail** become?

Some hailstorms can drop staggering quantities of hail. And since ice floats, the hail is then swept along with the runoff, piling up in huge drifts. In Colorado and Wyoming it is not at all uncommon to call out snow plows in summer to clear main highways of hail drifts. In Orient, Iowa, hail drifts in August 1980 reached six feet in depth and required days of plowing and melting before they were gone.

How big are **hailstorms**?

Hail often falls in swaths or streaks beneath its parent thunderstorm. These swaths are typically five miles long and a half-mile wide. But one 1968 superstorm in Illinois spawned a hail streak some 19 miles wide by 51 miles long with 82 million cubic feet of ice. The largest hail swaths are typically on the order of 10 by 100 miles.

Can hailstorms be **controlled**?

People have certainly tried. In fourteenth-century Europe, people attempted to ward off hail by ringing church bells and firing cannons. Modern versions of hail cannons are still in use in parts of Italy. After World War II, scientists worldwide experimented with "cloud seeding" to reduce hail. In Soviet Georgia, scientists fired silver iodide into thunderclouds from the ground. The idea was to make more, but much smaller and less damaging hailstones. There is little evidence that any of these techniques work.

What are some of the more **bizarre things** to fall from the sky?

In December 1933, huge hailstones were reported to have fallen around Worcester, Massachusetts—containing fresh, frozen ducks inside.

A thunderstorm that occurred in Germany on 9 August 1892 resulted in a heavy fall of rain . . . and hundreds of fresh water mussels.

A thunderstorm near Vicksburg, Mississippi, in 1930 resulted in some rather interesting hailstones. One had a solid piece of alabaster three-quarters of an inch across. Another was an eight-inch-long gopher turtle, entirely encased in ice.

Bucharest, Romania, reported a rainstorm on 25 July 1872. Along with the rain fell numerous small black worms that littered the streets.

On 17 July 1841, a shower of heavy rain and hail in Derby, England, was accompanied by hundreds of small fish and frogs, many of them alive.

Scientific American, in 1877, reported that a rain of snakes, some up to 18 inches long, fell over the southern part of the city of Memphis, Tennessee.

In 1687, hail fell in England containing the seeds of ivy berries.

In 1953 the town of Leicester, Massachusetts, was deluged by a fall of toads. Children were able to gather them up by the bucketful.

The term hailstone is a bit of a misnomer since no rocks are involved. However, in Sweden in 1925 a large chunk of limestone did fall from the sky, smashing into myriad pieces upon impact.

In 1961, a portion of Shreveport, Louisiana was pelted by small, unripe peaches during a passing thunderstorm.

In Spain, in 1902, a rain shower was observed in which the rain drops, upon touching the ground, gave off a crackling noise and emitted electrical sparks. The

event lasted for less than ten seconds.

What are some typical **consequences** of severe storms in recent U.S. history?

The following are typical of the many thousands of reports of severe weather received each year in the United States.

Dane County in southern Wisconsin was raked by severe thunderstorms on 7 July 1991. Winds rose from 2 mph to 80 mph in just thirty seconds. A hangar containing 20 aircraft collapsed at the Dane County Regional Airport, with damage at the airport alone in the $5 million range. And just to add insult to injury, a lightning strike drilled a hole in the airport runway.

On 21 June 1991, thunderstorm winds flipped nine aircraft at the Downtown Airport in Kansas City, twisting some like pretzels, causing damage up to $250,000. The winds were so strong that airport employees abandoned the tower.

Farmers complained mightily about the unusually wet weather in Iowa and southern Minnesota during May and June 1991. One farmer was reported to have taken his son waterskiing across a flooded field.

As up to seven inches of rain fell near Atlantic, Iowa, on 13 June 1991, creeks began to overflow. A local petting zoo was flooded, forcing rescuers to save 15 pygmy goats.

Stockton, Minnesota, population 500, was the scene of a flash flood from heavy thunderstorm rains on 22 July 1991. After only a three-minute warning, a wall of water 12 feet high smashed through the small town, causing major damage to 250 homes. Fortunately no one was killed or injured.

Some of the most severe thunderstorms ever to strike southern lower Michigan ripped across the state with wind gusts over 80 mph during the second week of July 1991. At one point, over 900,000 homes and businesses were without power. One utility reported that a total of 760 transmission lines and 1,770 secondary distribution lines were disrupted. Over 250 wooden utility poles were replaced. Repair crews came from as far as Chicago to help get the state plugged back in. One death occurred when a woman was crushed by a falling tree. The other four fatalities were electrocutions due to contact with fallen power lines.

LIGHTNING AND THUNDER

What is **lightning**?

Lightning refers to one of the several forms of visible electrical discharge produced by thunderstorms. The primary forms of lightning discharges are cloud-to-ground (CG), cloud-to-cloud (CC), in-cloud (IC), and cloud-to-air (CA). There are other rare forms, such as ball lightning. Lightning appears to be very bright because it is—its optical output is equivalent to approximately 100 million light bulbs going on and off.

Why does lightning appear to **flicker**?

When you see a flash of lightning, it often appears to flicker. If you were to videotape that flash and play it back in slow motion, you would find that it is composed of a succession of multiple strokes that usually follow the exact same path as the initial stroke. Each actual stroke lasts on the order of millionths of a second, separated in time by several tens of milliseconds (thousands of a second). The entire flash can sometimes last for a quarter-second or more. Some flashes have a very large number of strokes—the world record is thought to be 47 strokes in a single flash.

Does lightning **strike up or down**?

It goes both ways. Lightning is essentially a giant spark, but it occurs in a rather complex manner. There is often an initial breakdown within a cloud. Then a faintly luminous stepped leader emerges from the cloud and progresses in a series of jagged motions towards Earth at an average speed of around 200,000 miles per hour. As the **139**

One of the 100 or so lightning discharges that flash around the planet every second.

stepped leader approaches the ground, an upward streamer emerges from the object about to be struck. When the two meet, this completes the path to the ground, and the cloud is short-circuited with a brilliant, luminous, 60,000-mile-per-second return stroke from the Earth to the cloud. If the flash has more than one stroke, a dart leader emerges from the cloud and follows the same path to ground without branching, and as it approaches the object to be hit, another upward streamer emerges, resulting in the next return stroke. Occasionally the stepped leader originates at the ground (more likely from a tall building or tower) and moves upwards. The branching of the first stroke would then look like an upside-down tree. This could be considered true "upside-down lightning."

What are the **electrical currents** within lightning discharges?

They are highly variable. The average lightning stroke has a peak current on the order of 30,000 amps. But some discharges, especially those that are totally within the cloud, are only a few thousand amps. On the other hand, superbolts do occur, occasionally reaching 300,000 amps or more. The electrical potentials involved in lightning discharges can reach up to 200 million volts.

How many kinds of **cloud-to-ground flashes** are there?

There are basically two distinct types of CG flashes—negative CGs and positive CGs—though they generally look the same to the naked eye. The negative CG is by far the most common (perhaps 95 percent of the total); it lowers negative charge to the ground. The less common positive CG lowers positive electrical charge to the ground. The positive CGs typically have stronger peak currents, are less likely to have multiple strokes in a flash, and often have a continuing current that lasts many tens of milliseconds and that allows the struck object to be intensely heated, causing ignition. The positive CGs are thought to be the chief forest fire starters.

How **hot** is lightning?

The air in the core of a lightning bolt has been estimated to be heated to as much as 54,000°F. That happens to be about six times hotter than the surface of the sun.

How often does **lightning** occur on our planet in a year?

There are probably 1,500 to 2,000 thunderstorms active around the world at any given time. It has been estimated that lightning flashes approximately 100–125 times per second on a global basis. Lightning therefore flashes approximately three billion times each year across the whole planet.

Can you survive a **lightning strike**?

Yes—in fact the majority of the people struck do not die. Estimates of the mortality rate from lightning strikes range from 5 percent to 30 percent. But your odds of survival are significantly enhanced if you are struck in the presence of someone who knows CPR. The basic rule to follow if several people are struck by lightning is to "treat the dead first." Often lightning victims appear dead but are in fact in cardiac arrest. The immediate application of CPR can bring them back to "life." More often than not, people who have been struck and are screaming and howling, while obviously in pain, are usually in less imminent medical danger. Many people think that someone who has been struck by lightning is still somehow electrically charged. This is not true. There is no danger in touching a lightning strike victim.

More people are killed by lightning every year than by rattlesnake bites. You have less than 3 chances in 100 of actually dying from a rattler bite.

Can you **survive a lightning strike** more than once?

The answer is yes. In fact, at least one person has been hit by lightning seven times and survived. American park ranger Roy Sullivan was struck seven times between **141**

1942 and 1977. Known as the "Human Lightning Rod," Mr. Sullivan survived being struck even though his hair was set on fire twice and he suffered burns on various parts of his body.

Though lightning is hot, the discharge usually lasts such a short time that serious burns do not necessarily occur. And electrical current usually prefers to stay on the outside of an object. Unless the discharge goes through the heart or spinal column, a victim will often survive.

Can lightning **strike twice**?

Absolutely. As mentioned, in a typical lightning flash, several strokes often hit the same spot in rapid succession. Tall structures and buildings—such as the Empire State Building in New York City, the Hancock Building in Chicago, and the CN Tower in Toronto—are hit many times each year and are used in gathering information for lighting research. In general any object struck by lightning is more likely to be struck again than something that hasn't been zapped.

In odd cases, lightning even appears to strike twice with some malevolence. On 8 August 1937 three persons were killed by a bolt that struck the Jacob Riis Park Beach in New York. On 7 August 1938, almost one year to the day later, lightning struck the same beach and again killed three beachgoers. A home in Arvada, Colorado, was struck during a summer thunderstorm and sustained considerable damage. Just as repairs were nearing completion six weeks later, the house was hit again, resulting in $30,000 additional fire damage.

Can people have a **genetic propensity** for being struck by lightning?

Not likely, but there are oddities. A Midwest woman was struck by lightning in 1995. Not long before, her nephew had been struck and suffered temporary blindness. Her cousin was dazed in the 1970s when lightning struck her unfolded umbrella. The same woman had been struck once before—in 1965. Her grandfather was killed by lightning on his farm in 1921. And his brother was fatally struck by lightning while standing in the doorway of his house in the 1920s.

How many people are **killed and injured** by lightning each year?

Contrary to the popular belief that getting hit by lightning is a low probability event, it actually isn't all that uncommon. Over 7,000 Americans have been killed by lightning in a recent 34-year period. Your chances of being struck by lightning in the United States are between 1 in 250,000 and 1 in 400,000 in a given year. The chances

In this overhead view of a violent thunderstorm,
shadows are cast by two boiling turrets marking updrafts so strong they punch through into the stable stratosphere.

increase, of course, if you happen to be golfing, swimming, or boating (or just outside, for that matter) during a thunderstorm. In the United States alone, between 75 and 150 people are reported killed by lightning each year, with 5 to 30 times that number suffering injuries. And these statistics are thought to be underestimates of lightning casualties. It is possible that many lightning victims' cause of death is listed as burns or cardiac problems.

The deadliest month for lightning fatalities and injuries in the United States is July. The large number of thunderstorms combined with numerous vacation trips and other outdoor activities yields this deadly total.

Do lightning strike victims **burst into flames**?

In spite of the popular conception that this happens, those struck by lightning do not become fully engulfed in flames. Some of their clothing may be singed and smoking and they may have burn marks, but flames do not occur.

What is a **fulgerite**?

A fulgerite is fossilized lighting. It forms when a powerful lightning bolt melts the soil into a glass-like state. Several years ago in Michigan, a record-large specimen of fulgerite was found that measured 15 feet in length and had a white-green-gray color.

Visitors to the Great Sand Dunes National Park in Colorado get to see examples of fulgerites extracted from the nearby sand mountains. And in order to make a point to the many hikers who take off for a long stroll through the dunes, the last sign they see warns them to take cover in thunderstorms and "not to become fulgerites."

How safe are you from lightning **inside an airplane**?

Commercial airliners are generally quite safe during electrical storms. A commercial plane is, on the average, struck by lightning twice per year. Not to worry—the metal skin of the plane conducts the current along the outside of the aircraft, and fuel tanks are now designed to prevent entry of electrical charges.

The last major U.S. commercial airliner crash caused by lightning was more than 35 years ago and was the worst lightning-related aviation disaster ever. It occurred over Elkton, Maryland, on 8 December 1963. Lightning struck and penetrated the reserve fuel tank of the plane, igniting the vapors and causing a crash that killed 82 people. This tragedy resulted in numerous design changes in aircraft to prevent such accidents.

Flying through a thunderstorm can be a bouncy and sometimes unnerving experience. But while the up- and downdrafts can be a potential hazard, you really don't have to worry too much about lightning. If a plane is struck by a bolt, passengers might see a flash and hear a bang, but the strike is usually no worse than that.

Exceptions do still occur, however. On 4 August 1992, a DC-10 flying from Denver to Minneapolis flew into a thunderstorm and took a direct hit. In addition to burned out electronics, some of the rivets on the fuselage were damaged. The plane managed to land safely. In 1995, a lightning bolt struck an MD-80 aircraft as it was pulling away from the boarding gate in Phoenix. None of the 131 passengers on board were injured, but three airline workers outside were hurt seriously enough to require hospitalization. The plane made its scheduled trip to Chicago without incident.

How safe are you from lightning **inside your car**?

You are generally safe from lightning while inside a car. The rubber tires provide some shielding, but it is the metal body on the outside of the vehicle that provides the protection by offering a safe path for the current to flow to the ground.

Being struck in a vehicle can be terrifying, however. Lightning struck and severely damaged a pickup truck in Blue Earth County, Minnesota, on 7 August 1994. The occupants were uninjured. A car traveling on I-35 near Des Moines, Iowa, was struck directly by a bolt of lightning. The car stopped dead in its tracks, but the startled driver was uninjured. The car had major electrical damage, many small holes in its body, and all four tires eventually went flat. The roadway beneath the car had a one-yard-wide, several-inches-deep crater. The driver's first name was Rod and the most enduring effect of the incident was that his friends started calling him Lightning Rod.

In Michigan, two motorists had an accident during a rainstorm. While huddled under an umbrella watching the tow truck hoist their wrecked car, lightning struck the umbrella and hit the motorists and then jumped over and shocked the tow truck driver. None was seriously hurt.

How safe are you from lightning **inside your home**?

Being inside a building during a lightning storm is generally quite safe, as long as you are not talking on the phone (cordless or cellular excepted), holding on to plumbing fixtures, or working with electrical appliances.

There are some exceptions, however. A lightning bolt struck a house in Denmark, went down the chimney, knocked plaster off the living room walls, ripped curtains to shreds, and smashed a clock to bits, all while leaving a caged canary inches away unfazed. It continued on, breaking 60 window panes and all mirrors, blasting through the door into the back yard, and finally killing a cat and a pig before burrowing into the ground. When a house in Iowa was severely damaged by a lightning bolt, a pile of twelve dinner plates was found to have every other one broken.

On 24 October 1991 a resident of Chicago Heights, Illinois, was comfortably sleeping in bed when lightning hit; it traveled through a cable television line into the house, where it struck and set fire to the bed. The person was treated for shock.

Is lightning a sign from a **higher power**?

Many cultures throughout history have believed that it is. Ancient Romans saw Jove's thunderbolts as a sign of condemnation and denied burial rites to those killed by lightning. Some cultures have made medicines from stones struck by lightning. Roman, Hindu, and Mayan cultures all held the belief that mushrooms arise from spots where lightning has hit the ground. In ancient Greek mythology, Zeus was the great bearer of rain, thunder, and lightning. Spots struck by lightning were frequently fenced in by Athenians and consecrated to Zeus.

Does lightning ever have any **positive effects**?

It does make for some fantastic visual displays and stunning photographs. It also fixes the nitrogen in the air, which is used by plants. And there are some unusual incidents in which lightning had beneficial results.

According to an article published in *Scientific American* in 1856, an intense lightning discharge hit the ground in Kensington, New Hampshire, and made a hole about one foot wide and thirty feet deep, forming a well that soon filled with good water. **145**

A Greenwood, South Carolina, man (by profession an electrician) survived a direct strike by lightning 28 years ago. Since that time, he has never been cold. He can stay outside for hours in sub-zero temperatures, wearing summer clothing, without discomfort.

There are several stories of blind people regaining their sight after being struck.

It is reported that lightning once struck a house in Minnesota, setting it afire. The bolt then leaped across the street, striking a fire alarm box. The power surge resulted in an alarm being sounded, and the fire department responded promptly and put out the house fire.

There is a published claim of improved intelligence on psychological testing after a lightning strike. A woman in southern Illinois believes she became psychic after being hit.

Can there be lightning during a snowstorm?

Lightning is usually associated with thunderstorms, and therefore is thought to be a spring and summer event. Yet lightning does occur during winter, and even during heavy snowfalls and blizzards. Winter lightning appears to be unusually powerful, associated with loud and long thunderclaps. Sometimes snowfalls associated with lightning can reach up to three inches an hour. A man was struck by lightning during a blizzard in Minneapolis in March of 1996.

If a lightning flash takes only a fraction of a second, why does thunder last so long?

While we see the flash virtually instantaneously, the beginning and end points of lightning might be five or more miles apart. Due to the fact that sound travels more slowly than light, it takes more time for the shock wave to reach our ears. If the lightning channel was two miles long, and assuming it started directly overhead, it would take at least 10 seconds for the rumbling to stop.

What is the fair weather electric field?

If you had a "volt meter" measuring device, you could demonstrate that even in the fairest of weather the atmosphere has a potential gradient. Near the ground the electric potential in the air is roughly 100 volts per meter in the vertical. The total potential through the entire depth of the atmosphere is around 300,000 volts.

Are thunderstorms the only source of lightning?

Lightning is usually associated with thunderstorms. On a few occasions, though, it has been observed within giant steam and debris clouds from erupting volcanoes. Lighting and miniature tornado-like vortices were observed during the spectacular volcanic birth of the island Surtsey near Iceland. Giant plumes of smoke from large forest fires also have been known to produce lightning, although these smoke clouds were probably in the process of turning into regular thunderstorms. In the western United States, most forest fires are started by lightning. Sometimes the heat from the intense fires trigger new thunderstorms, which in turn can produce more lightning. This is called a feedback loop.

Can you make lightning indoors?

Easily, although on a rather small scale. When the relative humidity indoors is very low, which it often is in the winter, static electricity on your shoes and clothing can generate notable electrical discharges. It can also result in static cling on clothing. Each inch of spark represents a potential difference of 40,000 volts. So a three-inch discharge represents a 120,000-volt potential difference. That is why you want to protect your computer from static electricity.

Does lightning give off radiation besides light?

In 1895, William Roentgen discovered X-rays. Recently, atmospheric scientists were surprised to find that thunderstorms can produce X-rays during lightning discharges. Of course, lightning also radiates radio energy over a broad range of frequencies. Some of this energy is within the AM broadcast band, producing the familiar static heard on many summer afternoons and evenings. Another name for this static is sferics, as in atmospherics.

Where does lightning like to strike?

Lightning strikes most portions of the globe sooner or later, but it does have its favorite locations. Weather satellites suggest that the vast majority of lightning strikes to the planet occur over land areas, even though land constitutes only about a quarter of the Earth's surface. Not surprisingly, the tropics receive two-thirds of the lightning bolts. But some mid-latitude storms, such as those that roam the interior of the United States during summer nighttime hours, can produce prodigious amounts of lightning.

The lightning rod is still widely used to conduct lightning strikes harmlessly down cables into the ground, protecting buildings and homes.

What **sporting activities** are prone to lightning danger?

Virtually anything you do outside during the spring and summer involves a lightning risk. Swimming, boating, hiking, golfing, soccer—all present dangers. Most lightning deaths in the United States occur (in descending order)in open fields or ball fields; under trees; while boating and fishing; near tractors and heavy equipment; on golf courses; and on telephones (but not cellular or cordless ones).

A young man fishing in Indiana in 1993 was struck when the fishing rod he was carrying over his shoulder was hit by lightning as he walked away from a pond. He was hospitalized but did recover.

During a frisbee match in Nashville, Tennessee, on 10 April 1994, one person was killed and eighteen injured by a lightning strike.

Golfers are not struck by lightning significantly more often than other outdoor sports participants. However, there are many stories of lightning strikes on golf courses. Golfers Lee Trevino and Jerry Heard were both struck by lightning during the 1975 Western Open in Chicago. Both recovered after hospitalization. Over the years, hundreds of other golfers who failed to leave the course after lightning was spotted have been far less fortunate. In Minneapolis on 13 June 1991, one spectator was killed and five others injured while taking shelter under a tree during the U.S. Open Golf

Tournament. Then on 29 June four people were injured when lightning struck at a course in nearby St. Paul.

A 37-year-old man was killed by a bolt while golfing near Louisville, Kentucky. Two others were injured. All were standing under a cluster of trees. The safety rules are the same for all outdoor activities. If lightning threatens, get inside—and *don't* seek shelter under trees.

Outdoor recreation during thunderstorms is risky, and so is working outside. Postal employees, construction workers, and many others need to exercise caution. Five miners were killed in Texas in May 1985. They were all taking a lunch break sitting under a 35-foot oak tree. Only a single lightning flash was seen in a classic "bolt from the blue" scenario.

Cowboys and farmers are also at risk. One Utah cowboy was literally blown out of his saddle when struck by a lightning bolt in August 1993. He found a hole in his felt hat, his hair was melted in several spots, and he had numerous burn marks on his torso. His faithful steed did not survive.

Four amusement park workers were injured after being struck by a bolt of lightning while they were dismantling a ride in Warrick County, Indiana, on 2 September 1991.

Do lightning rods work?

Lightning rods, invented by none other than Ben Franklin, neither attract nor repel lightning bolts. They do, however, provide a safe path to ground for the flash. Indoor plumbing, which includes pipes buried deep in the ground and vents extending above the roof, has long served as a surrogate lightning rod for homes. However, the trend toward using PVC pipe instead of metal means lightning protection has vanished from many newer homes. To be effective, lightning rods must be properly grounded, and there should be no sharp bends in the cable leading from the air terminal (the pointed rod) to the grounding rod.

What role does lightning play in **forest fires**?

Lightning is the leading cause of forest fires in Alaska and the western United States. In the past decade, over 15,000 lightning-induced fires have burned over 2 million acres of forest nationwide. Arsonists and careless people who do not watch their camp fires are still a major cause of fires, especially near large cities.

149

Does it need to be raining for **lightning to strike**?

It is a myth that if it is not raining there is no danger of being struck by lightning. Bolts can and often do strike as far as 10 miles outside of the rain area of the parent storm. Recent research on lightning deaths found that most fatalities occurred in the period when the storm appeared to be ending. During the height of most thunderstorms, people are inside seeking protection from the rain. For approximately 10 minutes after the rain ends, and even after the sun comes out, lightning is still a threat. Remember, if you can hear thunder, you are close enough to the storm to be struck by lightning. Move at once to a sturdy building or vehicle.

How can you tell how far away a **lightning bolt** was?

How close was that lightning bolt you just saw hit the ground? Count the seconds between the flash of the lightning and the bang of the thunder, and divide by five, and you have the answer in miles. Sound travels at about 1100 feet per second, so timing the interval between seeing the lightning and hearing the thunder is a good indicator of how close you were to the actual strike.

Where should you go for protection when a **thunderstorm** is approaching?

A car or truck (with windows closed) or the inside of a building are the safest places. Where not to go? Avoid standing under trees, or near fences, railroad tracks, tents, hilltops, or golf carts. Also avoid holding onto telephones, electrical appliances, or plumbing. And stay out of the water! Open-sided rain shelters are not particularly good protection from lightning, either.

Could we **harness the power** of lightning?

First of all, "catching" lightning is not the easiest thing to do. But even if we could capture and store a bolt, there is less energy there than you might think. Though very powerful while it lasts, the typical stroke only lasts for millionths of a second. If the total energy of a single lightning flash were captured, it would run an ordinary household light bulb for several months.

Lightning is one reason why U.S. electricity bills are not lower—lightning strikes destroy more than $100 million worth of utility power transformers each year.

What if you are **caught outside** in a lightning storm?

If you are caught outside in a lightning storm and can't make it to a car or building, then get away from isolated trees, tall objects, and hilltops. (Being deep inside a grove of trees is safer than being exposed in the open.) Do not be the highest point around. Avoid direct contact with other persons (in a group, don't hold hands or hug each other), get into a ditch or shallow depression if possible, crouch down with feet together and with your hands on your knees. Remove metal objects such as belts and jewelry.

Is **lightning** related to **rainfall**?

Lightning is produced inside thunderstorms as a result of the formation of precipitation particles. For each lightning bolt that hits the ground, on the average about 200,000 pounds of rain are also formed.

Do we understand everything about **atmospheric electricity**?

Not yet. In 1991, two young girls near Bristol, England, were playing frisbee. Suddenly the disk was hurled back at one of the girls by some unseen force. Then both were enveloped in some sort of "yellow bubble." They received slight electric shocks, were thrown to the ground, and had problems breathing. Eventually they freed themselves and ran home, terrified by this inexplicable experience.

What is a **bolt from the blue**?

Quite literally, it means a lighting bolt that came from an area of blue sky. Lightning bolts can on occasion jump 10 or more miles out from their parent storm cloud and appear to strike in a region with blue skies overhead. Such was the case in 1995, when lightning struck a playing field near Miami, Florida, injuring 10 children and a coach. The skies were clear save for a line of clouds to the distant northwest.

Does lightning have **gender preferences**?

Lightning does seem to be picky about what it hits. Some studies suggest that it preferentially strikes oak trees over other species. And it certainly seems to be sexist, striking and killing men far more often than women. In Great Britain, over a two-decade period, 85 percent of lightning fatalities were men. In a recent study of lightning fatalities in Florida, 87 percent of the people struck by lightning were males. It is notable, however, that of all the people in this study that were struck by lightning, 44 percent of the women were killed by the lightning, while only 34 percent of the men died from the blast.

Lightning is six times hotter than the surface of the sun, and can have a peak current of more than 300,000 amperes.

What is **ball lightning**?

Ball lightning is one of nature's most mysterious phenomena. Usually seen during violent thunderstorms, the spheres of glowing light are typically the size of bowling balls or basketballs. They can last from a few seconds to many minutes. The spheres can simply vanish into thin air, but can also pass through window glass and screens, leaving burn marks behind. Not every scientist is convinced the phenomenon even exists. But there are numerous credible reports of balls of fire floating through the air, often after nearby lightning strikes. They usually do not cause much damage and even seem playful. They have been known to roll down the aisles of airliners or pass through an open window into a startled resident's bedroom.

On 8 June 1972 a two-inch diameter hole was punched through the window of an empty office in Scotland during a thunderstorm. Since the glass was melted and fused around both the inside and outside of the pane as well as on the circle of glass found on the nearby floor, it was presumed that ball lightning had passed through the window. Coincidentally, the office was in the University of Edinburgh's Department of Meteorology.

In Wales, on 8 June 1977, a brilliant yellow-green transparent ball bounced down a hillside. It was visible for about three seconds, and what was most remarkable about the ball was that it reportedly was nearly the size of a bus.

In 1996 in Gloucestershire, England, ball lightning entered a factory. Blue, white, and orange, it traveled along girders and machinery around the building, sending off a shower of sparks. The ball then hit a window and disintegrated. The incident lasted just two seconds and only damaged the company's phone system.

How often does lightning strike the ground in the United States each year?

Lightning detection networks suggest that bolts hit the ground some 25 to 30 million times per year. The lightning hot spot of the United States is central Florida. In a typical year each square mile of central Florida is struck approximately 10 times. Most parts of the country east of the Rockies are hit at a rate one-tenth to one-half that of Florida.

Does Florida always have the most lightning?

During the summer of 1993, when torrential rains caused disastrous flooding in the Midwest, the national lightning hot spot was located in Missouri, not Florida. That was the first time Florida did not have that honor since lightning detection networks were created.

How wide is a bolt of lightning?

A big lightning bolt might seem to be hundreds of feet across when it hits the ground, but in actuality the current channel is generally not much thicker than a pencil.

How far away from the lightning bolt can you hear the thunder?

Generally thunder cannot be heard much more than 10 miles from its source. In a city, where ambient noise levels are high, thunder is often audible only when the lightning strikes a mile or two away.

Can lightning come from ice?

As strange as it sounds, lightning, with a temperature hotter than the surface of the sun, only forms in clouds with large quantities of ice. Electric charge is generated during freezing and melting processes in the presence of snow and supercooled water droplets.

Are **lightning deaths** on the increase?

In the United States, lightning deaths per million citizens have declined some 70 percent since the 1950s, although the number of serious injuries has only dropped slightly. The reduction in deaths is probably the result of several factors, including widespread knowledge of CPR, which can be used to revive lightning victims.

But with more and larger outdoor gatherings and concerts taking place—outdoor crowds during summer pose a special hazard—the chances of a major lightning strike disaster are increasing. In July 1991, at least 22 people were injured when lightning struck a crowded beach in Potterville, Michigan. None of the victims was in the water.

If you are not **hit directly by lightning**, are you safe?

No. Lighting can travel through the ground for a considerable distance from where it strikes. It can easily enter your body through your feet. Four-legged creatures like cows and horses are even worse off, because their feet are further apart than ours, a fact that increases the electrical potential difference. On 8 June 1993, 10 cows were killed by lightning in Trempealeau County, Wisconsin.

Lightning could hit a tree, and then a phenomenon called side flash can cause a streamer to jump sideways to a nearby object. Grim photographs exist of a person being directly struck by a bolt, followed by a side flash leaping to the person standing several feet away. Both were killed. You could also be on the telephone telling your friends what an impressive thunderstorm you are having when lightning could strike the phone line and travel through the phone to you. Several people are killed this way each year. Cordless or cellular phones are safe to use during lightning storms.

Does **carrying an umbrella** increase your chances of being hit by lightning?

Probably. Even more risky would be standing next to a boat mast, leaning on a conductive object such as a metal fence, sitting on a railroad track, or swimming.

On 31 August 1991, lightning struck two recreational craft in the Gulf of Mexico off Panama City, Florida. The toll: 2 killed and 9 injured. Any time you are exposed in a thunderstorm, in particular near a tall, attracting object such as a tree or a boat mast, there is a real risk of being struck by lightning.

How long can a **lightning discharge** be?

If it is a "cloud-to-ground" bolt, then travel distance is limited by the distance from the interior of the cloud to the ground, which is rarely more than 10 miles. But cloud-to-cloud or intracloud flashes may reach for 100 miles or more in rare cases.

Are mountain climbers at special risk from lightning?

Yes. Falling off a cliff isn't the only risk for climbers. Between 1980 and 1991, lightning killed at least 50 people in Colorado—20 of them were killed while climbing or hiking. Mountain hikers should plan their climbs early in the day before most storms start. If caught in a thunderstorm, don't stay mounted on horseback. If in a group, spread out. Cars (if you can find one) are a safe haven. Taking shelter under an isolated tree can be deadly, but the cover of a dense forest canopy is relatively safer.

What was the **billion dollar lightning strike**?

In July 1977, a bolt took out a major power line in upstate New York, resulting in a massive 24-hour blackout in New York City. The resulting losses from looting were estimated at over one billion dollars. There was also a surge in the birth rate nine months later; maternity costs were not tabulated.

Do **cities** affect **lightning**?

Perhaps. Recent research suggests that the frequency of lightning strikes over and downwind from a number of midwestern cities can be 10 to 20 percent higher than that of surrounding areas.

What is **keraunophobia**?

The irrational fear of lightning is known as keraunophobia. The fear of thunder is termed brontophobia.

What is the most **dangerous time** of a lightning storm?

Near the end. A study of Florida lightning strike casualties found that the largest number occurred just as the storm was ending, not during its most intense part. Apparently people were too quick to declare the storm "over," and wandered outside from their protective shelters, only to be hit by the storm's last flashes.

155

If you are outside during a thunderstorm and your hair stands on end, leave the area: lightning may be about to strike. Minutes after this woman left, lightning struck the spot where she was standing.

What are some examples of **bad luck** during a storm?

Lightning struck a Waterford, Wisconsin, barn on 10 July 1992. Inside the barn were 6,000 bales of hay that were destroyed by the ensuing fire.

In July 1995, in Miller City, Ohio, lightning struck a poultry farm and 68,000 chickens were roasted.

In 1926, the Navy's largest ammunition depot was located in Lake Denmark, New York. The storage buildings were sturdy, and equipped with lightning rods. That wasn't enough, however. On 10 July, lightning struck, blowing up depth charges and TNT bombs.

On 2 July 1992, in the Chicago area, five people were killed by lightning and thousands were left in the dark for up to two days due to widespread, storm-caused power outages.

The summer of 1980 was a bad one for lightning strikes in Ohio. In Wickliffe, the entire high school football team was knocked down by a bolt during practice, and one player was injured. Then, 26 people were injured in Tuscora Park, Ohio, with one fatality. And a lone man was struck and killed—while digging a grave in a cemetery.

Lightning struck near a house in upstate New York on 17 July 1988. Among the unusual results: the tires on cars parked in the driveway went flat and the hubcaps were blown off. Also, the homeowner's contact lens actually popped out of his eye.

What is **heat lightning**?

Heat lightning is not a special form of lightning. It is simply the reflection of regular lightning off atmospheric dust layers from distant thunderstorms below the horizon.

Can you **outrace lightning**?

The electrical breakdown of the atmosphere during a lightning strike takes place at speeds on the order of 100,000 miles per second.

What if your **hair stands on end**?

If you are outside when a thunderstorm is nearby and your hair starts to stand on end, or a fishing line literally hangs in the air after casting, or a plastic rain coat suddenly begins lifting into the air, lightning may be about to strike. These phenomena are caused by an extremely high electric field in the atmosphere. Seek shelter immediately. If caught in the open, crouch down as close to the ground as possible without having your hands touch the ground.

What is **triggered lightning**?

On 26 March 1987, the U.S. Air Force launched a rocket from the Kennedy Space Center carrying a communications satellite into low-hanging rain clouds. Forty-eight seconds after launch, the rocket was "struck" by lightning—apparently triggered by the ionized exhaust plume trailing behind the rocket. The cost: $162 million. A similar thing happened to the Apollo moon landing launch vehicle during lift-off, though the mission continued without incident. Scientists routinely fire small rockets trailing copper wires into electrically charged clouds, often triggering a lightning strike in a predetermined location. This capability is being used to study the effects of lightning on electrical equipment and materials, as well as to conduct atmospheric research.

157

Can you get paid to be **struck by lightning**?

Some research pilots actually did. Their job was to fly heavily instrumented F106 aircraft through thunderstorms, deliberately trying to be struck by lightning. They were very good at their job, getting hit many times. This seemingly odd occupation was pursued to improve aircraft lightning safety features and to develop better forecasting tools for in-route lightning avoidance.

What is **thunder**?

Thunder is the sound emitted as a result of the rapidly expanding gases along the heated lightning channel. When lightning strikes very close by, one sometimes hears a tearing sound. This is believed to be produced by the stepped leader, which precedes the first stroke in a flash. The sharp crack heard at very close range, just prior to the main thunder crash, is caused by a ground streamer ascending to meet the stepped leader of the first stroke.

Can you have **thunder without lightning**?

By definition, thunder cannot exist without lightning. But you might not always be able to see the parent lightning discharge. Especially during the daytime, lightning discharges deep within a cloud are difficult to see. More than 80 percent of all lightning discharges remain inside clouds.

How **loud** is thunder?

 Sound intensity can be expressed in decibels (sometimes abbreviated dBA). A clap of thunder can typically register about 120 dBA, which is ten times louder than a garbage truck, chain saw, or pneumatic drill. In comparison, sitting in front of the speakers at a rock concert can expose you to a nearly continuous 120 dBA, which can seriously harm your hearing.

What are the **loud booms** that occur at beaches?

One enduring mystery has been reports by coastal residents of "water guns," loud thunder-like booms that seem to emanate from somewhere out over the sea. The reports have been made for centuries, so sonic booms would be ruled out. Natural gas explosions are one possibility.

Does thunder make **milk go sour**?

An old wives tale says that thunder causes milk to go sour. Neither thunder nor lightning can have that effect. The saying probably originated because the thunderstorm season comes with heat and humidity, and if not properly refrigerated, milk will turn sooner than usual in hot weather, thunderstorms or not.

OPTICAL PHENOMENA

What is the **solar spectrum**?

If one were to break up the sun's light into its spectrum of colors, the shortest wave lengths are found in the blue light and the longest wavelengths are with the red colors. The rainbow is the result of rain drops, acting as myriad tiny prisms, which are illuminated by the sun.

How fast does **sunlight travel**?

Sunlight, just like any other electromagnetic radiation, travels at 186,282 miles per second in a vacuum. The light from the sun takes 8.4 minutes to reach the Earth. The nearest star beyond our own sun is Alpha Centauri. Its light takes about 4.3 years to reach Earth. A light year is the distance that light travels in one year's time, 5.87 trillion miles. To get some idea of how far that is, imagine you were riding in a taxi that charged one dollar per mile. If you could spend an amount equal to the entire U.S. national debt (which is now over five trillion dollars) you would travel one light year.

What are **crepuscular rays**?

Bands of sunlight shining through breaks in clouds on the horizon create crepuscular rays, an optical effect sometimes called "the sun drawing water" or "Jacob's ladder." The diverging pattern of shadows and rays are made visible by the haze or smoke in the atmosphere. Similar phenomena can also be seen when the sun has set below the

Bands of light from the setting sun shine through a cloud bank, creating crepuscular rays.

horizon, but there are some distant clouds partially blocking the sunlight that still shines into the illuminated atmosphere above.

Why does the moon appear larger at **moon rise**?

The moon appears larger when it first rises in the sky than when it is higher largely because of an optical illusion.

Sometimes the moon looks larger than at other times because its orbit around the Earth is not circular, varying by about 13,000 miles in either direction from its mean distance. Thus, at its closest, it appears some 12 percent larger in the sky than when at its furthest point.

How high must you go for the **sky to turn black**?

You can begin to detect a distinct darkening of the blue in the sky once you reach the lower stratosphere, as do high-flying commercial jets. Once you get above 100,000 feet, the sky turns increasingly dark.

How far away is the **horizon**?

162 The question is deceptively simple. But for an observer at sea, what appears as the horizon (in miles) is equal to 1.317 times the square root of the height of the vantage point (measured in feet). Thus if you are 10 feet above the water's surface, the horizon is 4.16 miles away.

Why is the sky blue?

This is probably one of the toughest questions parents are asked, though the answer is actually fairly simple. The blue sky is a result of a phenomenon called Rayleigh Scattering. The air molecules scatter the colors of the sun's light differently, depending on wavelength. The blue portion of the solar spectrum predominates in the light we see in the sky. If you were a Martian parent answering that question from your offspring, you would be responding to, "Why is the sky orange?" On Mars, vast amounts of dust in the atmosphere produce scattering of the longer (red) wavelengths of light, rather like a perpetual sunset.

What is St. Elmo's fire?

In technical terms it is a corona from an electrical discharge typically found on tall, pointed metal objects that are grounded such as lightning rods, chimney tops, and ship masts. It usually occurs during thunderstorms and indicates a very high local value of the atmospheric electric field. Molecules of gas are becoming ionized and thus they glow, often with a blue color. The phenomenon is often seen on ships, and sometimes aircraft wings, and can take the shape of spear-like or tufted flames. The name derives from St. Erasmus, the patron saint of sailors, who popularized his name to St. Elmo. The phenomenon was often considered a good omen. However, if you are outside and nearby pointed objects start hissing and giving off St. Elmo's fire, don't count on good luck: get inside, because a lightning strike is likely.

What is diamond dust?

It is a form of precipitation composed of very small, unbranched crystals of ice that fall so slowly they seem to be floating in the air. They can fall either from a cloudy or clear sky. In the latter case, when illuminated by the sun or moon, the crystals sparkle like diamond dust. The effect, usually seen in direct sunlight, can also be created by artificial light beams, and often produces vertical sun or light pillars.

163

What are sun dogs?

Sun dogs (or at night, moon dogs) are bright spots that appear on either side of the sun or moon when there are thin layers of high ice crystal clouds causing unusual optical phenomena.

Sometimes people are startled to look into the sky and see what appear to be three suns, but this effect is just a manifestation of a sun dog, or solar parhelion. This optical phenomenon is caused by hexagonal ice crystals falling with the longest axis in the vertical position. The sun dogs, also called mock suns, appear as two brightly colored luminous spots, usually at 22-degree angles on either side of the sun.

Who first explained rainbows?

The cause of the rainbow was discovered by a German monk named Theodoric in 1304. By observing what happens to sunlight passing through a large, water-filled globe, he illustrated the reflection and refraction processes that occur as the sun shines through myriad raindrops.

What is the correct way to draw a rainbow?

The color scheme on the rainbow should have the red colors on the inside of the arc and the blue on the outside. The sequence should be red, orange, yellow, green, blue, indigo, and violet, inside to outside. An informal check of rainbows portrayed in various artworks over the centuries found that nearly half of the representations got it backwards.

Where do you look for a rainbow?

Three conditions must be met in order to see a rainbow. First, there must be raindrops, preferably from a heavy thundershower. Second, the sun must be shining. Third, the observer must be between the sun and the rain drops.

The rainbow will always be in the sky directly opposite the azimuth of the sun. The lower the sun is in the sky, the higher the arc of the rainbow. The best rainbow watching is looking east near sunset in the part of the sky directly opposite the sun.

Can a rainbow appear at night?

Yes. The lunar equivalent of a solar rainbow can occur. The nighttime rainbow is very rare and occurs when the moon is bright enough and positioned properly with respect to a falling mass of rain to produce the nocturnal relative of the sun's most spectacular optical effect. Generally, however, the colors are very muted, and often appear white.

What is **cloud-to-space lightning?**

Since 1886, scientific literature has periodically been publishing reports on observations of strange luminous phenomena occurring above thunderstorms. These phenomena, which seem to appear in a wide variety of shapes, sizes, and colors, most often do not start with the cloud (though there are exceptions), do not go all the way to space, and, while related to lightning, are not *true lightning*. In the past, little attention has been paid to these reports, but recently that has changed. Over the past few years, thousands of cases of at least three classes of phenomena have been identified and captured on tape using special low-light video cameras. These three types are sprites, elves, and blue jets.

What are **sprites?**

Sprites are huge "blobs" of light that occur briefly above thunderstorms. Rather like an auroral curtain that flashes on for just a blink of an eye, they can extend to 55–60 miles high (the ionosphere) and can be 100 or more miles wide. Most are invisible to the naked eye (you would need a "night scope" type of camera to see them) but some can be detected by the naked eye.

Sprites have been reported in the scientific literature since 1886, but since no one had a picture of one, mainstream science paid the reports little heed. Then, quite by accident, University of Minnesota scientists Professor John Winckler, Robert Franz, and Robert Nemzek captured one on video as they pointed a low-light camera at the sky while conducting a test for an upcoming rocket flight. Soon sprites were found to have been captured, again unintentionally, in low-light video taken from the space shuttle. The first successful attempt to intentionally capture a sprite on video occurred on 7 July 1993 by Walter Lyons, who taped more than 240 sprites above a thunderstorm complex in the U.S. High Plains. Scientists from the University of Alaska, Davis Sentman and Eugene Wescott, captured more sprites from a high-flying NASA aircraft less than 48 hours later. Since then thousands of sprite pictures have been obtained.

What are **elves?**

Elves, an acronym for emissions of light and Very Low Frequency (VLF) perturbations from ElectroMagnetic Pulse (EMP) sources, are another type of strange lights in the night sky resulting from lightning within thunderstorms. These bright flashes are like giant expanding donuts, which can spread out for over 200 miles, and are centered at about 55–60 miles altitude. The thin upper atmosphere gives off a brief glow as the electromagnetic pulse from a lightning bolt passes through on its way into outer space. Elves are too fast to be seen with the naked eye, lasting less than one-thousandth of a second.

165

What are **blue jets**?

The blue jet appears as a blue column of light that squirts out the top of thunderstorms at speeds of 62 miles per second, sometimes reaching a height of 25 to 30 miles. What exactly causes them is unclear, nor is it certain whether they may pose a threat to any vehicle that might be flying above cloud tops.

What are the best conditions to **observe sprites and blue jets**?

It must be nighttime and you need to be away from the lights of the city. There must be a very large thunderstorm system on the horizon, ideally between 100 and 300 miles away (check the radar display on the late news weather report) and there should be no intervening clouds to prevent a clear view of the region above the storm, nor should there be a bright moon. Let your eyes adjust to the darkness, usually over five to ten minutes. Then look in the sky above the thunderstorm. Try to judge the height of the storm and look to eight times the distance above. Shield your eyes from the lightning that may be flashing in the clouds below. The sprites tend to occur every few minutes in active storms. They appear to the eye rather like a reddish curtain of light lasting less than the blink of an eye. Some people perceive the color as orange, white, or green. Blue jets, which are extremely rare, appear to the eye as a blue flame or column of light shooting directly out of the cloud top. They may be more likely to occur from storms producing very large hail.

What is a **blue moon**?

This term is used when a full moon occurs twice in the same calendar month. It happens once every 2.7 years and never in the month of February because there are not enough days. The moon does not actually turn blue, though in late September 1950, the moon in eastern North America actually did appear blue, due to smoke from forest fires in western Canada. Any fine aerosol in the atmosphere can have similar effects. The 1991 eruption of Mt. Pinatubo in the Philippines also resulted in reports of blue moons around the world.

What is a **blue sun**?

Once in a blue moon is a common expression for a rare event. How about once in a blue sun? This expression could be used to indicate an extrememely rare event. The result of atmospheric particles from forest fires, sandstorms, and volcanic eruptions, a blue-looking sun is a truly infrequent, but very real phenomenon.

Why are **eclipse forecasts** so accurate?

Astronomical bodies such as the moon and planets obey the laws of celestial mechanics, which were formulated as equations by Sir Isaac Newton. The rotation of one body around another under the influence of gravity is extremely regular and repeatable. Once such basic parameters as the mass of an object and its distance from other objects are known, orbits can be predicted with precision. It does require a lot of arithmetic to forecast when one object will appear to pass in front of another (the moon blotting out the sun, for instance), but the calculations are pretty straightforward.

Do **solar eclipses** influence the weather?

Most definitely. Beneath the path of a solar eclipse, a giant shadow more than one hundred miles wide and perhaps several hundred miles long races across the Earth at speeds sometimes reaching 5,600 mph. As the sun's shadow (the umbra) passes, it produces a premature sunset. And just like at night, temperatures can fall. In the dry air of Baja California, during the eclipse of 11 July 1991, the midday temperature fell from 90°F to 74°F as the eclipse proceeded. With the day's second sunrise six minutes later, things started to heat up again. The lake breeze in Chicago that day nearly stopped as the land suddenly cooled in response to the dimming sun.

How can climatology help plan an **eclipse viewing expedition**?

Climatological predictions can give a good idea of what the weather might be in a given area on a given day, but it is not foolproof. On 11 July 1991, tens of thousands of eclipse watchers traveled to the Big Island of Hawaii to view the total solar eclipse, which promised to be one of the longest and most spectacular ever in the United States. Most converged on the northwest coast between Keahole Point and Kohala where climatology promised a 90 percent chance of seeing the sun. Unfortunately, on that morning, clouds also converged on the same spot, eclipsing the eclipse for many.

What is the longest possible period of totality for a **solar eclipse**?

The longest the totality in a solar eclipse can last is 7 minutes, 31 seconds. The total eclipse of 11 July 1991 over Hawaii, Mexico, and South America was a near record setter. It passed over the capitals of four countries and may have been seen by more people than any previous eclipse (thanks to passing directly over Mexico City.) Along portions of its path, totality durations as long as 6.5 minutes occurred, the tenth longest to be seen during the next five centuries. If you missed the big total solar eclipse of **167**

1991, hang in there. The next one that will pass over the United States will be on 21 August 2017. The occurrence interval of a solar eclipse at a given point on the Earth is on the order of once every 400 years.

What is the green flash?

The green flash is one of the more interesting atmospheric optical phenomena. You are sitting at the beach watching a beautiful sunset over the ocean . . . and suddenly you see a green flash at the point where the sun just sank beneath the horizon. You've just witnessed the relatively rare green flash, an optical effect caused by atmospheric prismatic dispersion, which results in the last rays of the sun appearing as green, rather than the usual white or pollution-tinged red. For an instant the sun has a brilliant green color due to refraction of that color in the solar spectrum as the sun's disk goes below the horizon. Normally lasting less than a second, the all-time green flash record may have been noted by Admiral Byrd in the Antarctic. As the sun rolled along the horizon, a green flash was recorded that lasted about thirty-five minutes. The green flash can also sometimes be seen at sunrise. There is a much rarer form of the phenomenon that shows up as blue light, called the blue flash.

Are there **double rainbows**?

Yes, they do happen. Some rain showers produce two rainbows, seen as concentric arcs. The inner (and brighter) rainbow has the red on the top and the blue on the bottom side. The outer (and dimmer) rainbow has the color scheme reversed. The technical name for the second arc is a supernumerary rainbow.

Do we all see **the same rainbow**?

No one sees the same rainbow. Everyone who looks at a rainbow sees a slightly different collection of light beams originating from the sun's light reflecting and refracting millions of raindrops.

What is **irisation**?

When a thin cloud of water droplets passes across the sun, it may cause a complex

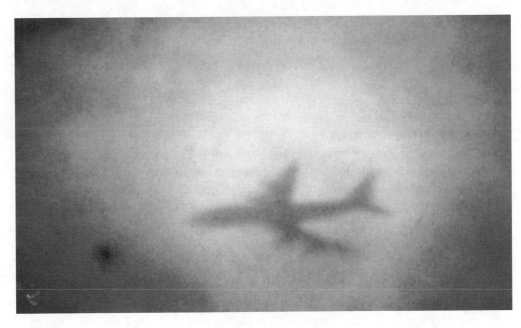

The shadow of a jet aircraft on the cloud underneath is surrounded by a colored ring of light, called a glory.

optical phenomenon called defraction. To the observer the thinner parts of the cloud can take on all kinds of colors. The mother-of-pearl-like effect often has many red and green highlights. This effect is most common in altocumulus clouds.

What is a **bolide**?

A bolide is an especially brilliant meteor. It usually ends its fiery descent with an explosion, sometimes called a detonating fireball. The color is often green, but many other shades can be present. They are frequently bright enough to be seen during the daytime.

Is an **all-red rainbow** possible?

Although extremely rare, there have been reports of red rainbows. All-white rainbows have also been noted on a few occasions.

What is a **glory**?

Sometimes called an anticorona or Brocken bow, the glory appears as concentric rings of color surrounding the shadow of an object cast by the sun on a cloud of water **169** droplets. It is often seen by passengers on airplanes who look down and see the dark shadow of the aircraft surrounded by brilliant-colored rings. It appears directly opposite the sun or moon from the observer.

What is **Hevelius's Parhelia**?

Hevelius's Parhelia refers to an optical phenomenon seen in the sky when ice crystals in high cirrus clouds play tricks with the sun's light. The parhelia appear as bright, white spots halfway around the sky from the sun to the antihelion (the area of the sky opposite the sun). They may result in part from the sun's light passing through randomly oriented, bipyramidal, hexagonal ice crystals.

Why can't the **end of the rainbow** be found?

Because the rainbow is an optical illusion, you just can't catch up to it. When you move, so does it. If you literally try to find the end of the rainbow, you'll have a long walk. The rainbow does not end in a physical place.

What is a **corona**?

There are several processes that produce a "ring" around the sun or moon. One such process results in the corona, which is produced by the sun or moon shining through a thin cloud of water droplets. The circular band of light has blue on the inside and red on the outside, the result of defraction of the light by numerous water droplets. The ring has a much smaller angular diameter than the relatively common halo. The diameter of the corona is influenced by the size of the cloud's water droplets, with smaller droplets making for a larger corona.

What is a **halo**?

The halo is a ring around the sun or moon produced by refraction of light through a thin cloud of ice crystals. The halo has the red colors on the inside of the ring shifting to blue on the outside. There are many forms of halos, depending on the shape of the ice crystals in the clouds, but a 22-degree ring is seen most often. In folklore, haloes are said to foretell precipitation, and in fact, the high icy cirrus clouds that can produce haloes are often the forerunners of storm systems.

What is a **mirage**?

Technically, a mirage is a refraction phenomenon in which the image of some object appears displaced from its true position. A very common mirage is the appearance of

A thin veil of icy cirrus clouds creates a halo around the sun.

"water" some distance down the highway on a hot summer day. In effect, the very hot layer of air near the road's surface refracts light rays from the sky on the distant horizon upward into your field of view. The mind interprets this as reflection from a water surface. It is the same thing that causes desert travelers to "see" lakes that are nothing but sand.

What is a **superior mirage**?

This doesn't refer to a high-quality mirage, but rather a spurious image of an object formed above its true position by abnormal refraction conditions caused by temperature layers in the atmosphere. This often occurs when there are low-level temperature inversions, such as warm air flowing over cold water. Sometimes people standing on beaches are startled to see ships, upside down no less, apparently floating in the air over the water. At night, distant lights can produce some amazing effects during these conditions.

What is the **Fata Morgana**?

Some mirages are so famous they even have their own names. The Fata Morgana was named by Italian poets who compared this strange apparition, often seen near the **171**

Strait of Messina, to a mythical submarine crystal palace called Fata Morgana. The Fata Morgana is a complex mirage that manifests itself by multiple distortions of images, generally in the vertical, so that distant objects such as cliffs and cottages are magnified into shapes resembling fantastic castles.

Have mirages ever affected **military battles**?

A military battle was "canceled" by a mirage that occurred in Mesopotamia in April 1916. Because of the mirage, neither the British nor Turkish gun batteries could see each other. So, rather than firing blindly, they all went home for the day.

What are **sun pillars**?

Sun pillars are luminous vertical streaks of light, white or sometimes slightly reddish in color, extending from above and below the sun. They often extend to 20 degrees above the sun and end in a point of light. They are the result of the light being reflected by the mirror-like surfaces of ice crystals. They are fairly common around sunrise and sunset, and often when the sun is below the horizon, the shafts of light appear to be suspended in air in the twilight sky. Rarely, on cold nights when there are thin ice fogs, the same type of phenomenon can be seen above bright lights.

What is a **parhelic circle**?

Also called a mock sunring, it is a halo consisting of a faint white circle passing through the sun and running parallel to the horizon. It is caused by reflection from the prism faces of hexagonal ice crystals falling in a certain orientation. Sometimes parhelic circles occur at the same time as light pillars, forming a cross shape. In times past, religious significance was often accorded these events.

How far can you **see through the atmosphere**?

The limit to visibility in a perfectly clear atmosphere (no dirt or pollution) is approximately 200 miles during the daytime. It is possible to see bright objects at night for more than 500 miles near the horizon.

What are some possible **scientific explanations** for what appears to be a UFO?

It could be a star or planet, which under certain atmospheric conditions can appear extremely bright and twinkle in a variety of colors.

It could be the landing lights of distant aircraft. These can appear extremely bright at ranges of 100 miles or more from airports.

An unusually bright flash or streak in the sky could have been due to a large meteor or bolide. Several such bolides, videotaped soaring across the sky, brilliant even in full daylight, have been passed off as "alien space craft."

It could be an optical phenomenon caused by ice in the air, such as a sun dog.

Have clouds been reported as UFOs?

Frequently. There is a well-known cloud formation, altocumulus lenticularis (meaning "lens shaped"), which is sometimes dubbed the "flying saucer cloud." Clouds are normally thought to move with the wind currents, but they do not always do so. Air flow over mountainous terrain often produces "standing waves," rather like those in a fast-moving stream as water flows over a rock. Moisture condenses on the upwind portion of the cloud and evaporates as it exits downwind. The wind may be blowing 150 mph, but the cloud stands still. These clouds can take many shapes, but they often look like disks or plates, sometimes stacked one upon another. During one of the many flying saucer flaps in the 1950s, *Life* magazine actually published a picture of a "UFO hovering over Holloman Air Force Base in New Mexico." It was a great picture of a perfect specimen of altocumulus lenticularis. There are many regional names given to the distinctive clouds formed by airflow over mountain barriers that are as varied as the clouds themselves: lenticularis, cap cloud, crest cloud, banner cloud, Moazagotl, foehn cloud, the table cloth, Chinook arch, rotor cloud, and Bishop wave.

What is syzygy?

Syzygy occurs twice a month, when the sun and moon are lined up, either with the new or full moon. In both cases, the gravitational effects of these astronomical bodies reinforce each other and the range between high and low tides increases.

TORNADOES

What is a **tornado**?

Technically defined, a tornado is a violently rotating column in contact with the ground that is pendant from a parent cumulonimbus cloud. On a local scale it is the most destructive of all meteorological phenomena. Many tornadoes only last for a few minutes and are on the ground for a few miles. But others can persist for hours, travel more than a hundred miles, and have paths that are a mile wide. The word tornado may have originated from a merging of the Latin *tornare,* which means to turn, and the Spanish *tronada* (thunderstorm).

What is the difference between a **funnel** and a **tornado**?

Both are violently whirling columns of air spawned by thunderstorms, but funnel is most often used to refer to a vortex that is still aloft, while a tornado has actually descended to the surface. If the condensation funnel has dropped more than halfway from the cloud base to the surface, it is generally called a tornado because that is the point at which damaging winds are likely to be occurring on the ground. Sometimes the term funnel refers to the visible portion of the vortex whether or not it has reached the surface.

What **causes tornadoes**?

Conservation of angular momentum is the technical explanation. Everyone has seen ice skaters doing spins—they spin slowly when their arms are extended, but as soon as **175** they tuck their arms in close to the body, their rate of rotation increases dramatically. Approximately the same thing happens in the atmosphere to cause a tornado. An air mass always has a certain amount of "spin," or vorticity. As that air is converged into the strong updraft of an intense thunderstorm, the rate of spin increases, much like that of the skater. But while the basic principle is recognized, the meteorological mechanisms are still poorly understood. Do tornadoes form deep within the cloud and extend downward? Or does the intense rotation form beneath the cloud and extend upwards? Or both? A major research program conducted in 1995 called VORTEX (Verifying the Origins of Rotation in Tornadoes Experiment) involved hundreds of meteorologists driving or flying near tornadic thunderstorms to probe them electronically and photograph them from every conceivable angle. But unlike the movie *Twister,* where the "answer" was immediately obvious, it will take years of data analysis to fully understand the information gathered from the VORTEX project.

What is **Tornado Alley**?

Tornado Alley refers to a broad belt of land in the United States that is hit by more tornadoes than any place in the world. It includes much of the middle part of the country, starting in Texas and working northwards into many of the Midwestern states, including Oklahoma, Missouri, Kansas, and Ohio. There is no exact geographic boundary of Tornado Alley, and tornadoes are certainly not confined to that region, since they have been spotted in every state in the United States.

Are there other places of **high tornado frequency** besides Tornado Alley?

Tornadoes actually can occur almost anywhere in the nation. And while the middle part of the country has the largest number, other regions are far from immune. Located outside of Tornado Alley proper, Harris County, Texas, over the past 20 years, has had the greatest chance of having one or more twister touchdowns. Better reporting in recent years has shown that central Florida has far more tornadoes than once thought. Alabama has the distinction of leading the nation in percentage of tornadoes rated of significant intensity, reporting four F5 tornadoes over the last two decades. In any given year, tornadoes can take "detours" and strike one area of the nation with unusually great frequency. Several years ago, Weld County, Colorado, far from Tornado Alley, had more twisters than any other U.S. county. Most of these, fortunately, were weak and short-lived.

How many tornadoes reach **"violent" status**?

Tornadoes of F4 or F5 on the Fujita scale are classified as violent tornadoes. Only two percent of tornadoes in the United States reach violent intensity, yet those few result

A tornado skipping at rooftop level toppled trees in this residential neighborhood in Minneapolis.

in 70 percent of all tornado deaths. Their winds exceed 200 mph and they can stay on the ground for an hour or more.

The strongest tornado wind speeds are estimated to be in the 260–318 mph class, which corresponds to F5 on the Fujita Scale. During the last decade, less than a dozen twisters were estimated to reach this kinetic milestone.

What **time of year** do tornadoes occur?

While the tornado season is generally considered to be March through August, twisters can occur in any month of the year. During the winter months, however, they generally occur only in the southern and southwestern parts of the country. Statistically speaking, May is the month when the most tornadoes occur, though the highest rate of tornado deaths occurs in April (storms in April seem to be more violent than those in May).

The tornado season never really ends—it just slows down and occasionally springs back to life during the winter months. On 22 November 1992, 13 confirmed tornadoes swept across Indiana, injuring at least nine people. On 9 December 1991, a tornado smashed through McLean County, Illinois. About 20 empty railroad cars were blown off the tracks near Bloomington. A camper was picked up and hurled 100 feet into a tree.

What is the **Fujita Scale**?

The Fujita Tornado Scale, or F-scale, measures the power and destructiveness of tornadoes. It is based upon the amount of damage caused, and classifies twisters into six categories of estimated wind speed (F0 through F5), ranging from 40 to 300-plus miles per hour. The Fujita Scale was devised by Professor T. Theodore Fujita of the University of Chicago, whose pioneering research on tornadoes serves as the basis of today's rapidly improving understanding of severe storms.

The Fujita tornado intensity scale is set up this way:

Category	Wind speed	Level of damage
F0 (light damage)	40–72 mph	Some damage to chimneys; twigs and branches broken off trees; shallow-rooted trees pushed over; signs damaged; some windows broken.
F1 (moderate damage)	73–112 mph	Roof surfaces peeled off; mobile homes pushed off foundations; outbuildings demolished; moving autos pushed off roads; trees snapped or broken.
F2 (considerable damage)	113–157 mph	Roofs torn off frame houses; mobile homes demolished; houses with weak foundations lifted and moved; large trees snapped or uprooted; light-object missiles generated.
F3 (severe damage)	158–206 mph	Roofs and some walls torn off well-constructed houses; trains overturned; most trees in forest uprooted; heavy cars lifted off ground and thrown; weak pavement blown off roads.
F4 (devastating damage)	207–260 mph	Well-constructed houses leveled; structures with weak foundations blown some distance; cars thrown and disintegrated; trees in forest uprooted and carried some distance.
F5 (incredible damage)	261–318 mph	Strong frame houses lifted off foundations and carried considerable distance to disintegrate; automobile-sized missiles fly through the air in excess of 300 feet; trees debarked; incredible phenomena occur.

No tornado is believed to have exceeded F5, and only a few of the 700–800 tornadoes that occur annually in the United States reach that destructive level.

What is the peak time of day for tornadoes?

Tornadoes are creatures of the late afternoon, when most thunderstorms are in progress. Over 40 percent of U.S. twisters are reported between 2 and 6 P.M. However, they can and do strike at any hour if conditions are right. Nighttime tornadoes, which are common along the Gulf Coast of the United States, tend to be even deadlier since they are obviously harder to see, and because most people are sleeping and unaware of warnings. Early morning tornadoes are unusual, but not impossible.

How many touchdowns are there in a **normal tornado season**?

In a normal tornado season, about 780–800 confirmed twisters touch down. The average number of fatalities over the past 30 years has been 82 per season. Usually, more Americans die from lightning strikes than tornadoes.

What **shapes** can tornadoes take?

Most people have a mental image of a tornado that is roughly like an elephant's trunk reaching out of the sky, but twisters do come in a variety of shapes and sizes. Sometimes they appear as roiling billows of smoke, other times as a twisting rope or a barely visible swirl of dust. Some extremely violent ones appear as several "snake-like" vortices whirling around a common center. Some especially large funnels, which can be a mile wide, are called wedges. A tornado often goes through a life cycle, starting as a classic funnel shape, then broadening and widening in its mature stage, occasionally looking like billows of smoke on the ground. The twister then enters the dissipating stage, also called the rope stage, where it becomes thinner, longer, and often very distorted. But even when a tornado is "roping out," it can still cause extensive destruction on the ground.

What are the **strongest winds** in a tornado?

Once they were thought to be over 500 mph, perhaps even supersonic. But research on damage patterns, as well as recent measurements made by portable Doppler radars, suggest the highest winds are in the 280–300 mph range. The strongest wind speeds in a tornado vortex are thought to occur about 300 feet above the ground. It is estimated that fully 90 percent of all tornadoes fail to register winds above 113 mph.

What are **suction vortices**?

The ability of a passing tornado to destroy a building, yet leave something only a few feet away totally unscathed, has long puzzled experts. Professor Fujita has made detailed studies of tornado films and videos, as well as airborne surveys of tornado damage. His studies have revealed that many funnels break down into smaller vortices rotating about a common center and moving forward with the mean speed of a tornado. These smaller vortices, only tens of feet across, cause much of the major damage and "freakish" destruction associated with tornadoes. These suction vortices, as they are called, have dramatically higher wind speeds than their parent funnel cloud. Thus, when a main funnel passes over a house, whether or not it is severely damaged may depend on if it is hit by one of the smaller, more destructive suction vortices. Some multivortex funnels do not look like the commonly held image of a tornado, but rather appear to be swirls of smoke and multiple snake-like columns whirling about a common core.

What is a **dust devil**?

Dust devils are atmospheric whirlwinds that superficially resemble small tornadoes, but form by totally different mechanisms. These swirling columns, which can rotate either clockwise or counterclockwise, form on sunny, hot days with relatively light winds, often over plowed fields or expanses of dirt or pavement. They result from superheated air near the sunbaked ground rising into the cooler air aloft. These whirling dervishes, unlike tornadoes, are not associated with thunderstorms. In some cases, wind speeds can easily surpass 50 mph. In a few cases, dust devils have extended to more than 5000 feet above the ground.

Most dust devils just swirl across an open field and kick up a few puffs of dust and a handful of leaves. Some exceptionally large and long-lived examples, however, if traveling over loose soil, can sweep upwards of 50 tons of dust and debris skyward.

What are some examples of **unusually strong dust devils**?

One of the biggest ever formed occurred over the Bonneville Salt Flats in Utah. It grew to 24,500 feet in altitude and traveled for over 40 miles. One huge dust devil was seen swirling in place for over an hour in a trucking terminal near Des Moines, Iowa. A large dust devil in New Mexico developed at the edge of a railroad embankment and excavated over one cubic meter of sand per hour for longer than four hours. It stopped only when it came across a bulldozer that was parked nearby; apparently the bulldozer

This mini-vortex was formed from a steam plume emanating from a power plant.

interrupted the airflow into the vortex. Some dust devils have been known to overturn trailers and even down power lines in Arizona.

An unusually powerful dust devil swept a 300-yard-long path near Minong, Wisconsin on 6 May 1995. It damaged the roof of a house, tore through a snow fence, and caused a small fire when it downed a power line.

In March 1995, on a beautiful sunny day in upstate South Carolina, a covered porch at a rural residence was smashed by a sudden gust of wind, with wreckage scattered over several acres. The best explanation offered for the destruction was that a large dust devil swirled up from the hot ground and destroyed the porch.

What is a **steam devil**?

A steam devil is essentially the same thing as a dust devil. The difference is that it occurs when extremely cold Arctic air passes over very warm water, as during a late autumn cold wave over the Great Lakes. The whirlwinds that rise up from the surface layer are marked by steam fog and not dust.

Are there other **tornado-like vortices** in the atmosphere?

Aside from dust devils and steam devils, intense atmospheric whirls can be found near any unusually strong heat source. Forest fires can induce intense fire whirls that can **181** exceed 100 mph and extend more than a mile upwards. Similar vortices have been seen near volcanic eruption clouds, the Kuwaiti oil fires during the 1991 Gulf War, and sometimes even in the exhaust plumes of large power plants.

What are storm chasers?

Storm (or tornado) chasers are a mix of professional scientists and amateur storm enthusiasts who try to guess where tornadoes are going to strike so they can be present to witness them. Professional tornado researchers have been chasing storms—driving to areas where they believe storms will form and then setting up for observations—for more than two decades. The chaser's goal is to get close enough to gather good data (whether video, radar, or other measurements) while staying far enough away to avoid being hurt or killed. There are numerous attendant hazards including giant hail, lightning, downburst winds, and rain-slick roads.

The movie *Twister* glamorized storm chasing but was highly unrealistic. No serious meteorologist would attempt getting as close to the funnels as the scientists portrayed in the movie did.

Do **rivers protect** you from tornadoes?

Definitely not. While some people believe rivers are a safe haven from tornadoes, there are numerous reports of twisters crossing major rivers such as the Mississippi and Missouri, damaging towns on both sides. At least 30 significant tornadoes have crossed the Mississippi River. In fact, more than one river boat has been struck and sunk by twisters.

Are you safe from tornadoes in **mountainous areas**?

Mountainous areas are safer than plains, but still not totally safe. There is more than one picture of a tornado flying past Pike's Peak, Colorado. In general, meteorological conditions are not often favorable for tornadoes in the mountainous western United States or in the smaller mountains found in the eastern United States, but tornadoes can occur in these areas. During the 1974 super tornado outbreak, some tornadoes traveled up and down 3000-foot mountains in the Appalachians. The "highest" tornado touchdown ever was recorded when a large twister left a 23-mile-long path of destruction in the Teton wilderness region of Wyoming. Much of the path was located above 10,000 feet.

Many communities believe that some local hill, or an Indian burial mound, somehow protects them from tornadoes, but this is not the case. In fact, the city of Tulsa, Oklahoma, was devastated by a tornado decades ago, and the massive funnel formed right over an Indian burial mound that was believed to ward off twisters.

Where do tornadoes form with respect to the **parent storm cell**?

Tornadoes have a strong tendency to form near the right rear portion of a thunderstorm cell, particularly if it is an isolated storm called a supercell. The apparent mesocyclone, out of which many tornadoes form, is often centered in this updraft region of the storm. Heavy rain and hail often fall to the north and east of this point, but the mesocyclone (which often appears as a hook-shaped echo on radar) tends to feature large, powerful updrafts that create a rain-free base. Thus, many tornadoes occur in areas with little rainfall, though large hailstones may be falling. It is for this reason that many tornado pictures show clear skies in the background.

What is a **wall cloud**?

A wall cloud is a feature that accompanies and precedes many tornadoes. It is an often abrupt lowering of the rain-free base of the cumulonimbus cloud into a roughly circular, low-hanging cloud one to four miles in diameter. Severe-storm spotters know that the cloud is a visible manifestation of the tornado mesocyclone. Wall clouds often have rapidly rising elements and rotate in the same direction as the tornado, but at a much slower speed. The "collar cloud" is a ring of clouds that sometimes surrounds the upper portion of the wall cloud. The rotating wall cloud has been known to appear as much as an hour before the actual touchdown of a tornado.

What is a **beaver tail**?

Severe-storm spotters have all manner of names for the cloud features that occur in severe storms. One feature, called the beaver's tail, is a broad flat cloud that swirls inward, often from beyond the storm area, into the wall cloud region. Its name derives from the fact that it looks like the tail of a beaver.

What is the **bear's cage**?

Storm chasers try to avoid this part of the thunderstorm. The bear's cage refers to the region of intense rainfall and widespread hail that usually falls north and east of the mesocyclone region containing the tornado. If you wish to see a tornado, it is much safer to approach the cell from the southeast through west quadrants. If you instead **183** try to "core punch" from the north, you're said to be in the "bear's cage." Aside from the danger of giant hail, intense lightning, downburst winds, and flash floods, you are likely to be driving your vehicle through a wall of rain and are very likely to come face to face with a twister having had little or no warning.

Do **hurricanes** spawn tornadoes?

Yes, and quite frequently. Some tropical storms that hit land have spawned dozens of small twisters, most frequently on the outer fringes of the advancing storm. It is by no means unusual to be under both a tornado and hurricane watch or warning at the same time. Hurricane Beulah (1967) generated as many as 115 tornadoes upon making landfall in the United States.

What are the **tornado death tolls** for the United States?

Over 10,000 Americans have been killed by tornadoes in this century. Since 1880, there have been at least 31 tornado outbreaks that claimed more than 100 lives each. Between 1961 and 1990, the state with the highest annual average tornado fatality rate has been Mississippi (10 per year), followed by Texas (8), Indiana (7), Alabama (6), and Illinois, Ohio, and Arkansas tied at 5 per year. The tornado death toll has been steadily decreasing, due in large part to improved warnings, communications, emergency preparedness, and possibly by the population shift into urban areas.

What was the **single deadliest tornado outbreak** in the United States?

The worst single tornado tragedy in the United States was the Tri-State (Missouri, Illinois, and Indiana) tornado outbreak of 18 March 1925. Several tornadoes claimed a total of 747 lives with an additional 2,027 persons being injured. The largest of these twisters killed 695 persons, the worst single tornado death toll ever.

What was the **worst tornado disaster** in the world?

The United States may have the largest number of tornadoes, but it may have lost the dubious distinction of hosting the deadliest tornado. A twister struck about 40 miles north of Dhaka, Bangladesh, on 26 April 1989. At least 1,109 were killed, 15,000 injured, and 100,000 left homeless. Other twisters have struck that region and may have taken even greater tolls.

What were the **deadliest years** for tornadoes in the United States?

The deadliest year for tornado deaths in the United States was 1925, when 794 died, many in the infamous Tri-State tornado outbreak in Missouri, Illinois, and Indiana. In more recent years, 1953 saw the grim total of 519 fatalities. Tornadoes in May and June of that year were especially deadly. On 11 May, 114 died in Waco, Texas. On 8 June, 116 perished in Flint, Michigan. The next day the weather system spawned a tornado in Worcester, Massachusetts, resulting in the loss of 90 more lives.

The year with the fewest reported U.S. tornado deaths is 1986, when only 16 people perished.

What is the **longest tornado path** on record? The **shortest**?

A tornado on 18 March 1925 was believed to have stayed continuously on the ground for 215 miles. The Great Tri-State tornado swept across portions of Missouri, Illinois, and Indiana. In 1917, a tornado was reported to have traveled from Missouri to Indiana—a distance of 293 miles—while staying on the ground for seven hours and 20 minutes. Without a detailed damage survey, however, it is difficult to know if the tornado really stayed on the ground the entire time or if the track instead resulted from several successive tornado touchdowns. At the other end of the spectrum, one tornado was reported to be on the ground for a mere seven feet.

What is the **average path length** of a tornado?

In the United States, the average tornado has a path length of about 5 miles and a width of 160–170 yards. The total amount of area involved in a typical twister is about one square mile. However, the feared maxi-tornado can exceed many times these numbers.

What the biggest recorded **tornado outbreak** ever?

During this century, the largest single tornado outbreak occurred on 3–4 April 1974. Over a span of 16 hours, some 148 tornadoes slammed into 13 states east of the Mississippi, plus the province of Ontario, Canada. Casualties were extensive—315 people were killed, 5,484 persons were injured, and property damage was approximately $500 million. The total path length for the 148 twisters was 2,598 miles, with each twister having a mean path length of 19 miles. Six of the 148 funnels reached F5 intensity. Six towns were struck not once, but twice, in one day. The City of Xenia, Ohio saw both good fortune and misfortune. While many people were killed and the high school was largely destroyed, most of the students had departed for the day. A group practicing a

185

play in the gymnasium saw the funnel approaching and was able to reach a safer hallway just in time—a school bus was hurled through the roof onto the stage where they had been practicing.

What are some other **infamous tornado outbreaks**?

Palm Sunday (11 April) 1965 ranks as perhaps the second greatest outbreak of tornadoes in the United States. Some 37 tornadoes killed 271 people in the Midwest. The injury toll was greater than 5,000.

On 21 November 1992, some 94 twisters blasted 13 states, ranging from Mississippi to Indiana to Maryland, leaving 26 dead, 641 injured, and causing $291 million in damages. On 28 March 1984, 22 tornadoes killed 57 people, injured 1,248, and left behind $200 million in damages in South and North Carolina. Ohio, Pennsylvania, and Ontario, Canada, were ravaged by 41 tornadoes on 31 May 1985, killing 75 (in the United States), injuring 1,025, and causing $450 million in damage.

It is notable that none of these outbreaks occurred in the area that is normally thought of as Tornado Alley—large outbreaks can and do occur almost any place east of the Rockies. One outbreak that did occur in Tornado Alley was one of the largest in recent years. On 26–27 April 1991, in an outbreak spanning from Texas to Iowa, 54 tornadoes touched down, killing 21 people, injuring 208, and causing $277 million in damage.

What are your chances of being **struck by a tornado**?

Even if you were to camp out permanently in the heart of Tornado Alley, statistically speaking you would most likely have to wait 1,400 years before being struck by a twister. Thus the "point probability" of direct strike by a tornado is quite small. During the peak of tornado season, however, at least five a day touch down somewhere in the United States. Many tornado researchers have never seen a tornado. Professor T. Theodore Fujita, developer of the Fujita scale for measuring tornado intensity, had to wait more than 30 years before seeing his first live tornado.

Are there tornadoes in **California**?

Twisters are relatively uncommon in California, but the state does average four per year, and southern California has seen a growing number of tornadoes. The peak season is January through April. A new record was set in 1991 when 16 twisters were recorded; more than two dozen more occurred during the winter of 1992–93. Twisters are not that uncommon in the Los Angeles basin and tend to occur when strong winter storms slam ashore. And, as more real estate is urbanized, there will be more for the tornadoes to strike. Tornadoes are also not that uncommon in the Central Valley, including the greater Fresno area.

Each year almost 1,000 tornadoes touch down in the United States.

What state **leads the nation** in tornadoes?

There is no one answer to that question, since there are different ways to define "leading" the nation. Here are some of the leaders in different categories:

Texas—number of total tornado deaths (regardless of size)

Oklahoma—number of significant and violent tornadoes per square mile

Florida—number of tornado touchdowns per square mile, of any intensity

Arkansas—highest number of killer tornadoes per square mile

Kansas—highest number of F5-intensity tornadoes since 1880

Iowa—greatest number of F5 tornadoes per square mile

Tennessee—highest percentage of tornadoes that cause fatalities

Alabama—highest percentage of tornadoes rated of significant intensity

Kentucky—highest percentage of all tornadoes ranked as violent (F4 or F5)

Indiana—highest rank on scoring system measuring all aspects of tornadoes

Mississippi—highest percentage of population killed by tornadoes

Delaware—highest concentration of tornadoes causing injury

What is a **waterspout**?

A tornado over water is often called a waterspout. It is generally weaker than its land-based cousin. Peak winds are typically in the 50 to 100 mph range. Some waterspouts can move inland, where they can cause damage and injuries. Waterspouts usually are associated not with severe storms, but with average rain showers and weak thunderstorms. They are sometimes found over the Great Lakes and even over the Great Salt Lake in Utah during periods of intense cold air outbreaks over the warm water. The area around the Florida Keys is prime waterspout territory. When a land-based tornado happens to move offshore (for example, crossing a lake or a river) it may be technically called a waterspout at that point, but it is really still a tornado.

Two men were reported missing on 18 June 1993 after a waterspout swept through Chicago's harbor. One man was blown off a pier while the other, a wind surfer, was nowhere to be found after the vortex passed.

Are there any **patterns** to tornadoes?

Certain meteorological patterns are known to produce tornadoes, but as for picking a certain day of the week or month, there is no reason to suspect any pattern. Oddly enough, however, according to the *Storm Tracker* newsletter, in 1994, there were fatal tornadoes in the United States on the 27th of each month from March through August, as well as in November. In fact, 85 percent of the tornado fatalities in 1994 occurred on the 27th day of the month.

Is it possible to **survive a tornado**?

Yes. In fact the odds are very much in your favor if you follow the basic safety rules. During the monster outbreak of 3–4 April 1974, over 30,000 people were directly in the path of tornado vortices. While 315 people did perish, a full 99 percent of those directly affected by the tornado outbreak survived.

Who is **most at risk** during a tornado?

People in mobile homes and automobiles are in the most danger, while the elderly, the very young, and the physically or mentally impaired also face increased risk. Also at risk are those who have a language barrier and don't understand a tornado warning, as well as anyone else who doesn't take precautions when a tornado threatens.

What can you do before a tornado to **improve your survival odds**?

Have a storm plan in place—everyone in the home, school, or office should know in advance where to go in case of a tornado. You should consider having drills. Be sure you know what county you live in or are visiting, since warnings are often issued on a county basis. Have a National Oceanic and Atmospheric Administration (NOAA) weather radio with a warning alarm to automatically receive warnings. Also monitor commercial radio and television. Have a supply of batteries for flashlights and the weather radio as well as a first-aid kit.

What should you do if a **tornado threatens**?

All your planning should pay off if a tornado threatens your community. If there is warning, or if skies appear threatening, do not panic but be prepared to move immediately to a previously designated shelter, such as a basement or an interior room. Remember, at night or during heavy rainstorms, waiting to see the funnel before taking action can be a fatal mistake.

If an underground shelter is not available, move to an interior room or hallway on the lowest floor and get under a sturdy piece of furniture. Bathrooms, with their reinforcing plumbing, are often relatively good shelters. The bathtub itself is a good option. Stay away from windows at all costs—flying glass is a major hazard. If caught outside, lie flat in a ditch, culvert, or ravine, or under a sturdy bridge. The goal is to avoid being struck by flying debris. Mobile homes, even if tied down, offer little protection from tornadoes, and should be abandoned. Protect your head from flying debris. Wearing a football or motorcycle helmet is not a bad idea. The scene in *Twister* in which the stars survived being lashed to a well pipe while being engulfed in an F5-level tornado made no sense whatsoever. They would have been blasted by high-speed debris and would have been killed or seriously injured.

Is your car a **safe haven** in a tornado?

Definitely not. Under no circumstances should you wait out a tornado in an automobile. And do not try to outrun a tornado in your car. If caught you are likely to be rolled over or go airborne and have the equivalent of a 100 mph head-on collision.

In 1967, a woman driving down a street in Chicago was intercepted by a tornado. The car became airborne and was hurled a considerable distance. This very lucky woman landed dazed, but otherwise unharmed. Reporters asked her what she did while she was whirling around up there. "Well," she said, "I stepped on the brakes and blew the horn."

Automobiles are among the most dangerous shelters during a tornado. During a 1989 tornado in Huntsville, Alabama, 12 of the 21 people killed by the storm were seeking refuge in cars. Of the 42 killed in the 1979 twister in Wichita Falls, Texas, 25 were in cars, and of those, 16 were actually trying to outrun or outmaneuver the funnel.

A truck driver was struck by a twister that crossed a highway about 30 miles east of Buffalo, New York, on 3 September 1993. The truck was picked up and hurled through the air, into the path of a tractor-trailer rig. Both drivers were killed.

Should you open windows or doors during tornado warnings?

No—this is an old wives' tale. Creating an opening for the wind may actually increase damage, and in any case going to a window to open it exposes you to deadly flying glass. It used to be thought that much of the damage from tornadoes resulted from unequalized air pressure as the funnel moved over a building, but it is now clear that wind effects are the primary cause of most structural damage. Any opening, including a garage door, that allows for the wind to enter a structure and begin acting on walls and ceilings increases the chance of damage.

Is a **train** a safe place to be during a tornado?

A train is better than an automobile, but it has its risks. The passenger train "The Empire Builder" encountered a large tornado on the tracks near Moorhead, Minnesota, on 27 May 1931 and didn't fare well. Eight coach cars, each weighing more than 80 tons, were derailed. One shifted more than 80 feet off the tracks. One passenger was killed.

What are the signs that a **tornado is imminent**?

Even with today's advanced radars and spotter networks, tornadoes can form suddenly and without warning. Your own senses then become your most important defense mechanism. A thunderous, roaring noise growing ever louder should be cause for immediate alarm, especially if there have been severe thunderstorms present in the area. Many tornadoes occur outside a storm's rain area, so don't assume that the lack of rain is a good omen. In fact, tornadoes can even strike when the sky is clear and blue. Large hail is often present in the vicinity of funnels.

Are tornadoes **unique** to the United States?

No. Tornadoes occur in many countries around the world, though three out of four twisters that do touch down occur in the United States. They tend to occur at mid-latitudes in regions experiencing strong surface fronts and jet streams aloft. Tornadoes are rare in the tropics. Australia may be the closest to the United States in terms of tornado potential, but since they tend to touch down unwitnessed by humans in the sparsely populated outback, it's difficult to determine the total number of twisters there. Other countries experiencing a significant number of twisters include New Zealand, South Africa, Argentina, and in the Northern Hemisphere, much of middle Europe from Italy north into England and Russia. England tends to get very severe weather—at least 50 tornadoes were reported in an 82-year period ending in 1949. October is the peak month for British twisters. Japan, eastern China, northern India, Pakistan, and Bangladesh have all experienced tornadoes, and Bermuda and the Fiji Islands are also at risk.

Though the frequency of actual touchdowns is much lower than in the United States, other countries also experience killer tornadoes. On 24 June 1904, a killer tornado swept through portions of Moscow, Russia, claiming at least 24 lives. On 19 March 1978, a tornado struck New Delhi, India. Seventeen people perished and 700 were injured. In 1995, hundreds of people were reported killed in a tornado in Pakistan.

What is the first known **recorded movie** of a tornado?

While spectacular tornado shots made with home camcorders are now almost commonplace on the nightly news, getting such footage in the old days wasn't so easy. The first known film of an actual tornado was made in 1933. This is nearly two decades earlier than the 1951 Corn, Oklahoma, tornado film once believed to be the earliest. And no, *The Wizard of Oz* did not use actual tornado footage—it was a Hollywood special effect.

What tornado almost changed **American pop culture**?

A tornado almost changed the course of popular music in the twentieth century. On 5 April 1936, a devastating tornado ravaged Tupelo, Mississippi, killing more than 200 residents. One of those who survived the devastation was an infant by the name of Elvis Aaron Presley.

Have any **well-known buildings** been struck by tornadoes?

The path of a tornado knows no bounds. A killer tornado that once struck Atlanta, Georgia, included the governor's mansion as one of its targets.

Historic buildings haven't fared much better. An 1875, one-room schoolhouse in Cuyahoga Falls, Ohio, was destroyed by a twister in 1992.

On 27 March 1991, a tornado raised the roof off St. James Catholic Church in Sag Bridge, Illinois, and damaged old grave markers in the churchyard. The church, built in 1837, was previously a French fortress. Efforts are underway to preserve the church.

A strong tornado smashed into the Oregon Correctional Center in Wisconsin and destroyed 20 buildings while 76 inmates were eating lunch on 17 June 1992. Much to the prisoners' chagrin, however, the fences held fast.

This marquee shows an attempt at gallows humor after the sign above it was destroyed by a tornado.

Has anyone even seen the **inside of a tornado** and lived to tell about it?

On 3 May 1943, McKinnet, Texas, farmer Roy Hall survived being trapped inside a 150-yard-wide tornado funnel, where he was able to look straight up into the rapidly whirling vortex for over a thousand feet. It had small mini-tornadoes spinning off the smooth wall of clouds, and these were illuminated by almost constant lightning.

Can tornadoes **pluck chickens**?

One of the more interesting bits of tornado folklore is that chickens can be de-feathered as a result of the low pressure from a tornado. That may not actually be the case, but many feathers can be blown off by the high winds.

How many tornadoes actually **kill people**?

Only between one and two percent of tornadoes in the United States result in human fatalities. During 1993, there were 1,167 confirmed twister touchdowns. Of these only 16 resulted in deaths. The low death toll of recent years is a testament to the ongoing improvement in the timeliness and accuracy of the National Weather Service's tornado watches and warnings. It also suggests that once warned, it is relatively easy to find a safe haven even if a tornado is bearing down on you.

Do tornadoes have **eyes**?

The eye of a hurricane is the often clear, nearly calm center about which the violent storm rotates. But does the much smaller tornadic vortex have a similar calm core? Very possibly. Both theoretical and observational evidence suggest that the inner portion of the tornado funnel may be largely cloud free and have relatively light winds.

Do **cities** attract tornadoes?

This is not easily answered. Population density significantly affects the probability for a tornado being detected and reported. In many rural areas there is nothing for a tornado to hit, nor anyone there to see it. Moreover, in rural areas in the past, people have been less prone to report tornadoes to authorities. Thus, the statistics, which reflect more tornado activity in urban areas, may not be accurate. Some have suggested that tornadoes tend to avoid cities, but there seems to be little evidence of this. Some scientists believe that cities might even attract tornadoes. Scientists at Texas A&M University found that urban and suburban counties are more likely than primarily rural counties to experience tornadoes.

What is the largest **nonhuman death toll** from a tornado?

On 26 April 1994, a tornado swept through a turkey farm in Barron County, Wisconsin, destroying the farm and killing approximately 10,000 birds.

What is the **record for tornadoes** in July?

July 1993 set an all-time U.S. tornado record for that month with 234 twisters touching down. A normal July total is 94, but July 1993 was hardly a month for normal weather—the great floods along the Mississippi and Missouri Rivers also occurred that month.

Are tornadoes attracted to **trailer home parks**?

One might think so, given the frequency with which they appear in storm damage reports. The reasons are twofold. First, there are many more trailer home parks in the nation than people realize. Second, the construction of these homes is such that they provide far less resistance to strong winds than conventional homes. If not properly tied down, they can begin to "fly" at wind speeds of just over 70 mph. Trailer parks

193

should have storm shelters. Never try to ride out a tornado, or even a severe thunderstorm, in a trailer home.

A tornado ripped through a southeastern Tennessee trailer park on 26 April 1994. A man was killed. His daughter and grandson were in a nearby trailer that was also destroyed; the daughter landed in a pile of rubble 300 yards away and the 9-month-old infant was tossed 100 yards. Both the woman and the child were injured but survived. On 5 August 1992, a South Carolina tornado rolled a mobile home over three times, but its lucky occupant managed to walk away without serious injury.

Can you have a tornado when the sun is shining?

Yes. Many tornadoes occur very close to the edge of the violent thunderstorm cloud from which they are spawned. It is entirely possible, especially in the hour or so just before sunset, that the tornado itself could be in full sunlight for much of its life cycle.

Are the number of tornadoes increasing?

The number that occurs each year is not increasing, but the number of twisters that are actually spotted and reported has definitely increased. In the 1950s, there were 4,793 tornadoes reported in the United States. In the 1980s, there were 8,194. The higher number is not a reflection of a dramatic rise in the incidence of tornadoes. More people now live in or travel through the most tornado-prone parts of the nation, so tornadoes are more likely to be spotted. There have also been tremendous improvements in communications and means of reporting severe weather.

Are tornado fatalities increasing?

No. In fact, there has been a long-term downward trend in tornado deaths in the United States, in spite of a steadily increasing population. In 1993, a then-record number of 957 severe thunderstorm and tornado watches were issued by the National Severe Storms Forecast Center (now named the Storm Prediction Center), but the death toll from severe thunderstorms and tornadoes dropped to 33, well below the 30-year average of 82 fatalities. In the pre-forecast days in the 1930s, tornadoes alone killed almost 200 people each year.

What date is least likely to see a tornado in the United States?

If recent history is any guide, the 16th of January is a good bet. It is the only day of the year in this century (through 1996) on which a tornado touchdown has not been reported somewhere in the United States. There is at least one tornado touchdown somewhere in the United States on at least half the days of a typical year. More than 10 tornadoes on a single date occur only 2.4 percent of all days, or nine times per year.

What is the "green sky" effect?

During severe thunderstorms, a definite green tint in the dark clouds is often a sign of nearby hail and perhaps tornadoes. A satisfactory explanation for this effect has yet to be agreed upon by meteorologists. Some believe it may be due to optical effects caused by the large amounts of ice suspended aloft in the storm's powerful updrafts.

Was 1992 a "good" year for tornadoes?

1992 was a bumper year for U.S. tornadoes, with a record total of 1,293 reported. On 15–16 June 1992, a massive outbreak of 123 twisters was the second largest in history. The 399 reported touchdowns in June was the most ever recorded in a single month. In 1992, seven states broke their records for tornado touchdowns: California (19), Delaware (6), Louisiana (77), Maryland (13), New York (25), Ohio (61), and Colorado (81).

Is there **double jeopardy** in tornado strikes?

Since tornadoes are relatively rare, at least in certain regions of the country, it would seem that you are unlikely to be caught in a tornado more than once. There are many cases in which the unlikely happened, however. On 2 December 1982 a Missouri man was injured when his home was flattened by a tornado. He moved into a trailer home for temporary shelter—and three weeks later was killed when a second tornado came through town.

A house in Collinsville, Illinois, was destroyed by a tornado on 18 May 1883. The rebuilt structure was hit again on 30 March 1938.

The possibility of a building being struck by a tornado is rare, typically less than once per 1000 years. But one church in Guy, Arkansas, has been damaged by twisters three times during the last century.

Statistically, you will have to wait hundreds of years to be struck by a tornado standing at any one spot in Tornado Alley. But on 16 March 1942, Baldwin, Michigan, was lashed by a tornado. Before the clean-up could even begin, another twister passed through town 25 minutes later.

A Spanish-American War veteran survived the war and returned home to Nebraska in 1899, only to have his house destroyed in a twister. He then moved to Omaha, where his house was destroyed by a tornado in 1913.

Why do the residents of **Cordell, Kansas,** get nervous every May 20th?

May 20th is a day on which the residents of Cordell, Kansas, probably watch the sky with more than the usual interest. In 1916 on that date the area was swept by a tornado. It happened again on 20 May 1917, and yet again on 20 May 1918.

Is the **southwest corner** of the basement the safest?

Contrary to weather folklore, the southwestern corner of a basement is not the safest place to hide during a tornado—in fact it may be more dangerous than other parts; research has shown that the southwest corner may actually accumulate more debris than other parts of a basement. Your best chances to avoid injury are to be under a stairwell or heavy table that can protect you from flying or collapsing debris.

How are **tornadoes detected**?

The new nationwide network of NEXRAD Doppler radars being installed is designed specifically for tornado warning purposes. While the NEXRAD cannot detect the actual tornadic vortex, except under unusual circumstances, it often can spot the larger scale rotation (the tornado mesocyclone) out of which many tornadoes form. NEXRAD does not always detect conditions leading to smaller tornadoes, but it is highly likely that it will provide warning, often with many minutes of lead time, of most major twisters.

Do tornadoes strike cities in the **northern United States**?

When severe thunderstorms and tornadoes are mentioned, one tends to think of locales like Kansas and other parts of the Great Plains. But severe tornadoes have been known to strike far north of that region. The worst tornado outbreak ever to hit the Twin Cities metropolitan area in Minnesota occurred on 6 May 1965. The death toll reached 13, and 371 were injured. At least 321 homes were destroyed and 1,196 more were damaged. One of the most spectacular series of tornadoes ever to hit the United States roared through the Fargo, North Dakota, area in 1957. And tornadoes have been spotted well north into Canada and Alaska. In Manitoba and western Ontario, on 18 July 1991, small tornadoes and downburst winds gusting faster than 90 mph flattened up to 30 million trees in less than an hour, the worst such loss in Canadian forestry records. A major tornado has also struck Edmonton, Alberta.

What is the pressure inside a **tornado vortex**?

No one is sure, except that it is very much lower than that of the surrounding atmosphere. One recording barometer that survived a near miss by a passing funnel dipped to around 24.00 inches (the average home barometer doesn't even go that low).

What is a **gustnado**?

The Weather Service doesn't issue gustnado watches or warnings, but such creatures do exist. Formed in the strong wind shears along the gust front marking the strong outflowing winds at the leading edge of intense thunderstorms, gustnadoes look very much like small tornadoes, and in fact they are. While rather short-lived in nature, they can cause some damage.

Was the invention called **"Dorothy" in the movie** *Twister* **based on a real experiment?**

In *Twister,* Dorothy was an instrumented package that a team of meteorologists were trying to place in the path of an F5 tornado in order to obtain the first measurements inside a monster twister. Far from being pure fiction, it was an obvious take-off on TOTO—the totable tornado observatory developed by NOAA and university scientists. Actual measurements of pressure and winds inside a tornado are hard to come by. Scientists designed this portable wind-sensing system that was meant to be dropped in the path of an oncoming twister in the hope of getting invaluable measurements. Due to the fact that the real scientists weren't as reckless as the characters in the movie, TOTO was never deployed with complete success. But the fundamental idea is a good one.

Can tornadoes **rotate the "wrong way"**?

In the Northern Hemisphere the vast majority of tornadoes rotate in a counterclockwise direction, though a small percentage do spin in the opposite direction. Southern Hemisphere tornadoes rotate in a clockwise direction. Smaller vortices in the atmosphere, such as dust devils, tend to have a larger percentage spinning in an atypical direction.

How does **television news** cover tornadoes?

In July 1986, a television station traffic helicopter in the Minneapolis/St. Paul area gave the viewers of the 5 P.M. newscast something more than a traffic jam to watch. A spec- **197**

Debris from a house demolished by a tornado comes to rest in a grove of trees.

tacular tornado slowly moved across the northern part of the metro area, all the while being broadcast "live." The pilot did start getting nervous when large parts of trees began falling out of the sky around him. Since then, television news choppers across the interior of the United States have vied with each other to get the "live" tornado scoop on the air. So far none has crashed—either from the tornado or into each other.

What are some of the larger objects that have been transported by a twister?

Severe weather hit Tazewell County, Illinois, on 9 May 1995. The winds were so strong that a 40 x 80-foot farm building was blown a quarter of a mile across a field.

A tornado struck a church. It ripped off its steeple and carried it some 15 miles. Another whirlwind sucked up an ice chest weighing some 800 pounds and transported it over three miles.In 1973, a Nebraska tornado struck a house and carried a 500-pound baby grand piano over a quarter of a mile through the air.

The Lubbock, Texas, tornado of 1970 was famed for its destructive force. A 41-foot-long fertilizer tank, weighing 13 tons, was blown or rolled three-quarters of a mile away from its original position.

On 7 February 1988, in Lancashire, England, the winds, which had been in the 10–20 mph range, suddenly gusted to 106 mph. The gust tossed a 150-pound sheep-feeding trough more than 15 feet, but otherwise had little impact.

What sorts of debris have been **lofted by tornado funnels?**

Tornadoes and waterspouts often act rather like a vacuum cleaner, sucking almost anything from the surface high into the atmosphere. But what goes up does come down, often causing great surprise in the process.

A gentleman living near London in 1984 heard loud thumping sounds on his roof one night while watching television. The next morning he found a half-dozen flounders and whitings up to six inches in length on his roof. Other locals found fish deposited in their gardens. A waterspout was considered a likely source of the fish fall.

In the year 1578, a large number of yellow mice fell from the sky in the city of Bergen in western Norway.

In Wolverhampton, England, there occurred a violent thunderstorm in 1860 in which the heavy rain became mixed with small, sharp, black pebbles. Enough fell in some places that they had to be swept away.

Pennies and half-pennies fell out of the sky near Bristol, England, in 1956. And in Germany in 1976, two ministers were surprised to find some 2000 marks worth of bills fluttering earthward.

In Maryland, on 7 January 1969, literally hundreds of badly injured ducks fell from the sky, apparently having been hurt while in flight. The cause for their injury is unknown.

The smelt run on the Great Lakes is a springtime fisherperson's delight. But there was one report in 1986 of smelt literally falling from the sky, by the thousands, near Alpena, Michigan.

Near Stoke-on-Kent, England, on 21 March 1983, severe weather produced numerous weird effects, including ball lightning, several tornadoes, and a shower of seashells from the sky.

A small boat floundering in 8-foot waves during a spring 1986 storm near Lake Huron's Thunder Bay had more than capsizing to worry about. Thousands of small silvery fish fell from the sky like a silver rain.

During a thunderstorm in Charleston, South Carolina, on 2 July 1843, thunder was heard, rain fell, and an alligator fell, right on Anson Street.

In England, after a 1983 thunderstorm, there were numerous reports of chunks of coal, some several inches across, scattered throughout several towns. Some were found inside melting hailstones.

How far can **tornadoes carry debris**?

For hundreds of years there have been reports of strange objects (fish, turtles, and the like) falling from the sky, presumably after being lofted by some distant tornado. Now researchers at the University of Oklahoma's School of Meteorology have begun compiling such reports, particularly of objects that can be traced back to their point of origin, as part of a study of severe storm dynamics. Recent research has collected a long list of items that have fallen from the skies. Some of the more interesting things that have been found: plate glass, land title papers, a music box, trousers, coats, a wedding gown, a tie rack, dead ducks, a cow . . . and an airplane wing. A jar of pickles was swept away and landed 25 miles downwind, unbroken.

On 25 April 1880, a tornado that went through Noxubee, Michigan, carried an entire bolt of cloth some eight miles—without it unraveling.

On 13 June 1953, Emily McNutt of South Weymouth, Massachusetts, found a wedding gown in her backyard. It was dirty but otherwise in good condition. Some detective work traced the gown back to a woman in Worcester, Massachusetts, some 50 miles away. The gown was just one of the many pieces of debris flung over eastern Massachusetts by the passage of a powerful tornado.

On 15 April 1979, canceled checks began fluttering out of the sky in Tulsa, Oklahoma. Their source: a bank struck by a tornado earlier in the day in Wichita Falls, Texas. They were carried by the thunderstorm for over 200 miles.

A tornado destroyed a motel near Broken Bow, Oklahoma. The motel's sign was later found in Arkansas.

COLD AND WINTER STORMS

Is it true that **no two snowflakes** are alike?

No two snowflakes are exactly alike. But all snow crystals are six-sided, at least at the microscopic level. Some snow crystals can appear to the eye as long needles, bullets, or capped columns, as well as the six-sided star variety. An ice crystal develops a needle shape, for instance, if the six-sided crystal is, say, 200 times thicker (longer) than it is wide. The uniqueness of each snow crystal was demonstrated to the world by the efforts of amateur Vermont meteorologist W. A. Bentley, who spent nearly 50 years of his life making microphotographs of thousands of individual snowflakes. Over 2000 of these were published in 1931 by the American Meteorological Society. The earliest known attempt at portraying the delicate crystal nature of snow was a 1555 woodcut published in Rome and made by Olaus Magnus, Archbishop of Uppsala, Sweden.

What causes the different types of **snow crystals**?

Snow is not frozen raindrops. Snow crystals are the result of water vapor depositing on a speck of ice or some minute particulate, often a piece of mineral dust. The deposition of water vapor into the solid form results in a crystal, the characteristics of which depend very much on the temperature and water vapor content within the cloud. The shapes can be in the form of plates, needles, feathery dendrites, or capped columns. It is possible to predict the general characteristics of ice crystals formed in clouds if you have enough information about the temperature and moisture conditions. Snowflakes are actually aggregates of many individual ice crystals.

One of thousands of snow crystal microphotographs taken by amateur meteorologist W. A. Bentley.

How quickly can **winter** start?

On 12 September 1993, the temperature soared to 94°F near Ft. Collins, Colorado. By 10 A.M. the next morning, the temperature was 33°F and there were three inches of snow on the ground.

Which is warmer, the **North Pole** or the **South Pole**?

The North Pole is warmer than the South Pole. The Arctic Ocean does have at least some breaks that help warm the overlying air mass a little. The South Pole, by contrast, sits high atop a mountainous glacier and is often 50°F colder during the dead of winter than the Northern Pole. More than 95 percent of the Earth's snow-covered land lies in the Northern Hemisphere.

How much **water equivalent** is there in snow?

On the average, ten inches of snow melts down to about one inch of water. There is, however, tremendous variability from storm to storm. In general, snow that falls during low air temperatures is much "fluffier"—that is, less dense—than that which falls when the temperature is near freezing. Thus, 25 inches of Colorado champagne powder might melt down into one inch of liquid, whereas just three inches of Central Park snow might yield an inch, a testament to the soggy nature of many eastern seaboard snowstorms.

What is **sleet**?

Frozen raindrops are called sleet, or more technically, ice pellets. Snow is not frozen raindrops. Snow crystals form directly in clouds and do not melt or refreeze on their way to Earth.

202

Ice storms are one of winter's worst hazards.

What is **freezing rain**?

Freezing rain is exactly what its name implies. Very often during winter, precipitation forms in warmer layers aloft that override the dense cold air near the surface. When rain falls, or snow melts while falling through an intermediary warm layer, rain falls into the shallow cold layer near the ground. There, either the air temperature is below freezing and/or the ground and objects are still below freezing after an extended cold snap, and the liquid rain freezes on contact with colder surfaces. This precipitation is also called glaze, or an ice storm, and it is usually much more slippery than snow.

What is an **ice storm**?

A little freezing rain is very slick and a lot can be a catastrophe. Ice storms are often winter's worst hazard in a broad belt from Nebraska, Oklahoma, Kansas, and eastward into the Middle Atlantic states, though these storms have occurred over most of the nation that experiences winter weather. By some measures the ice storm from 28 January to 4 February 1952 was the most devastating on record, killing at least 22 persons from Louisiana, Mississippi, and Arkansas.

Glaze usually does not accumulate more than an inch thick during most ice storms. But deposits of eight inches in diameter were reported on wires in northern Idaho in January 1961. Six inches coated wires in northwestern Texas in November **203**

1940 and in New York in December 1942. In Michigan in 1922, the ice coating on a one-foot-long piece of number 14 telephone wire weighed 11 pounds. It has been estimated that an evergreen tree 50 feet high with an average width of 20 feet may be coated with as much as five tons of ice during a severe ice storm. Needless to say the impact on traffic is enormous, with over 85 percent of all ice storm deaths related to traffic accidents.

In February 1994, an ice storm in the southeastern United States resulted in nine fatalities and over three billion dollars in economic losses. From Texas to Louisiana to North Carolina, up to four inches of ice and sleet accumulated at some locations. Some utility customers were without power for a month.

What is a **degree day**?

Degree day is a computation devised by the utility companies to help predict the approximate rate at which you will spend energy dollars to heat or cool a building. It is assumed that neither the furnace nor the air conditioner is on when the outside temperature averages 65°F. To compute the number of *heating* degree days, take the average temperature for a day (the high plus the low, divided by two) and subtract it from a base of 65. If the daily average was 20°F, that day had 45 heating degree days. To compute *cooling* degree days, calculate the daily average temperature, and then subtract the base of 65. Therefore, a day with an average temperature of 80°F had 15 cooling degree days. Your heating/cooling energy bills rise and fall along with the accumulated heating/cooling degrees days over a season. Your energy bill will often have a comparison of the current month's heating/cooling degree days compared with the prior year or the long-term norm.

What is the **largest snowflake** ever observed?

In 1888, in Shirenewton, England, it snowed and made things white in a hurry. Giant snowflakes, 3.75 inches across and a quarter-inch thick, fell and covered the ground to a depth of two inches in just two minutes! That is not a world record, however. On 28 January 1887 it snowed in Ft. Keough, Montana, and some of the flakes were measured at 15 inches in diameter! Some snowflakes fell that made splotches in the fields "larger than milk pans."

Snowflakes, as opposed to individual snow crystals, may consist of clumps of many crystals, over 100 in extreme cases.

What are some record **24-hour snowfalls** in the United States?

In 1995, Buffalo, New York, had a 24-hour record snowfall (for a large town or city) of 37.9 inches. In U.S. weather history, only Valdez, Alaska, has recorded a greater one-day dumping in a large city: 47.5 inches in January 1990. On 26 December 1947, New York City had its biggest one-day burst of snow: over 25 inches.

How much ice exists in **Greenland and Antarctica?**

The Greenland ice cap contains three million cubic kilometers of ice. If it were all to melt, the global sea level would rise more than 7.5 meters (24 feet). The Antarctic ice cap contains twenty-nine million cubic kilometers of ice. If it were all to melt, the global sea level would rise some 65 meters (210 feet).

What is the **coldest temperature** possible?

On the Earth's surface, air temperatures in the -130°F range are found near the South Pole. But absolute zero, the coldest it can get anywhere, is -460°F (-273.16°C). The coldest thing you are likely to personally be near is liquefied helium, at -452°F.

Is **wet ice** more slippery than dry?

Ice covered with water is more slippery than "dry" ice. The reason why an ice skater can glide so effortlessly across a frozen pond is that a thin film of water forms between the blade and the ice. Recent research shows it is not the high pressure of the thin blade pressing on the ice, but more likely the friction of the moving blade that heats up the ice to form the liquid water. And during periods of winter freezes and thaws, it is common for melted snow to pool over ice, greatly increasing the number of "slip and fall" accidents. Cold, "dry" ice is often much less slippery underfoot.

Can alcohol **warm you up** on a cold winter day?

You're lost in a howling blizzard, curled up in a snowdrift. Suddenly a big warm tongue slurps across your face. You're saved by a St. Bernard! But before you chug the contents of the keg, check to be sure it doesn't contain alcohol. A shot of cognac might give you a momentary "warm glow," but on the whole alcohol greatly increases your body's sensitivity to cold. If you are planning to be outside for an extended period of time in cold weather, do not rely on alcohol to "warm you up." Try drinking a warm, non-alcoholic beverage, like hot chocolate. Alcohol, in addition to reducing the body's tolerance to cold, impairs judgment, leading to weather-related injuries and accidents.

What is the **windchill factor?**

The windchill factor (or windchill index) is a number, often expressed as an equivalent temperature, that expresses the cooling effect of moving air at different temperatures. **205**

The lower the windchill, the more calories of heat are being carried away from the exposed surface of the body. The concept was originally developed in 1939 by Paul A. Sipel, an Antarctic explorer and an expert on cold climate issues. Many of the experiments originally conducted measured the time it took a small container of water to freeze as a function of air temperature and wind speed. Other factors such as sunshine, your own metabolism, and of course protective clothing, will determine how cold you feel outside. The windchill is a useful guide to quantify the impact of the two major factors (air temperature and air motion) on your bare skin. There is increasing discomfort but little serious danger of frostbite for properly clothed persons down to about -20°F windchill. Between -20 and -70°F, danger of frostbite increases rapidly unless special precautions are taken. Below -70°F, exposed flesh can freeze within 30 seconds and Arctic survival gear is an absolute must. And keep in mind, skiers and snowmobilers create their own wind. Cruising through the woods on a calm zero degree day, riding a snowmobile at 23 mph, exposes the passengers to a -45°F windchill.

A temperature of 23°F with a 50-mph wind has the same effect as a temperature of -10°F and a 7-mph wind. Both result in approximately a -20°F windchill factor. In other words, the effect of the wind causes the body to lose heat in both cases comparable to standing in calm air having a temperature of -20°F.

Do you have to **antifreeze your car** to the lowest windchill?

You should antifreeze your car according to the owner's manual. Just in case you aren't sure—you need only antifreeze your car down to the coldest expected air temperature, not the lowest windchill value expected. Check your city's all-time record low temperature and use that as a guide. But if you live in a low-lying area away from the warmth of a city, you may need to subtract another 10 or 20 degrees. There is no need to add enough antifreeze to account for the windchill. Even though the windchill might hit -100°F, you do not need to protect your car beyond the air temperature itself. Neither your car's engine nor anything else can be cooled below the air temperature; it will just cool down to that temperature much faster if the wind is blowing strongly.

What can be done with **unripe tomatoes** when an early first frost is forecast?

The first frost or freeze of the oncoming winter is good news and bad news. It kills the mosquitoes, but also your tender garden plants. You can pick your remaining green tomatoes, put them in a paper bag in a cool dark place, and they will ripen in five to eight days.

What are the **coldest temperatures** ever recorded in the United States and Canada?

The coldest temperature ever recorded in the 48 contiguous states was in Rogers Pass, Montana: -70°F on 20 January 1954. A U.S. national record of -80°F was observed at Prospect Creek, Alaska, in the Endicott Mountains of northern Alaska on 23 January 1971.

Canada's coldest temperature was -81°F at Snag in the Yukon Territory. The only problem with that measurement was that the thermometer scale only went to -80°F and the observer had to make a pencil mark and later guess how cold it actually was.

What is the coldest temperature recorded in the **Northern Hemisphere**?

Siberia justly deserves its cold reputation. Temperatures of -90°F have been observed at both Verkhoyansk and Oymyakon, Siberia. And these were strictly air temperatures—they did not include windchill.

What is the coldest temperature ever **recorded on Earth**?

The coldest temperature ever recorded on the surface of the Earth was measured at the Russian research station at Vostok, Antarctica. The air temperature reached -128.6°F (-89.6°C) on 31 July 1983. And that was without any windchill. How cold can it get at the surface of the planet Earth? Some estimates claim that in the calm polar winter night, above the surface of the snow, temperatures of less than -140°F are possible.

Are **cold winters** followed by cold springs and summers?

In general there is little correlation in large-scale weather patterns from season to season. The winter of 1994–95 was among the warmest on record in many parts of the United States. In the Twin Cities local street departments used 8,000 fewer tons of salt and 30,000 fewer tons of sand on area highways, for a savings of $750,000. But the following spring was decidedly colder than normal, so much so that many lakes in Minnesota still had part of the winter ice cover as late as the second week of May.

December 1989 was the fourth coldest month on record in the United States. But with a shift in the jet stream winds aloft, the following January was the warmest ever. In many parts of the United States, the coldest winter and hottest summer both occurred during the Dust Bowl year of 1936.

What are urban ice slabs?

An April 1995 ice storm in Chicago caused the usual travel headaches for pedestrians and drivers. But in a city with so many high-rise buildings, there is another hazard. When warm temperatures return, giant sheets of ice peel loose from the tall structures and fall to the ground. Michigan Avenue has been closed to traffic for up to several hours as giant shards of ice threatened those below.

Have U.S. winters been getting milder?

There have been some very mild winters in the United States in the past decade. During this century, the warmest winter in the 48 contiguous states was 1991–92. Close behind were the winters of 1953–54 and 1994–95. There have been some very cold winters as well. During the February 1996 Midwestern cold wave it was so cold that a zoo in Minnesota reported polar bears were leaving bloody paw prints because their feet froze to the ground! On 16 January 1996, temperatures plummeted in northern Minnesota. It was -57°F at Embarrass, MN. But it was International Falls that was really embarrassed; the town near the Canadian border had to cancel its annual macho cold festival, Icebox Days, because it was too cold even for them. Things then got colder: Groundhog Day 1996 was a day for neither man nor beast to be out checking for shadows over much of the nation. Tower, Minnesota, logged the coldest temperature ever recorded in a state known for its brisk winters: -60°F (and that was without windchill). Meanwhile, those who elected to visit Mickey and Goofy in Disney World on that day were 145°F warmer, as Orlando sizzled in 85°F weather.

Is there a difference between a frost and a hard freeze?

A frost refers to ice crystals forming on a surface—the proverbial frost on the pumpkin. However, the air temperature does not necessarily need to be below 32°F for frost to be found. Many surfaces such as grass, metal, and glass can cool many degrees below that of the air temperature as it is recorded by the National Weather Service (NWS) in a shelter some five feet above the ground. That is why you may need to scrape your car's windshield even when the weather announcer is informing you that the latest temperature is 37°F.

Frost flowers on a cold window pane in winter.

A hard freeze is a term used by the NWS to indicate air temperatures of 26 degrees or less for at least four consecutive hours. It usually means the end of most annual vegetation.

What is a **blizzard**?

By National Weather Service definitions, a blizzard occurs when there are wind speeds of 35 mph or more with considerable falling and/or blowing snow causing poor visibility, which can frequently be much less than a quarter of a mile. Although the term used by the general population may suggest otherwise, a blizzard does not require new snow to be falling. Sometimes the phrase ground blizzard is used to describe a storm in which all the airborne snow has been re-suspended from that which had previously fallen.

What is a **whiteout**?

A whiteout results from extreme blizzard conditions in which blowing or falling snow reduces visibility to the point where the sky, the air, and the ground become indistinguishable— everything is white.

Persons caught in a whiteout, aside from the hazard of frostbite, can become completely disoriented and lose their way, often with dire consequences.

209

Which is coldest, **Moscow or Minneapolis**?

In Moscow, Russia, the average January temperature is about +14°F, about the same as southern Minnesota, where Minneapolis is located. The warmest month is July, with a pleasant mean of 66°F in Moscow and 74°F in Minneapolis. The Twin Cities' all-time coldest temperature was -41°F, while Moscow's official all-time low is -27°F.

Where are the **snowiest places** in the United States?

In the lower 48 states, the snowiest location in the western United States is Blue Canyon, California, which annually averages 241 inches of snow. In the eastern United States, Marquette, Michigan, gets 129 inches in a typical year. The snowiest large city in the lower 48 is Syracuse, New York, which plows some 112 inches each year.

The Mt. Shasta Ski Bowl in California received 189 inches of snow from 13 to 19 February 1959, the record for a single U.S. snowstorm. Silver Lake, Colorado, holds the 24-hour snow dump record for the western U.S. mountains, with 76 inches on 14–15 April 1921. Paradise, on Mount Rainier in Washington State, recorded 1,224.5 inches from 19 February 1971 to 18 February 1972 to establish a one-year record. The deepest snow depth ever measured was 37.5 feet in March 1911, at Tamarac, California.

What is a **Nor'Easter**?

Deep low-pressure systems form over the eastern United States, and once obtaining a feed of warm, moist air as they approach the Atlantic, they can grow explosively while moving offshore, parallel to the coastline. Since the winds rotate counterclockwise, most of the snow or rain in coastal cities is accompanied by strong northeast winds, hence the name Nor'Easter. Most of the major snowstorms in the highly populated mid-Atlantic and New England states arise from Nor'easters.

Predicting snowstorms is one of the most difficult aspects of a forecaster's job. Over a half billion dollars a year is spent on preparation for snow removal for storms that are predicted but never materialize, but far more than that can be saved by accurate predictions of the amount and timing of snowfalls affecting major metropolitan areas.

In December 1992, a slow-moving Nor'Easter battered the northeast U.S. coast. Nineteen people died, and damage approached two billion dollars.

A nearly stationary Nor'Easter storm off the U.S. East Coast produced some very topsy-turvy weather in January 1956. Temperatures rose to nearly 50°F from New Jersey to Maine. Meanwhile, freezing weather gripped Florida and ice formed on the tropical island of Jamaica.

In February 1969, records were set for snow depth throughout much of New England. The snow lay 70 inches deep at Rumford, Maine. At higher elevations, Mt.

Mansfield, Vermont, recorded 114 inches and Pinkham Notch, New Hampshire could boast of 164 inches (almost 14 feet).

What is the **rain/snow line**?

In many large winter storms there is often a distinct boundary between the areas receiving snow and those getting rain. This boundary is a major forecast problem for cities such as Boston, New York, Philadelphia, and Washington, D.C. A 30-mile shift in the rain/snow line can mean the difference between a soggy commute and a winter nightmare. The rain/snow line in storms in the northeastern United States can be influenced by many factors, including proximity of the warmer ocean. On 2–3 March 1969, Port Jarvis, New York, was dumped with 33 inches. Yet in Liberty, New York, just 30 miles to the south, mostly rain fell and only one inch of snow was recorded.

Strong local differences in snow can result from other factors. In 1992, 21 March was the first full day of spring. But a localized snowstorm pelted northern Chicago, including O'Hare airport, with 8–10 inches of snow. Three jets slid off runways and 300 passengers were stranded overnight, depending on the Salvation Army for dough-nuts and the airlines for blankets. All of Chicago's 238 plows were out in force. Over 3,000 homes lost electrical service. Just 30 miles south of O'Hare, however, it was a bit more spring-like, as only one inch of snow fell.

What is **lake effect snow**?

On cold fall or winter mornings, "steam fog" can often be seen lofting from the still-warm surfaces of rivers and lakes. Over the Great Lakes, this massive evaporation of lake water results in clouds and the infamous "lake effect" snow squalls on the down-wind shoreline.

Bennetts Bridge, a small town about 30 miles east of Oswego, New York, can claim one of the greatest 24-hour snowfalls ever in the eastern United States. On 17 January 1959, 51 inches fell in 16 hours.

During the 1976–77 winter in Buffalo, New York, it snowed for 40 straight days. And then a massive blizzard struck, with 70 mph winds, leaving 30-foot drifts and 29 dead, some frozen to death in their vehicles. Almost 20,000 people were stranded away from their homes for days.

What was the **Storm of the Century**?

Depends on where you are. There have been dozens of "Storms of the Century" in the twentieth century. In several states, the Storm of the Century occurred in 1975. The great blizzard of 10–11 January 1975 in the upper Midwest was one of the worst, with winds reaching 90 mph in Iowa, windchills of -80°F in North and South Dakota, trains

Snow mounds in a snowbelt downwind of the Great Lakes.

stranded in snowdrifts in Minnesota, 19 inches of snow in Nebraska. At least 80 people perished and 55,000 head of livestock were lost.

The East Coast has had several "100-year storms" during this decade alone. But the Blizzard of 1993 was certainly worthy of the extensive press coverage. The total death toll on land was at least 270, and 48 more were lost at sea. Fatalities were reported from Florida to Maine. Economic losses were between three and six billion dollars. For the first time, every major airport on the East Coast was closed at one time during the storm.

Over three million customers lost electrical power during this colossal storm. Hundreds of roofs collapsed due to the weight of the snow. At least 18 homes fell into the sea on Long Island, and 200 homes on North Carolina's Outer Banks were severely damaged. Up to six inches of snow fell as far south as the Florida Panhandle, and 100+ mph winds over the Gulf resulted in a 12-foot storm surge that killed seven people in Taylor County, Florida.

Wind gusts above 75 mph were commonplace along the entire eastern seaboard. The Dry Tortugas, west of Key West, Florida, peaked at 109 mph. Flattop Mountain, North Carolina, gusted to 100 to 107 mph, and 144 mph winds were noted on Mt. Washington. At lower elevations, 89 mph on Fire Island, New York was not unusual.

The highest reported snowfall from the 1993 storm was 56 inches at Mount LeConte, Tennessee, and Syracuse, New York, logged 43 inches. There were 25 inches at Pittsburgh, 9 inches in Boston, 13 inches in Washington, D.C., 17 inches in Birmingham, Alabama, and 4 inches in Atlanta. Freezing temperatures plunged as far south as Daytona Beach, Florida. It was estimated that the volume of water that fell was equivalent to 44 million acre-feet, the equivalent of 40 days' flow of the Mississippi River past New Orleans.

The Great Blizzard of 1993 was so powerful a storm that it set record-low barometer readings over a dozen states. New all-time record-low barometer readings were set in Columbia, South Carolina; Raleigh, North Carolina; and Richmond, Virginia—wiping out records established in "lesser" storms like Hurricanes Hugo and Hazel. A record-low barometer was reported at 28.43 inches in White Plains, New York. If the Blizzard were a hurricane it would have been rated as a category three storm on the Saffir/Simpson sale for hurricane strength. It affected 26 states and 50 percent of the nation's population.

What are some low points from the nasty winters of 1992–93 and 1993–94?

The Great Blizzard of 1993 was just a great exclamation point in a series of major winter events in two of the worst winters in memory. While that blizzard in March 1993 in the eastern United States garnered a lot of publicity, it was preceded by the "Great Nor'Easter of December 1992." Interior portions of the mid-Atlantic and New England states were dumped on by as much as 30 inches of snow, 23-foot waves slammed into the Massachusetts shore, and total damages were near two billion dollars.

The vicious cold wave during the latter part of January 1994 must have made all talk of global warming seem suspect. Temperatures dropped as low as -27°F in Indianapolis, and over 130 persons perished from the cold. Some have proposed a theory that ice ages are cyclical, and after a relatively long interglacial period, the start of the next ice age may in fact be overdue.

A cave is one place where you would think weather had little impact. Yet the extremely cold weather in Kentucky in January 1994 caused enough chilled air to infiltrate Mammoth Cave to cause thermal contraction of the rocks, loosening a 100-ton limestone slab (some 70 feet long and 20 feet wide). It crashed to the surface in the Rotunda Area, the most significant geological event in the cave since it opened in 1861.

On 1 November 1993, Marquette, Michigan, and Key West, Florida, had comparable weather: both cities set new record lows—two of over *five dozen* daily record lows set around the nation as the first really chilly air of that extremely cold season swept southward from Canada.

The Cold Wave of January 1994 was notable not only for its severity but for its longevity. Chicago had daily minimums of -10°F for a record 10 days.

The wicked winter of 1993–94 in the eastern United States was also deadly. At least 113 people died due to car accidents, heart attacks while shoveling, and exposure. By comparison the 1993 Great Midwest floods took 48 lives and Hurricane Andrew's toll in the United States was 56.

In Bowling Green, Kentucky, Halloween was upstaged on 30 October 1993 by six inches of snow, the earliest recorded fall of measurable snow for that area.

The coldest temperature reported in Alaska during January 1993 was -68°F, in the town of Coldfoot.

Some winters just don't know when to give up. On 1 June 1993, St. Cloud, Minnesota recorded its latest ever freezing temperature.

What were some of the impacts of the Great Blizzard of '96?

The Great Blizzard of '96 took at least 187 lives. It also cost quite a bit of money: insurance companies received over 300,000 damage claims totaling more than $595 million. Worse, over $700 million in flood losses in Pennsylvania alone were largely uninsured. The total price tag was over three billion dollars. Extremely accurate forecasts did allow many to take precautions, however. Airlines moved planes out of the storm's path so they weren't stranded in drifts and thus saved millions by restoring flight service early.

A band of snowfalls greater than 24 inches stretched from West Virginia into New York City and southern New England. Up to 31 inches fell on Long Island. The Washington, D.C., area ranged between 23 and 29 inches. Little Switzerland, North Carolina, received 30 inches. One inch fell at both Atlanta and Athens, Georgia.

What was the Blizzard of 1888?

Actually, there were two famous blizzards that year. The 11–14 March 1888 Nor'Easter lashed the eastern seaboard from Chesapeake Bay to Maine. Snowfalls averaged 40 inches or more over southeastern New York and southern New England. Over 400 deaths were recorded, more than 200 in New York City alone. The winds howled for four days, reaching 50 to 70 mph at times. In New York City drifts reached beyond second-story windows. The highest reported snowfall was 58 inches at Saratoga Springs, New York. The highest drift was 52 *feet* at Gravesend, New York.

Two months before, a blizzard had swept the High Plains eastward to Minnesota. The combination of gale winds, blowing snow, and extremely rapid temperature falls made it a very dangerous storm, and there was a large loss of life. Many people and thousands of cattle perished.

What was the **coldest winter this century** in the 48 contiguous states?

In the last 100 years, the coldest winter in the 48 states was the winter of 1978–79.

How dangerous are **avalanches?**

Avalanches are a major, unrecognized, weather-related hazard. A leading cause of winter recreation fatalities in the western United States, a typical avalanche unloosens some 100,000 tons of snow.

In 1910, a massive avalanche in the state of Washington pushed three steam engines and several passenger cars off a steep embankment. The plummet of nearly 150 feet killed over 100 people.

Avalanches are in general a major threat to life and limb in mountainous areas. If you are skiing, alpine or cross country, avoid long slopes greater than 30 degrees as they are most prone to let go in an avalanche.

How many **avalanches** occur in the United States annually?

The number of avalanches reported annually in the United States is between 1,200 and 1,800. The number of avalanches that actually occur, however, are probably 100 to 1000 times that number.

What U.S. state has the **highest annual death toll** from avalanches?

The state of Colorado leads the nation in deaths from avalanches, with an average of six to eight persons lost each season. Alaska and Montana also have high avalanche tolls. Most fatalities result from people, especially skiers and snowmobilers, traveling into areas that have been posted as off-limits.

In what month is an **avalanche** most likely to occur?

The deadliest month for avalanche victims in the United States is February. Since 1950, almost 90 persons, many of them skiers, have perished in snowslides. June, July, and August, incidentally, are not fatality free. Summer snowslides in high terrain sometimes catch hikers and campers by surprise.

What is permafrost?

Permafrost refers to frozen soil that remains unthawed even from the summer heat. Some permafrost zones have been frozen for centuries or even thousands of years. Permafrost can be found in regions where the annual air temperature is 23°F or less. Construction in polar regions is especially tricky due to permafrost; if proper precautions are not taken, heat from the building's foundation can melt the soil around the building and cause major structural problems.

Permafrost exists in regions such as central Alaska and Siberia, where the annual mean temperature is less than about 23°F and the summer's warmth fails to penetrate to the depth of the frozen ground.

What is the average depth of **frost penetration** in the lower 48 states?

While permafrost does not occur in the lower 48, soil does freeze to considerable depths during the heart of the winter. Frozen soil often extends down 40 to 80 inches over parts of Minnesota and North Dakota. Depths to 30 inches or more are common in New England. Even in the southern states, Oklahoma to North Carolina, soil occasionally freezes to a depth of several inches. These measurements are relevant to anyone who is planning to bury water or sewer pipes, or any object that could freeze or be subject to the heaving of the Earth that accompanies freeze and thaw cycles.

Has ice ever reached the **Gulf of Mexico** via the Mississippi River?

Only twice in recorded history (1784 and 1899) have ice floes drifted past New Orleans on the Mississippi River and entered the Gulf of Mexico. The cold wave of early February 1899 is considered by many to be the coldest of them all. The Mississippi River was frozen from its source to its mouth. The ice was two inches thick at New Orleans. That winter, ice chunks were observed passing into the Gulf of Mexico for the second time in history.

What happens to **farm and range animals** during blizzards?

People can go inside when the weather turns ugly, but many animals aren't so fortunate. The Blizzard of 1886 in Iowa, Nebraska, Kansas, Oklahoma, and Texas struck

Snowed in and plowed in, Chicago, Illinois.

with no advance warning. As many as 100 people in Kansas alone were believed to have frozen to death. The storm significantly changed the fortunes of many cattle barons, as up to 80 percent of all herds were wiped out in the storm areas. In more modern times, 100,000 head of livestock were lost in the Dakotas, Nebraska, and Minnesota in 1966 as winds gusting to near 100 mph caused snowdrifts over 30 feet.

In early November 1943, a major blizzard froze over one million turkeys as massive drifts covered farms in South Dakota.

Mild weather can have an impact also. During the summer of 1992, the mouse population over much of the United States soared as a result of the very mild winter season of 1991–1992. In Ohio, spring mouse populations were ten times that of a normal year.

How warm can it be and **still snow**?

The absolute record may not be known, but at LaGuardia Airport in New York City, flakes were once observed when the air temperature was 47°F. Snow falling from colder clouds above with near-surface air temperatures in the upper 30s is commonplace.

What are **ice flowers**?

These are not a species of plant, but huge, flower-like ice crystals that can sometimes form in quiet, slowly freezing bodies of water.

217

What are **snow rollers**?

Most residents of northern climes have made a snowman. You simply start with a ball of snow, roll it around, and it gets larger and larger. Under special situations, the wind can do the same thing. A small piece of blowing snow starts accumulating more and more snow, and soon a roll of snow perhaps a foot long and many inches in diameter results.

Who won the **1995 International 500** Snowmobile Race?

Mother Nature. On 11 January 1995, the International 500 Snowmobile Race in Minnesota was canceled because there was no snow on the ground. The lack of snow in neighboring Wisconsin was costing some snowmobile rental agencies up to $2000 a day.

How may people die each year due to **exposure to the elements**?

By ignoring nature's hazards, especially exposure to cold (which results in hypothermia), over 2000 Americans die each year when involved in outdoor recreation, not including weather-related traffic accidents.

Can there be **hurricanes in the Arctic**?

Winter storms called polar lows often begin to take on some of the characteristics of hurricanes, even including eye-like structures. Hurricane Andrew was an impressive storm, with a minimum central pressure at landfall of 27.22 inches. But intense winter storms in the North Atlantic can develop even lower barometer readings. On 10 January 1993 a North Atlantic howler had a central pressure of 26.86 inches. The most intense such winter storm was reported on 15 December 1986; it had a minimum pressure of 26.53 inches.

Do the **Great Lakes** ever freeze over?

While they rarely freeze totally, during severe winters such as 1993–94, most of the lake surface is solid. In 1979, all of the Great Lakes were frozen from shore to shore for one of the few times in recorded history. Even in milder winters there is enough ice to halt traffic of the huge Great Lake vessels from late January through late March. The ice can stay quite late in the season, especially on Lake Superior. On Memorial Day weekend in 1996, picnickers along Lake Superior's shoreline cooled their beers and sodas by using chunks of ice still floating in the lake.

How much **road salt** is used on highways?

One would not think Kentucky would be a leader in road salt usage, but the winter of 1993–94 was a rough one in the East. The state of Kentucky used record amounts of salt on highways—over 140,000 tons. For years the city of Chicago has stockpiled salt near the Chicago River, with the pile sometimes reaching heights of 50 to 100 feet.

Which state has the most **national temperature extremes**?

California is often called the land of extremes, and that includes temperatures. During 1992, the state recorded the hottest temperature in the United States on 148 days, and the coldest temperature (outside of Alaska) on 38 days. On a few days it had both.

In every month of the year, the warmest temperature ever recorded in the United States is 100 degrees or higher. In 9 out of 12 months, the record low is sub-zero. For June the record minimum is 2°F; for July, 10°F, and for August, 5°F.

Does it ever get **too cold to snow**?

Not really. But as temperatures fall, so does the amount of moisture in the air. Thus, at very cold temperatures, snowfalls tend to be lighter. The South Polar cap is certainly cold, and it does snow there, but many areas only have 10 or 20 inches per year—almost a "snow desert." The snow at the South Pole, however, never melts.

What is the best weather for **maple sugar production**?

The optimal conditions for maple sugar production are temperatures below freezing at night and mild daytime readings. Otherwise the sap just doesn't flow as well.

What is a **heavy snow warning**?

A "heavy snow warning" generally means that four inches or more of snow will fall in 12 hours, but not necessarily accompanied by high winds or very cold temperatures. A "blizzard warning" means sustained winds in excess of 35 mph will cause considerable blowing and drifting of snow, but need not be accompanied by heavy falls of new snow.

How do **snow fences** work?

Snow fences are installed near roads and highways to prevent snow from drifting over the roadway. They cause a reduction in wind speed and increase turbulence, resulting in snow accumulating downwind of the fence for a distance approximately ten times the height of the obstruction.

What is a **Norther**?

A Norther is a strong, cold wind that sweeps across the high plains of Texas, south into the Gulf of Mexico. They are sometimes called Blue Northers.

What is the **"snow eater?"**

The Chinook—also called the "snow eater"—is a very dry, warm wind descending off the lee slopes of mountains in the western states. When it arrives at the surface, temperatures can rise as much as 60°F in one day or less (sometimes much less).

The Chinook winds that occur along the Front Range of the Rockies in Colorado are famed for bringing spells of warm, windy weather during the depths of winter. But it's not all good news: wind gusts have been measured at up to 147 mph.

How cold does it get in **Florida**?

The cold wave of late December 1989 caused widespread misery and economic loss, including severe damage to the Florida citrus crop. Some citrus trees literally exploded: since the fruits are full of water, when the temperature falls below 20°F, they crack and split with a loud bang.

In Jacksonville, Christmas 1989 was the first with snow cover in over 100 years of record-keeping.

Up to four inches of snow fell on Milton, Florida, on 6 March 1954, a new single-storm record for the "Sunshine State."

One of the greatest historical cold waves occurred in February 1899. Snow fell as far south as Ft. Meyers, Florida, on 13 February 1899.

On 12 January 1982, the temperature fell to 23°F in Orlando, Florida. Even Miami has seen snowflakes—flurries have been reported that far south, but they have never made it to Key West.

Can it **snow in Arizona**?

There are parts of the state that are frequented by heavy snows. In January 1949, Flagstaff, Arizona, received a total of 105 inches of snow. In this high country near the rim of the Grand Canyon, heavy winter snows are commonplace.

On 16 November 1957, 6.4 inches of snow fell in Tucson. This was not only their heaviest snowfall ever recorded, but the first ever in November. Not too far from steamy Tucson, there are mountain peaks where you can actually do some winter snow skiing.

Is snow a major factor in traffic accidents?

B ad weather is often a factor in traffic accidents, but not as often as many suspect. Statistics compiled in 1991 suggested that a vast majority of crashes (86 percent) occurred when adverse weather was not a factor. When weather was poor, it was rain (71 percent), not snow (12 percent), that was most often the weather complication.

Have there been instances of **heavy snow on the Gulf Coast**?

On Valentine's Day, 1895, in Rayne, Louisiana, a snowstorm began that dropped 24 inches of snow for a state single storm record.

Has snow ever buried **Savannah, Georgia**?

Savannah, Georgia, is known for its temperate climate and mild winters. On 10 January 1800, however, 18 inches of snow fell on the city.

Which of the following cities has had **temperatures below 0°F: San Francisco, California; Tallahassee, Florida; Phoenix, Arizona; or Tucson, Arizona**?

The answer is Tallahassee. This city near the Gulf Coast is somewhat exposed to blasts of Canadian air coming south through the middle of the continent, and has recorded mercury levels as cold as -2°F.

How early in the year has it **snowed in Colorado**?

Being a mile high, Denver can turn from summer to winter in a hurry. Denver's earliest snow was 4.2 inches on 3–4 September 1961.

The 4th of July 1993 in Colorado was fairly cool, cool enough, in fact, that snow fell just 30 miles west of (and 5000 feet higher than) Denver. Snow plows were called out to clear roads on Mt. Evans. On the prior day, up to a foot of snow fell in parts of Grand Teton National Park in Wyoming!

The great early freeze on Halloween 1991 in Colorado killed more than 400,000 trees on Colorado's plains, a region short of trees in the first place. It typically costs $400 to remove a large tree and perhaps $150 to replant. Thus this freeze had hundreds of millions of dollars in economic impact on trees alone. The city of La Junta, Colorado, lost approximately 30 percent of its trees.

In the High Plains, unseasonable snows are not unique. The earliest a trace of snow has ever been recorded in Chicago is 25 September; it happened in 1928 and again in 1942. The latest measurable snow was 0.6 inches on 10 May 1923.

How thick is the **South Polar ice cap**?

The thickest part of the Antarctic ice cap is near Vostok, the Russian research station. It is estimated to be 15,670 feet thick. Vostok, Antarctica, is the coldest place on Earth. In August (during the Southern Hemisphere winter), the average high temperature is about -80°F with an average low of about -100°F. And that doesn't include windchill.

Are there **ice caps** in the Northern Hemisphere?

The Northern Hemisphere does provide some competition for the Southern Hemisphere. The friendly-sounding name of Greenland was selected by early Scandinavian explorers in hopes of attracting settlers. And while it was warmer in those days, Greenland wasn't and isn't very green. Around the edges, there are small habitations and some limited vegetation. But for the most part the giant island is a huge glacier. In the central part, around 100 inches of snow fall per year—but melting isn't an issue since the mean annual temperature is around -20°F. The coldest temperature ever was reported at Northice: -87°F. The depth of the accumulated ice sheet is over 10,000 feet in some places. Thule, Greenland, the fabled Air Force outpost, once reported a wind gust of 207 mph.

How much sunlight does **snow reflect**?

One reason very cold weather often follows a snowstorm is because fresh snow reflects more than 80 percent of the sun's energy directly back to space. It also efficiently radiates heat back into space during the night hours.

What is **Arctic sea smoke**?

Arctic sea smoke forms when bitterly cold air passes over cracks in the ice pack, which have exposed relatively warm seawater. The resultant fog swirls upward, looking very much like smoke escaping.

Can it snow from a **clear sky**?

A phenomenon called "diamond dust" appears to be just that. In extremely cold weather, ice crystals can condense directly from the air and sift earthward from an otherwise clear sky.

What is the **January thaw**?

A period of mild weather is said to recur each year in late January, especially in New England and other eastern states. The daily temperature averages (1877–1952) in Boston, for instance, show a well-marked peak on 20–23 January. The same peak can be found in other cities. The mild air generally results from the return southerly flow from a polar high-pressure system slowly moving eastward into the Atlantic.

January thaws occur elsewhere, too. In mid-January 1992, it was very cold with ground blizzard conditions in northwestern Iowa: windchills plunged to -60°F. Then came the thaw. Two weeks later it got so warm that the giant 150-foot tall ice palace built for the Saint Paul Winter Carnival started melting. It became so hazardous it had to be torn down early.

What should you do if **frost or freeze** are in the forecast for the first time of the season?

Be sure to drain your garden hoses and sprinklers to avoid freezing and damage. When water freezes, it expands with enough force to burst even strong pipes. Also, if you are about to turn on your heating system for the first time, be careful. It should be checked out. Bird nests built in chimneys during the summer could cause dangerous backdrafts of deadly carbon monoxide into the house. If you use a wood-burning fireplace, make sure your chimney is free of creosote, which could cause a nasty chimney fire.

What can farmers do to **save crops** threatened by frost or freeze?

In most agricultural areas, frosts and freezes often occur on clear nights with very light-to-calm winds. The coldest air layers are in the lowest few tens of feet above the surface, just where the sensitive vegetables and trees are. Several methods can be employed to save crops. Smudge pots, which produce huge clouds of dense smoke, help trap some of the Earth's heat by preventing it from escaping into space. They do, however, create major air pollution. Giant fans are often used to stir the warmer air, often found just aloft, down to the surface. Spraying the crops with water and allowing them to be coated with ice is very effective. This sounds like a cure that is worse than the disease. When the water freezes on the vegetation, however, the latent heat **223** released imparts just enough heat to warm the plant. The ice layer also acts as an insulator against the even colder air.

Even if your agricultural activities are a bit more modest than a corporate farm, you still might want to protect your plants. In spring or fall when frost threatens your tender plants and vegetables, place a sheet of newspaper, a bucket, or a glass jar over them. These methods can sometimes trap enough heat to protect them.

What are some examples of **spring and summer snows** that have occurred in the United States?

On 24 August 1992, up to a half foot of snow fell over the higher elevations of the Rockies, only an hour's drive from Denver.

Spring 1992 was cold in the East and South. Greenville, South Carolina, had its first ever May snowflakes, and they fluttered through the north Georgia skies for the first time in a century. But the mountains of North Carolina got feet, not flakes. Up to *five feet* of snow fell at some higher elevations on 7–8 May 1992.

The first two days of summer 1992 brought frost to parts of Indiana, destroying some 80,000 acres of field crops.

On 8 August 1882, a ship sailing in Lake Michigan reported that six inches of snow had accumulated on its decks. Flurries were also reported ashore at a few locations.

Memorial Day, 1992, was memorable in Minnesota for the wrong reason—it snowed up to two inches of wet snow, even in some of the southern counties. On Memorial Day weekend in 1996, parts of Nevada shared this fate as seven inches of snow canceled many picnics.

On 10 May 1990, Wisconsin received between 6 and 12 inches of wet snow over eastern portions of the state. In Waukesha, just west of Milwaukee, over 20,000 trees were damaged due to the weight of the snow.

Even in cold Duluth, Minnesota, snow during the summer doesn't usually happen. But on 18 September 1991, Duluth was dumped by a record 2.4-inch snowfall, breaking the previous September record of 1.5 inches.

On 2 May 1978, Elkhart, Kansas, had eight inches of snow and Amarillo, Texas, had 11 inches. Also, a "spring" snowstorm began on 3 April 1955 and dumped up to 25 inches of snow in portions of Connecticut.

The 1991–1992 "winter season" in the Twin Cities was a bit long, even by their standards. It started on Halloween with the biggest snowstorm in the area's history. And on the first full day of summer, 21 June 1992, frost was scraped from many windshields as overnight low temperatures approached the freezing point. The corn crop in southern Minnesota was damaged by the exceptionally late frost.

On the first full day of autumn in September 1992, much of the lower 48 states continued to enjoy summer-like weather. But Fairbanks, Alaska, made the jump to winter as temperatures dipped to 7°F, and stayed below freezing all day. The fall season then kicked off with nearly two feet of snow and a temperature that failed to raise above 40° for weeks.

Does hot water **freeze faster** than cold water?

Hot water will not freeze faster than the equivalent amount of cold water in the same type of container exposed to the same condition. However, a bucket of water that has first been boiled, and then allowed to cool to the same temperature as the bucket of cold water, may indeed freeze faster. The boiling process removes some of the air bubbles normally trapped in the water; air bubbles reduce thermal conductivity and can slow down freezing.

Can it snow on the **equator**?

If the elevation is high enough, it can snow on the equator. Snow falls regularly in the Andes of Ecuador, in fact. In the heart of central Africa, Mt. Kilimanjaro and Mt. Kenya are frequently snowcapped. Even in tropical regions, the freezing level of the atmosphere is rarely higher than 20,000 feet.

What is a **growler**?

Growler is one of the many names for sea ice. It refers to a small patch of floating ice, often originating from an iceberg, that floats low in the water and sometimes appears greenish in color.

What was the Twin Cities' **biggest snowstorm**?

Minnesotans endure long, hard winters, and they didn't have to wait long for the season to start in 1991. The last week of October was big for record setting in the Twin Cities. First the 1991 Twins became the first baseball team ever to go from the cellar to World Champions in one season. Then a few days later, the snowiest Halloween ever (8.5 inches) was followed by about 20 inches of additional snow the next day, making it the biggest snowstorm ever in this snowy part of the world.

What is the **coldest city** in the lower 48 states?

Butte, Montana, has more days with temperatures dropping below freezing each year (223) than any other major town in the lower 48 states.

Heading north to Alaska, Barrow has the coldest annual mean temperature in the United States at 9.3°F. In summer, the average is only 36.4°F. But the lowest winter (December–February) temperature average belongs to Barter Island, Alaska, with -15.7°F.

What is "heavy" snow?

According to National Weather Service usage, snowfall rate is considered "heavy" when visibility, without fog or blowing snow, is reduced to less than 5/16th of a mile. In a forecasting sense, a heavy snow warning is usually meant to imply four inches' accumulation in 12 hours or six inches' or more accumulation in 24 hours.

How many states are members of the "60 Below Club"?

Seven. Alaska, Colorado, Idaho, North Dakota, Minnesota, Wyoming, and Montana are the only states with recorded all-time minimum temperatures of -60°F or less.

Can you get frostbite in Hawaii?

One doesn't normally associate winter weather with that tropical paradise, but snow falls frequently atop the mountain peaks. Temperatures dropped to 14°F on 2 January 1961 on Mount Haleakala. Hawaii is the only state not to have a sub-zero all-time minimum; the 12°F recorded on Mauna Kea on 17 May 1979 was the closest they got.

Has the Hudson River ever frozen solid?

In January 1918 the Hudson River froze solid as far south as upper Manhattan Island in New York City. Some daring souls actually crossed the river on the ice pack to the New Jersey shore.

What is the typical annual range in temperatures in the lower 48 states?

During the 1990 calendar year, recorded temperature extremes in the lower 48 states spread over 180°F, ranging from -55°F to 125°F. But that was nothing unusual. During 1993, the coldest temperature recorded was -68°F on 22 January in Manly, Alaska. On 2 August it was 195°F warmer in Death Valley, California (a toasty 127°F).

What are the **growing seasons** for the eastern United States?

Gardeners of necessity are weather wise. The length of the growing season is determined by the dates between the last and first freezes. In warmer cities such as Los Angeles, San Diego, and Miami, years may pass without frost or freeze. But the average growing season in Duluth, Minnesota, is a mere 125 days. Lander, Wyoming, manages 128 days and Bismarck, North Dakota, can muster up 136 days without frost.

How **rapidly** can temperatures fall?

A cold front is the boundary between a warm air mass and the colder one that is replacing it. At Springfield, Missouri, on 15 November 1955, the temperature fell from 77°F to 13°F overnight. A cold front had gone through. The temperature at Browning, Montana, fell 100°F in 24 hours from 23 to 24 January 1916, tumbling from 44°F to -56°F, again due to a passing cold front.

The 11th of November 1911 had the warmest temperature for that date on record in Chicago, with one man becoming overcome by the heat. Yet, overnight the temperature plunged 61°F, and two Chicagoans froze to death on 12 November.

In Fairfield, Montana, on 24 December 1924, it was 63°F at noon. By midnight it was 84°F colder, -21°F.

On 12 January 1911 in Rapid City, South Dakota, the mercury fell 62°F in just two hours, from 49°F at 6 A.M. to -13°F at 8 A.M.

Topping that, the temperature plunged 58°F in 27 minutes in Spearfish, South Dakota, on 22 January 1943, falling from 54°F at 9 A.M. to -4°F at 9:27 A.M.

In just 15 minutes, a 47°F plunge occurred at Rapid City, South Dakota on 10 January 1911, from 55°F at 7 A.M. to 8°F at 7:15 A.M.

How can **snow** keep you **warm**?

Snow is an excellent insulator against cold. In chilly parts of the country, during winters with barren ground, frost depths can reach to over six feet. But even with sub-zero air temperatures, the ground can remain unfrozen if a thick blanket of snow is present to trap the Earth's heat. The ground you walk on can be used as a thermometer. If one were to dig down perhaps six to 12 feet (depending on soil types and other factors), the soil temperature very closely approximates the mean annual temperature of the overlying atmosphere.

Also, if trapped outside in a blizzard, don't try to fight the howling winds. Dig yourself a snow cave. By shielding your body from the blasts of the wind, you can greatly decrease the rate at which the body loses heat and greatly improve your chances of survival.

227

What is the record for **consecutive hours below zero** in the 48 states?

The month of January 1969 was one of the coldest on record in western North America. The city of Havre, Montana, had below-zero temperatures for nearly 400 consecutive hours.

What should you do if **stranded in your car** in a blizzard?

If stranded in your car during a blizzard in open country, stay with your car. Many have died of exposure after trying to get help and becoming disoriented in the blinding "white out." Keep the car running, but beware of snow blocking the exhaust pipe, allowing deadly carbon monoxide to build up in the car. Avoid consuming alcohol. Keep your car's hazard lights blinking, especially if you are on or near a traffic lane—snow plows can be dangerous. Carrying a cellular phone when driving in wintery weather is a smart move.

Do **cold waves** cause more deaths in the southern or northern United States?

On a per capita basis, there are more deaths from cold waves in the southern United States than in the northern plains. In the South, homes are not well insulated from the brief and rare extremes of cold. Poor insulation combined with the inexperience of the population often produces tragic, and unnecessary, casualties.

HEAT AND HUMIDITY

Which gets hotter, **Hawaii or Alaska**?

It would seem that Hawaii just has to be hotter than Alaska. The palm trees may make you think that Hawaii is the hotter of the two states, at least as determined by the state's all-time record high. But in Hawaii, the temperature is continuously modulated by the Pacific ocean. The interior of Alaska, during the long summer days, can become very hot indeed. On the average, Hawaii is much warmer than Alaska, but the all-time hottest temperature reading in Hawaii was only 100°F, on 27 April 1931, at Pahala. During summer, temperatures in Alaska's interior can occasionally reach above 90°F. The all-time state record is 100°F, on 27 June 1915, at Ft. Yukon. So Alaska and Hawaii are in fact tied when it comes to all-time record high temperatures.

Temperatures as cold as -1°F have been recorded in Alaska during the month of June. Barrow, Alaska, the northernmost permanent settlement in the United States, has an average summer temperature of only 36°F. In 1990, its average temperature during the month of July was 43°F. But you could find warmth that far north if you wanted it. Circle City, Alaska, peaked at 91°F on the 4th of July. In winter, Hawaii does have a slight edge over Alaska, though Alaska has seen readings in the 60s during December, January, and February.

In the same vein: which state is hotter, Florida or North Dakota? If your measure is all-time recorded high temperature, the latter wins the bet. Florida's top temperature is 109°F at Monticello, while the residents of Steele, North Dakota, once baked in 121°F readings.

San Juan, Puerto Rico, may be in the tropics, but temperatures are moderated by **229** the nearby ocean. When temperatures did reach 96°F on 28 March 1958, it set an all-time-high record.

How does the dew point relate to **summer comfort**?

Meteorologists use many measures to describe the moisture content of the air, including relative humidity, mixing ratio, specific humidity, wet bulb temperature, absolute humidity, and dew point. In some cities it is commonplace for the media to report the dew point, which is simply the temperature to which the air must be cooled in order for dew, clouds, and fog to form. The higher the dew point, the more moisture is in the air. In summer, dew points in the 50s or lower make for very comfortable conditions. With dew points in the 60s, things can sometimes feel a bit sticky, especially as it rises to near 70. Once the dew points are in the 70s, most people will start complaining about the humidity. Rarely does the dew point go above 80, but when it does, it can be extremely uncomfortable.

Can you have too little **relative humidity**?

When the winter heating season gets underway in most parts of the nation, don't forget the other component of your air—humidity. In northern regions, dry winter air, when brought inside and heated to room temperature without adding additional moisture, can result in relative humidities of less than five percent. This is as dry as the Sahara Desert. The average indoor relative humidity should be between 30 and 60 percent on the average. If your wintertime indoor relative humidity consistently averages less than 30 percent, you could experience cracked, dry lips and respiratory discomfort, and your furniture could dry out and age far more quickly than it should. "Indoor thunderstorms," discharges of static electricity, could also occur, signalling that the air is too dry.

Where is the **most humid place** in the world?

If you hate humidity, almost any tropical jungle is a place to be avoided, but there is one place where you want to be sure not to go. Parts of the Ethiopian coastline along the Red Sea can claim to be the world's dew point Capital. The dew point (the temperature at which the moisture in the air would condense if cooled at constant pressure) is an excellent measure of how humid the air feels. When the dew point passes 70°F, most people find the air oppressively sticky. Rarely does it top 80°F in the United States. But along the Ethiopian Red Sea coastline, the June *average* is 84°F.

Where are the **driest places** on Earth?

One list of top candidates would include the South Polar region, which is almost a

The prickly pear cactus, like all cacti, is physiologically adapted to cope with the extremely hot and arid conditions of the desert.

desert in terms of precipitation. Since cold air holds very little moisture, very little precipitation falls at the South Pole by comparison to most mid-latitude and tropical locations. Only a few inches of water equivalent—almost always in the form of snow—falls annually in many areas of Antarctica. The snow then becomes part of the permanent ice cap.

There are also two bands of deserts ringing the Earth that are associated with the belts of semi-permanent high pressure near the Tropics of Cancer and Capricorn. The Sahara, Sonoran, and Atacama Deserts are part of these belts where only a few inches of precipitation may fall each year, almost all of which is immediately evaporated back into the air.

231

Does it ever rain in the **Arabian Desert**?

Yes. In fact, in early November 1995, flash flooding in and around Jiddah killed five people and injured twenty-five. The worst rainstorm to hit in over 30 years was accompanied by wind gusts to 74 mph. On the whole, however, the region is quite dry. Aden, in South Yemen, averages only 1.8 inches of rain per year.

How can **hay bales** catch fire by themselves?

A fire occurring in a large stack of hay bales is not that uncommon. Aside from lightning, how might the weather cause this? The summer of 1995 was a hot one through much of the central United States. In Missouri, methane, emitted within big bales of freshly cut hay, began spontaneously combusting in locations all over the state.

How deadly are **heat waves**?

Heat waves, the force of which often goes unrecognized, are responsible for number-ous weather-related deaths. In a "normal" summer, some 175–200 Americans die from the effects of summer heat and too much sun. Between 1936 and 1975, tens of thou-sands of Americans died from problems related to heat and sun exposure.

During the brutally hot Dust Bowl summer of 1936, when many all-time heat records were established in the United States, and before air-conditioning was wide-spread, as many as 15,000 people may have perished. The hot summer of 1901 in the Midwest was especially deadly. At least 9,508 heat-related deaths were noted that year. Even in the modern "air-conditioned" era, over 1,250 died during the brutal Midwest heat wave of 1980. The 1988 drought and heat wave resulted in as many as 10,000 heat-related fatalities.

Most recently, 13 July 1995 was a day for heat extremes in the Midwest. Temper-atures and dew points had reached some of the highest levels seen this century. LaCrosse, Wisconsin, soared to 108°F. The dew points in Wisconsin were reported to be well over 80°F—higher than those found in steamy tropical jungles. The toll was especially severe on the elderly in Chicago, where over 500 perished before the heat wave was over. And cattle fared no better: at least 4,000 head of livestock were reported killed by heat stress in just one day.

What was the most **costly drought** in the history of the United States?

At least in the modern era, the 1988 drought in the United States was among the most severe on record. Economic losses exceeded $40 billion. The drought was very wide-spread, affecting over 200 million people. The hottest month on record in the Northern Hemisphere was July 1988. The drought that year was one of the three worst this century. In parts of the Midwest it was so dry that during thunderstorms rain evaporated before reaching the ground, and lightning set fire to people's lawns.

The devastation of the Dust Bowl of the 1930s is hard to calculate in economic terms, but the devastation of that period, unlike the 1988 drought, defined an epoch and completely altered the fabric of middle American culture.

Is **water** a scarce commodity?

Water is becoming an increasingly scarce commodity, especially in the western United States, with the increase in population and a tax on water supply usage. Conservation is very important and everyone can help save water. Consider this: if each home in the nation had one faucet leaking at the rate of one drop per second, a billion gallons per day would be wasted.

What month has the greatest **global temperature extremes**?

July is the month of temperature extremes. Both the hottest and coldest temperatures ever recorded at the surface of the Earth occurred during this month. The two temperature extremes spanned 265°F, between the deserts of Libya and the South Polar ice cap.

What place in the United States is the **hottest, driest, and lowest**?

Death Valley, California, is a place of extremes. It is extremely hot, the 134°F record-high being the nation's hottest (and number two worldwide). The Valley averages 1.8 inches of rain per year, the nation's driest. Its elevation is 280 feet below sea level—the nation's lowest point. Death Valley's 134°F was not an aberration. Twice during the 1980s, Death Valley reached 127°F, hotter than anywhere else in the country. From April to October, you can count on Death Valley to be one of the hottest places in the United States, and indeed the world. Typically the temperature tops 100°F between 140 and 160 days each year—and that's in the shade! In 1990, Death Valley was the first place in the United States to reach 100°F that year (on 10 April) and it had 97 consecutive days with high temperatures over the century mark, topping out at 125°F during July for the nation's yearly maximum.

Death Valley, one of the hottest places on Earth during the day, is sometimes broiling during the night, too. The overnight low on 26 June 1994 would have set a daily record-high temperature in many U.S. cities. After sizzling the day before at 124°F, the Death Valley nighttime minimum was 102°F.

What is the **lowest highest temperature** in the 48 contiguous states?

Since California is a state of extremes, one might guess it would contain the city having the lowest all-time maximum temperature, excluding some towns located high in mountainous regions.

Eureka, California, has the distinction of being the city with the lowest all-time maximum temperature in the 48 contiguous states, due to its proximity to the cold waters of the Pacific Ocean. On 26 October 1993, hot Santa Ana winds from the interior allowed Eureka to reach a new all-time high temperature—87°F—breaking its previous record by 2 degrees.

What are the **highest temperatures** in Asia, Australia, and South America?

Asia's highest recorded temperature is 129°F at Tirat Tavi, Israel. South America can claim 120°F at Rivadavia, Argentina. Australia can be a hot place, with 128°F maximum at Cloncurry, Queensland. Marble Bay, Western Australia, soared above 100°F for 162 consecutive days.

What is the highest temperature ever recorded in **Antarctica**?

It does actually get above freezing in Antarctica along the coast during the summer. It once reached a positively balmy 58°F in the output of Esperanza.

What is the highest temperature ever recorded on **Earth**?

Libya is the hottest spot in the world, meteorologically speaking. The temperature in the Libyan desert was once officially measured at 136°F.

What hazards do **dust storms** represent?

Snow and rain are not the only weather hazards on the highway. A dust storm in 1995 along the Arizona-New Mexico border caused havoc on the highways. At least eight were killed and 20 injured in highway accidents.

Are **forest fires** a hazard in the northern and eastern United States?

Forest fires are not just a problem in the western United States. On 1 September 1894, a searing heat wave in the Midwest resulted in one of the nation's worst fire tragedies. In a forest fire in the area around Hinkley, Minnesota, over 400 perished in the massive conflagration. Alaska also has a serious forest fire problem during its mild, dry summer. Lightning from summer thunderstorms is a major cause of arctic forest fires. During droughts in the past few years, major blazes have sprung up in the pine forests of suburban Long Island and New Jersey.

The western United States still has the majority of serious wildfires. On 5 July 1994, 13 firefighters were killed as a Colorado wildfire blew out of control on Storm King Mountain. Ironically, on the same date, at least 13 people drowned in Georgia flooding from the rains of Tropical Storm Alberto.

Forest fires are not exclusively a North American problem. The summer of 1992 was unusually dry in western Russia and the Baltics. At one point over 18,000 fires were reported burning in forests and marshlands.

How hot does the **surface of the Earth** get?

While the air temperature taken for weather reports (about five or six feet above ground) has never been officially recorded as hotter than 136°F (in the Libyan dessert), the actual surface of the soil can easily exceed 180°F. You can indeed fry an egg on some urban pavements and car tops on very hot days.

Which state has **natural air conditioning**?

In many areas, summers are cooled by breezes from adjacent bodies of water, but Hawaii may have the best claim to being naturally air-conditioned. Though closer to the equator than any other part of the United States, gentle sea breezes and trade winds cool the islands during the day. Temperatures only occasionally exceed 90°F near the beaches. And at night, cooling drainage winds (called the mauka breeze) pour down off the mountain slopes.

What is Indian Summer?

Indian summer, a lovely period of dry, warm days in middle to late autumn, usually after the first killing frost, is one of the more pleasant interludes on the American weather scene. It does not occur every year, but in some years may occur two or three times. The term is most frequently used in the northern United States and usage has been traced back to at least 1778. It may relate to the way Native Americans availed themselves of the nice weather to increase their winter food stores. In Europe, a similar weather pattern has been called Old Wives' summer, Halcyon days, St. Martin's summer, St. Luke's summer, and All-hallown summer.

What was the **Dust Bowl**?

The Dust Bowl is the name applied to the severe drought that affected large areas of the central U.S. during the mid-1930s. Several years of below normal rain, very high summer temperatures, strong winds, and poor agricultural and ranching practices devastated the fertile lands of this region. At least four major dust storms blew vast clouds of soil from the plains of New Mexico, Colorado, Oklahoma, Kansas, and west Texas into the eastern United States. "Black Sunday," 14 April 1935, saw a dust storm engulfing Stratford, Texas; the dust was so thick that some residents, in spite of wearing face masks, actually suffocated. The stream of drought refugees to California—many from Oklahoma—became known as the "Okies."

Conditions in some parts of the southwest in the 1950s were actually more severe, but improved farm and rangeland management practices reduced the dramatic clouds of dust to less spectacular proportions.

Why are the hot sticky days of August sometimes called the **"dog days"**?

There have been several theories. Perhaps the summer heat tended to drive dogs "mad." More likely the term results from the prominence in the sky of the bright star in Canis Major—Sirius, the Dog Star. During July and the first half of August, Sirius is the brightest visible star and rises in the east at about the time of sunrise. Ancient Egyptians believed the heat of this brilliant star added to the sun's heat to create the hot weather.

What are some examples of **extremely hot temperatures** in recent years in the United States?

In Sacramento, California, between 13 July and 20 August 1992, the daily high temperature topped the 90°F mark every day.

San Francisco is known for its almost bland climate, not too hot and not too cold—except on 17 July 1988. The City by the Bay roasted at 104°F.

During the last week of June 1994, the high temperature in El Paso, Texas, soared past 110°F every day.

According to the calendar, May is still spring. But at the end of May 1991, the thermometer indicated otherwise. Through the 30th, Philadelphia had seven straight days with temperatures exceeding 90°F; there were twelve such days throughout the month. Washington, D.C., sizzled at 98°F and on the 31st, peaked at 99°F.

During the summer of 1990, it was so hot in the southwestern United States that on a few occasions commercial jet traffic was halted because of fears the air was too thin in the 120°F heat to allow safe take-off on available runways.

What **economic impact** does a cool summer have?

Any major change from "normal" weather has an economic impact, usually positive for some, and negative for others. The cool summer of 1992 in the eastern United States limited some tourism and hindered crop growth, but consumers found it to be a boon for their pocket books. Milwaukee utilities reported eight percent less energy use and Kentucky gas and electric bills were down ten percent. Iowa residents used their air conditioners 65 percent less than the previous August. Since many utilities sell more energy during the summer (because of the nearly ubiquitous air conditioning in the United States), a cool summer hurts the power companies' bottom line.

What was the **Year without a Summer**?

Climate aberrations are nothing new. The year 1816 is often called the "Year without a Summer." It was so cold in portions of the eastern United States that crops were ruined, with few maturing north of the Potomac and Ohio Rivers. Frost and snows were common in June, July, and August. Heavy snow fell in New England between 6 and 11 June. On the 4th of July, the high temperature in Savannah, Georgia, was only 46°F. Crop failures were widespread not only in New England, but in Canada and Western Europe as well.

237

The year 1816 is also referred to as the "poverty year" and "eighteen hundred and froze-to-death." In Vermont, in June, ice formed an inch thick on some lakes. Almost the entire corn crop and many vegetables were killed throughout New England. On 7 June, snow drifts measured 18 to 20 inches in parts of Vermont and southern Quebec. At least one farmer was reputed to have frozen to death. Severe cold snaps continued throughout the summer, interspersed with periods of relatively mild weather.

This unusual summer weather was caused by an eruption of Mount Tambora in Indonesia in 1815 that ejected billions of cubic yards of fine dust more than 15 miles into the atmosphere. The dust layered out in the stratosphere, where it remained for a very long time. The result was considerable cooling worldwide. Other factors also played a role: In 1812, the volcano Soufrière erupted on St. Vincent Island in the Caribbean, and the Mayon Volcano erupted in the Philippines in 1814. At that time there were sunspots on the surface of the sun big enough to be seen with the naked eye. Thus, two factors (volcanic activity and sunspots) that scientists discuss in the ongoing debate about long-term climate change were in play at the time.

If 1816 was the "Year without a Summer" in the eastern United States, 1992 could be called "The Year without Much of a Summer." July was the third coldest in modern records with only 1950 and 1915 being colder. Some leaves started turning color in early August over parts of northern Michigan.

What are some of the planet's **temperature extremes**?

The range of recorded all-time temperature extremes in the United States is 214°F (-80 to +134); for the entire planet the range is 265°F (-129 to +136). By contrast, the temperature range on some tropical islands has never been more than 30 or 35 degrees.

During the decade of the 1980s, the average daily minimum temperature record for the United States (excluding Alaska) in January was -22°F. During July, the average daily maximum was a balmy 112°F. The actual recorded extreme readings over the 1980s: -55°F to 127°F.

In the United States the northern plains have the greatest annual variability in temperature. Far less variable are coastal cities in the extreme southwest (San Diego) and southeast (Key West). On a daily basis, average low and high temperatures may be less than 10 degrees apart in the Pacific Northwest but swing by more than 40 degrees in the dry western mountains.

How hot can it get in **Washington, D.C.**?

President Clinton's first July in the White House was the hottest month ever in the city, with an average temperature of 83.1°F. This average was raised by four days over 100°F and sixteen straight days above 90°F.

What is a **heat burst**?

The heat burst is one of the oddest of atmospheric phenomena. Thunderstorms are usually associated with cooling gusts of winds. But sometimes air is forced from 20,000 feet down to the surface, warming by compression all the way to the ground, and causing phenomenal temperature jumps. In Glasgow, Montana, on 9 September 1994, the temperature at 5:02 A.M. was 67°F. A heat burst from a nearby storm shot the temperature to 93°F by 5:17 A.M.—tying the date's record high. By 5:40 A.M., the mercury had subsided back to 68°F.

Another heat burst was reported to have occurred on 6 July 1949, in Portugal, when a meteorological observer saw the temperature soar from 100°F to 158°F in about two minutes. This reading was never certified as a world record.

The night of 15 June 1960, a heat burst struck northwest of Waco, Texas. Under largely clear skies, wind speeds gusted over 80 mph, and one resident claims his thermometer rose to 140°F. Damage to buildings and crops was also recorded.

What are some other unusual **short term temperature increases**?

On 26 April 1992, North Platte, Nebraska, shivered with a record-low temperature of 24°F. Four days later, North Platte sweltered in 98°F record heat. One day in 1993, Worland, Wyoming, set a record low of 41°F. The very next day they soared to a record high of 102°F. Not to be outdone, Pueblo, Colorado, set a record low of 52°F, and a record high of 101°F—all on the same day, 26 July 1993.

On 21 February 1918, the temperature in Granville, North Dakota, soared 83°, from -33°F in the morning to 50°F just twelve hours later. On 19 January 1892, Ft. Assiniboine, Montana, saw the mercury jump 42° in fifteen minutes, from -5°F to 37°F. In Kipp, Montana, on 1 December 1896, the temperature jumped 34° in seven minutes, and 80° in several hours. Some 30 inches of snow disappeared in a half-day. The Chinook wind is the cause of all these temperature gyrations, including a 49° surge in just two minutes in Rapid City, South Dakota, on 22 January 1943. At 7:32 A.M. it was -4°F, and two minutes later, it was 45°F. Car windshields that were still cold frosted over instantaneously all over town.

What is the **Bermuda High**?

The weather system that often dominates the eastern United States during summer is called the Bermuda High. This semi-permanent anticyclone residing over the western Atlantic results in warm, humid south and southwest airflow over much of the region for many days or weeks at a time.

How much **irrigation** is used worldwide?

In many areas of the world, natural rainfall is inadequate to grow crops. Currently over 3,000 cubic kilometers of irrigation water are used worldwide to spur production. In some areas, the resultant evaporation is large enough to begin affecting local climates. In northeastern Colorado, where irrigation is extensive, researchers suspect that the additional irrigation water moistens the atmosphere enough to trigger more intense thunderstorms. Other researchers have found that the contrast between hot, dry, non-irrigated regions and irrigated farms can actually create local wind systems.

What are the **sunniest cities** in the United States?

1. Yuma, Arizona
2. Redding, California
3. Phoenix, Arizona
4. Tucson, Arizona / Las Vegas, Nevada
5. El Paso, Texas

Between 1960 and 1969, measurable rain fell on only four days in the months of May and June in Yuma. Thus, the chance of rain during these months was less than one percent. Yuma has an average of 4,000 sunny hours each year.

What areas have the most and fewest **cooling degree days**?

"Cooling degree day," a term devised by the utility companies, refers to a unit of measure that helps predict how much money will be spent to air condition a building or home. For those who really enjoy the hum of the air conditioner, the regions having the most cooling degree days (more than 4,000 per year) are, not surprisingly, south Florida, south Texas, and the interior deserts of southwest California and Arizona. But the southwestern consumer, because of the very low humidity of the air, has the advantage of being able to use an inexpensive evaporative cooler, popularly called a "swamp cooler." By evaporating water into an air stream you can cool that air to what is called its "wet bulb temperature." The wet bulb temperature is the lowest temperature to which air can be cooled by evaporating water into it. It is one of several measures of the moisture content of the air. The drier the air mass, the greater the difference between the dry bulb (regular) temperature and the wet bulb temperature (this difference is called the wet bulb depression). Therefore the drier the air, the cooler it can be made using simple evaporation. Given the low wet bulb depressions—in other words, the slighter difference between air temperature and moisture content—found over Florida, conventional refrigeration techniques must be employed.

The areas where air conditioning is required less often (less than 500 cooling degree days per year) include areas of the West Coast that are very close to the water, the northern tier states, the northern Great Lakes, and northern and coastal New England.

What is the **Heat Index**?

The commonly heard expression says, it's not the heat, it's the humidity. Actually, it's usually both. High humidities greatly reduce the ability of the body to cool itself by its own evaporative cooler—sweat. In order to find an analog to the winter wind chill (wind and cold acting in tandem to produce discomfort and hazards to the body), various combinations of heat and humidity have been proposed. One in common use is the Heat Index, sometimes called the apparent temperature. When the heat index is between 90°F and 104°F, sunstroke, heat cramps, or heat exhaustion are possible with prolonged exposure and physical activity. When the index ranges between 105°F and 129°F, adverse physiological responses can become a real threat. If the value goes above 130°F, sunstroke, heat cramps, or heat exhaustion happen very quickly.

Heat Index Chart

Air Temperature and Relative Humidity versus Apparent Temperature

RELATIVE HUMIDITY(%)

Air Temp	0	5	10	15	20	25	30	35	40	45	50	55	60	65	70	75	80	85	90	95	100
140	125																				
135	120	128																			
130	117	122	131																		
125	111	116	123	131	141																
120	107	111	116	123	130	139	148														
115	103	107	111	115	120	127	135	143	151												
110	99	102	105	108	112	117	123	130	137	143	150										
105	95	97	100	102	105	109	113	118	123	129	135	142	149								
100	91	93	95	97	99	101	104	107	110	115	120	126	132	138	144						
95	87	88	90	91	93	94	96	98	101	104	107	110	114	119	124	130	136				
90	83	84	85	86	87	88	90	91	93	95	96	98	100	102	106	109	113	117	122		
85	78	79	80	81	82	83	84	85	86	87	88	89	90	91	93	95	97	99	102	105	108
80	73	74	75	76	77	77	78	79	79	80	81	81	82	83	85	86	86	87	88	89	91
75	69	69	70	71	72	72	73	73	74	74	75	75	76	76	77	77	78	78	79	79	80
70	64	64	65	65	66	66	67	67	68	68	69	69	70	70	70	70	71	71	71	71	72

Heat Index (or Apparent Temperature)

How many states are members of the "120 Club?"

To be in the 120 Club, a state must record an all-time high temperature exceeding 120°F. Nine states have reached or exceeded this temperature:

Arizona	127°F
Arkansas	120°F
California	134°F
Kansas	121°F
Nevada	122°F
North Dakota	121°F
Oklahoma	120°F
South Dakota	120°F
Texas	120°F

EARTHQUAKES AND VOLCANOES

What is the **Richter scale**?

The Richter scale for earthquakes is a numerical scale in which each unit increment involves a logarithmic increase in the size of the ground waves generated at the earthquake's source. An earthquake rated a 7.0 is 10 times more powerful than one rated a 6.0. The scale was devised by Charles W. Richter and his colleague Beno Gutenberg at the California Institute of Technology in 1935. A 1.0 earthquake is imperceptible and has the energy equivalent of six ounces of exploding TNT. A 5.0 (moderate) quake can cause some damage and has the energy equivalent of about 200 tons of TNT. A great quake (8.0) is equal to the energy of 6.27 million tons of TNT, or about 300 atomic bombs. The Great Alaskan Earthquake of 1964 was rated a stunning 9.2 on the scale. In simple terms, earthquake magnitude can be described in terms of potential effects:

Magnitude	Possible effects
1	Detectable only by instruments
2	Barely detectable, even near the epicenter
3	Felt indoors
4	Felt by most people, slight damage
5	Felt by all; damage minor to moderate
6	Moderately destructive
7	Major damage
8	Near total damage
9	Extreme destruction

Other scales have been proposed to quantify earthquake damage potential and energy release, one of these being the modified Mercalli scale, originally proposed in 1902 by Guiseppe Mercalli.

How many earthquakes occur **each year**?

There are probably millions of small quakes each year, many so small as to go unrecorded. Most of the larger earthquakes are registered by an international network of seismographs. In a given year the following number of quakes can be expected as ranked by Richter scale magnitude:

Magnitude	Number per year
8	2
7	20
6	100
5	3,000
4	15,000
3	100,000+

What is a **seismograph**?

The seismograph is a device for detecting and recording Earth movements, primarily those originating with earthquakes. Analysis of the data from several seismographs can determine the energy released by the quake and its approximate location (latitude, longitude, and depth below the surface).

Who developed the **first seismograph**?

The first crude seismograph was invented in China by Zhang Heng around 132 C.E. A rather fanciful-looking device, it was a copper-domed urn with dragon's heads circling the outside, each containing a bronze ball. Inside the dome was suspended a pendulum that would swing with the ground shock, and knock a ball from the mouth of one of the dragons into one of the open mouths of eight bronze toads below. The ball made a loud noise, signaling an earthquake. The epicenter (origin) of the earthquake could be determined based on which ball had been released. The device, perhaps more properly called a seismoscope, could apparently detect unfelt earthquakes that had occurred up to 400 miles away.

What is a **tsunami**?

Tsunamis, sometimes mistakenly called tidal waves, have nothing to do with the tides, but rather are caused by massive undersea earthquakes or landslides. The name comes from the Japanese term for "harbor wave," which is a rather appropriate term since it is in the shallow waters of harbors where they are most pronounced. After the quake, a chain of waves races across the ocean at speeds greater than 500 mph. Over the deep ocean, tsunamis are barely perceptible. But once they reach shallow waters, they can grow to monstrous heights. Generally earthquakes have to be stronger than 6.5 on the Richter scale in order to create significant tsunamis. Major tsunamis occur in the Pacific Ocean about once every six years. For some unknown reason, tsunamis are more common during the months of March, August, and November.

In 1883, an eruption in Krakatau in Indonesia spawned waves up to 130 feet high that leveled 165 coastal villages. The death toll was at least 36,000 people. In more recent times, an earthquake in the Sea of Japan struck Japan's Okushiri Island with a wave nearly 10 feet high, washing some 120 people to their death.

Do tsunamis strike **Hawaii**?

Hawaii is a target for tsunamis generated by earthquakes originating almost anywhere around the Pacific Rim. On 1 April 1946, the residents of Hilo looked seaward and couldn't believe their eyes. All of the water had drained from the three-mile-wide harbor, leaving thousands of flapping fish exposed on the sea floor. Then, suddenly, a giant wave came surging onshore, destroying the beachfront. The water then retreated again, after which two more giant waves crashed ashore. In Hilo, 96 people perished that day, with 63 more deaths in other parts of Hawaii. The source of the waves was a magnitude 7.3 earthquake in the Aleutians some 2,200 miles away. In May 1960, 61 people were killed in Hawaii by a tsunami. January 1988 saw 30-foot waves slamming into the state, though this time there was ample warning.

What happened to the **sand dunes** of New South Wales?

The lack of sand dunes near the oceanfront in New South Wales, Australia, has prompted a theory that they may have been washed away by a giant tsunami. Some 100,000 years ago a wave, possibly a quarter-mile high at its presumed origin in a Hawaiian earth slide, swept across the Pacific and scoured the coastline clean.

What is a **fault**?

A fault is a crack in the Earth's crust along which tectonic plates (or pieces of these plates) slide along, beneath or up and over each other. Earthquakes result from the sudden release of strain that has built up over a period of time as two blocks of the Earth's crust try to grind and slide along past each other. Seismic gap theory is one useful technique in earthquake prediction; it states that the portion of a fault that has not slipped (quaked) for the longest time is most likely to experience the next quake. **245**

The Los Angeles Coliseum suffered structural damage from the 1994 Northridge quake.

What is the **San Andreas Fault**?

The San Andreas Fault is the "master fault" of the intricate network of faults that cuts through rocks in the coastal region of California. It is a fracture in the Earth's crust along which two parts of the crust have slipped away from each other. The fault is about 600 miles long, extending almost vertically into the Earth to a depth of at least 20 miles.

The lateral shift along the San Andreas fault is about the same as the growth rate of your fingernail. This may seem trivially slow, but such movement adds up over time. If untrimmed, fingernails could grow about a foot a year. In 20 million years, Los Angeles will have slid north and will be located immediately opposite San Francisco.

Do earthquakes emit **radio waves**?

Quite by accident scientists at Stanford University discovered that strong low-frequency radio emissions preceded the 1989 Loma Prieta earthquake in California. There have been other reports suggesting ELF (Extremely Low Frequency waves) and even lower frequency electrical disturbances are associated with earthquakes. This is an area of potentially valuable research. Since radio waves travel 5,000 times faster than seismic waves, if a radio earthquake signature can be found, it would be possible to devise an early warning system. While the added time would not be sufficient to warn people, it would be enough to shut down critical systems such as gas and electricity networks, chemical plants, nuclear reactors, and computers, in order to automatically secure facilities that would be badly needed in the post-quake recovery period.

What are **earthquake lights**?

We tend to think of earthquakes as only affecting the ground and things on it. There are many confirmed reports, however, of the atmosphere glowing with a strange light before and during major earthquakes. The skies over the Andes mountains appeared to be on fire the night of the great Chilean earthquake of 1906. In the Cerro Gordo quake in California in the 1860s, reliable witnesses reported sheets of "flame" emerging from the rocky sides of the nearby Inyo Mountains. Sheets of what looked like lightning flashed across the sky just before the tragic Tangshan quake in China in 1976.

While there is little formal research available on the topic, it appears that earthquakes generate substantial electrical disturbances. Piezoelectric effects result from compression and strain applied to many rocks and minerals. The suspected low frequency electromagnetic waves emanating from earthquake zones might also stimulate the atmosphere to produce optical emissions in ways not yet understood.

What do **earthquakes and UFO reports** have in common?

A study found a significant correlation between reports of UFOs and seismic activity occurring within 150 kilometers of each other. It has long been suspected that powerful radio waves are emitted when rocks undergo significant stress. These in turn can cause luminous electrical phenomena in the atmosphere. Some people see these highly unusual luminous displays and attribute them to extraterrestrial activity.

How well can seismologists **predict earthquakes**?

One of the holy grails of geophysicists is the accurate prediction of earthquakes. There have been many promising starts, and many deep disappointments in this field. In 1966, China launched an ambitious program of earthquake research. They made use of many potential signs of impending earthquakes. These included small tremors increasing in strength and frequency, changes in the water levels in wells, tilting of the ground surface, changes in the electrical conductivity of the Earth, and anomalous behavior of animals. In winter 1974, in the Liaoning Province in northeast China, many of the signs were there. An evacuation of the cities of Haicheng and Yingkou was ordered on 1 February. On the 4th, a magnitude 7.3 quake struck. The cities were leveled and 300 people died, but hundreds of thousands of lives were saved. Many other quakes, though, have struck without any glimmer of warning, including the Northridge earthquake in 1994. Predictions of a large quake at Parkfield, California, were made by seismologists who had noted that a series of quakes at that location occurred like clockwork about every 22 years. The site was heavily instrumented, but the predicted Parkfield earthquake is now several years overdue.

Scientists have revised the chance of a major earthquake (magnitude 7.0 or greater) striking southern California by the year 2024 upwards from 60 percent to 85 percent. Most troublesome are the Parkfield and Imperial Valley regions as well as the zone extending from Santa Barbara across the Ventura Basin on to San Bernardino.

Is the Earth's **magnetic field** constant?

The intensity and orientation of the Earth's magnetic field is not at all constant. In fact, it has changed strength by over 50 percent during historical times. The magnetic polarity of the planet can sometimes abruptly shift—the North and South magnetic poles change places! These changes are the result of giant convection currents within the molten interior of the planet.

What meteorologist shook up the **geological world**?

A German meteorologist named Alfred Wegener published a book in 1912 entitled *The Origins of Continents and Oceans*. Like many others, he had noted how the coastlines of South America and Africa seemed that they would fit together amazingly well. He hypothesized that an ancient super continent had existed until about 200 million years ago. Continental drift then tore apart this massive landmass he called Pangaea; the pieces are the present continents, which are still drifting apart. The idea was met with scorn and ridicule from virtually the entire scientific community. But beginning in the 1960s, a series of discoveries led to the gradual acceptance of the theory of continental drift, which has profoundly altered our view of the Earth. The Earth is not a fixed, static orb. The Earth's crust is, in fact, constantly changing as various "plates" drift around on top of the fluid inner core. Today, continental drift is the cornerstone of our understanding of planetary climatic and geological processes, and it explains, among other things, the numerous belts of earthquakes and volcanoes worldwide.

What was the first evidence of **continental drift**?

Largely considered a pseudoscientific notion until the 1960s, Wegener's continental drift theories began to gain acceptance when the implications of the change in the magnetic polarity were realized. When molten rock cools and hardens, the magnetic particles in it become frozen and aligned along the Earth's current magnetic polarity. It was noticed that rocks on the floor of the Atlantic Ocean had alternating bands of positive and negative magnetic polarity. These strips were parallel to the great ridges that occur on the Atlantic sea floor. It was found that this pattern could be best explained if the sea floor were spreading outward away from these ridges. Thus, between the ridges, lava was upwelling from within the interior of the Earth. As it spread it was pushing the continental plates ever further apart.

Can animals foretell earthquakes?

It sounds rather like the stuff of pseudoscience or tall tales, but there is mounting evidence that some animals can sense impending earthquakes, and will exhibit changes in their behavior. Detailed studies in China documented marked behavioral abnormalities before the 1974 Haicheng and 1975 Tangshan quakes. Among the species that showed anomalous behavior were pigeons, chickens, fish, rats, dogs, pigs, cattle, cats, and weasels. Pandas were reported to have held their heads and screamed. Dogs became noisy. Horses stampeded and neighed. Earthworms emerged in great numbers, eels went upstream in many rivers, and snakes emerged from their underground dens in the dead of winter. Many animals exhibited their unusual behaviors for hours, days, or even weeks before the quakes.

What is **plate tectonics**?

The more technical name for continental drift is called plate tectonics. The Earth's outer shell is broken up into about 15 to 20 large tectonic plates. They are about 60 miles thick and made of lighter material that "floats" on the heavier molten core of the Earth's interior. Their extremely slow motion results from the slow circulations within the molten core of the Earth. In certain zones the plates are pulled apart, while in other areas they crash together, building mountains in the process. Continental drift occurring over eons of time is a major factor in determining active earthquake zones. Now satellites and laser ranging stations are precisely measuring the locations of seismic points on the Earth with an accuracy of one centimeter (0.4 inches). Therefore, over time the relative motion of land masses with respect to one another can be calculated. No more "lost continents." The origin of the term tectonic comes from the Greek word *tekton*, meaning builder, in this case of mountains.

Plate tectonics resulted in a headline in *Science* magazine on 8 December 1995: "Missing Chunk of North America Found in Argentina." This headline referred to a discovery that provided another confirmation of the continental drift theory originally proposed by Wegener. A 500-mile-long chunk of our continent broke off about a half-billion years ago and merged into what is now South America.

249

What is a **black smoker**?

Deep-sea explorations have resulted in the startling discovery of huge hydrothermal vents on the deep ocean floor near the point where the great plates are moving apart due to sea floor spreading. These vents, called black smokers, shoot massive quantities of superheated water and chemicals such as hydrogen sulfide into the ocean. Though miles below the surface and devoid of any sunlight, some of the vents are found to be teeming with life, including such strange creatures as tube worms.

What causes **rogue waves**?

On rare occasions, extremely large waves can appear out of nowhere, for no immediately obvious reason. A large freighter sailing during fair weather off the coast of Spain in the 1960s was suddenly struck by a huge 90-foot wave. The ship survived. The wave may have been generated by an undersea landslide.

A rogue wave swept the Florida beaches near Daytona Beach on 3 July 1992. The 18-foot wall of water injured 75 people and smashed hundreds of cars parked near the beach. First thought to be the result of an undersea landslide, it was later determined to be caused by the pressure effects of a fast-moving line of thunderstorms.

What were the **most intense earthquakes** in the 48 contiguous states?

A series of major earthquakes struck the Missouri Bootheel in the vicinity of New Madrid. They began on 16 December 1811 and lasted into 1813. At least three of the temblors were thought to substantially exceed 8.0 on the Richter scale, reaching magnitudes 8.4, 8.6, and 8.7. The quakes were felt over two-thirds of the United States. Chimneys toppled in Cincinnati and church bells rang in Boston. The Mississippi River was actually observed to flow backwards at times, and it changed its course in a number of regions, creating new lakes such as Missouri's Lake St. Francis and Reelfoot Lake in Tennessee. Given the sparse population in the area at the time, there were no known deaths. Quakes such as these that form in the middle of tectonic plates (intraplate quakes) are not well understood. They are known to occur far below the surface on faults formed perhaps some 500 million years ago.

Such major quakes are thought to occur every 550 to 1,100 years. However, quakes in the magnitude 6 to 7 range are likely to occur in the general New Madrid area every 90 or so years. Were the 1811–13 events to be repeated today, widespread devastation would be reported in cities such as St. Louis and Memphis.

What magnitude was the **San Francisco earthquake of 1906**?

The great San Francisco earthquake of 18 April 1906 has been estimated to have reached 8.2 or 8.3 on the Richter scale. The ground shook at 5:12 A.M. when the North American and Pacific plates slid sideways along a 250-mile-long segment of the San Andreas fault. The quake started with about 20 seconds of trembling; then more violent shocks lasted for 45 to 60 seconds. The official death toll has long been listed at 700, but recent research suggests that up to 3,000 of San Francisco's population of 400,000 perished. The entire business district was destroyed, and three out of five residences in the city crumbled or burned. More than 500 city blocks were totally or partially obliterated. Hundreds of thousands of people evacuated the city by train or ferry. One of those to leave was the famous Italian opera singer Enrico Caruso, who was sleeping in the Palace Hotel at the time of the quake after performing *Carmen*. Seen wandering around in the street wearing a fur coat over his pajamas, he swore he would never return to San Francisco, and he kept that promise.

In 1868, the Haywood Fault south of Oakland, California, caused a quake that also damaged San Francisco. Seismologists estimated that the San Francisco Bay Area has a 60–70 percent chance of a magnitude 7 quake by the year 2020.

What was the **Loma Prieta** earthquake?

This earthquake had the distinction of being broadcast live on national television during the 1989 World Series. A portion of the Bay Bridge collapsed, as did large segments of an elevated freeway in Oakland. The earthquake, centered some distance south of the Bay Area, registered a 7.1 on the Richter scale. A total of 67 people died. Property damage was estimated at $15 billion. Damage to buildings was especially severe in the Marina district of San Francisco. This area was built largely on unstable landfill (some of it debris from the 1906 quake), which greatly amplified the Earth's trembling.

What was the **Northridge earthquake**?

The greater Los Angeles area has had eight major earthquakes in the last 90 years. The Northridge quake, on 17 January 1994, was the most recent; it registered a 6.7 magnitude on the Richter scale. It killed 61 people, left 6,000 injured, and resulted in $15 billion in property damages. In comparison to the 1906 San Francisco quake, it was minor league, releasing only one-thirteenth of the energy of that historic temblor. Yet the location in which the earthquake occurred resulted in great losses. And had the quake not struck at 4:31 A.M., thousands of people may have been killed. Also unsettling was the fact that the Northridge quake occurred on a previously unknown fault.

The first floor of many apartment buildings collapsed due to the 1994 Northridge earthquake.

This fault may be one of many more remaining to be discovered within the Los Angeles basin.

What was the most intense large urban area earthquake in recorded history?

Though records are very sketchy, it appears that the massive earthquake that struck Lisbon, Portugal, on 1 November 1755 may have registered a magnitude of at least 9.0. The first series of shocks lasted six or seven minutes. The city was virtually leveled. Over 60,000 people perished. The quake was felt as far away as Sweden and the giant tsunami wave it generated swept much of the North Atlantic, with waves of twelve feet striking the West Indies.

What was the deadliest earthquake in the twentieth century?

At 4 A.M., 28 July 1976, a 7.8 magnitude earthquake destroyed the city of Tangshan in China. Official estimates said 250,000 people perished—one-quarter of the area's total population—but western observers suspected that the real toll was closer to 750,000, or three out of every four residents. It took less than a minute for 89 percent of the homes and 78 percent of the industrial buildings to be turned into piles of rubble. The city was located in the middle of a tectonic plate and not thought to be in a region of high quake potential, not having experienced a large quake in six centuries. Thus, the city's building codes were less stringent than in more earthquake-prone parts of the country. The main shock was calculated to have the energy of 400 Hiroshima-sized atomic bombs.

How did ancient societies interpret earthquakes?

Most ancient societies found religious explanations for earthquakes. The ancient Maya in Central America believed earthquakes were the gods' way of thinning out the population of humans when they became too numerous. Japanese myth said their island chain rested on *namazu*, a giant catfish that flopped about from time to time when the gods were displeased. According to Hawaiian myth, Pele is the Hawaiian volcano goddess who causes Kilauea to erupt whenever she has a temper tantrum. Vulcan was the Roman god of fire who kept his blacksmith's forge within a volcanic mountain off the coast of Italy. The catastrophic quake that destroyed Lisbon, Portugal, in 1755 resulted in a furious debate among European philosophers as to why God destroyed the city, the seat of the Holy Inquisition, during High Mass on All Saints Day.

What was **North America's** strongest quake?

On Good Friday of 1964, several seismologists attending a conference in Seattle were having dinner in the restaurant atop the Space Needle. They noticed that their soup started swashing in their bowls. They knew something bad had happened.

That "something" happened twelve miles beneath the coast of Prince William Sound, Alaska, as some 75,000 miles of Pacific Ocean crust slid past an equal amount of continental crust. The cataclysm caused the entire Earth to ring like a bell. The ground bulged two inches as the seismic wave passed Chicago. Water levels in South African wells oscillated. Around the greater Anchorage area, the Earth shook violently for up to four minutes. The Alaskan quake was rated a stunning 9.2 on the Richter scale and released some 5,000 times the amount of energy of the 1994 Northridge, California, temblor.

Collapsing buildings caused some deaths, but it was the tsunami wave that claimed most of the 115 lives. The walls of water, up to 40 feet high, slammed into the **253** coastline. In Seward, Alaska, the waves caught fire after an oil spill from a ruptured storage tank. Hours after the quake, giant waves washed into Crescent City, California. Though warned, people returned to their homes before the train of waves had finished pounding the coast. The last wave, which was 20 feet tall, swept 11 residents to their death. Four others died in Oregon. The sea wave ultimately struck the coastline of Antarctica.

What was the **strongest earthquake** ever measured?

A giant earthquake shook parts of Chile on 22 May 1960. It was estimated that this quake reached a magnitude of 9.6 on the Richter scale. The energy released by the snapping of the Earth's crust was equivalent to exploding 25 percent of the planet's nuclear arsenal. By another measure the energy equaled the entire annual U.S. energy consumption. Thousands of people died in Chile. The resulting tsunami claimed 61 people in Hawaii, 120 in Japan, and 20 in the Philippines.

What kinds of **shock waves** are generated by an earthquake?

In an earthquake, several types of waves radiate out from the epicenter. The first to arrive are the P (primary) waves. They alternately compress or expand the material through which they travel. They may be the source of the roaring noises that often signal the onset of large quakes. The slower S (shear) waves arrive next and they move the Earth both up and down and side to side.

R (Rayleigh) waves move across the surface like rolling ocean waves. On 23 January 1812, one of the greatest of the New Madrid earthquakes struck. An observer described the appearance of the Earth as the quake started: "The Earth was observed to be rolling in waves of a few feet in height, with a visible depression between. By and by these swells burst, throwing up large volumes of water, sand and charcoal." Similar two-foot ground waves were reported in the 1886 Charleston, South Carolina, temblor. There have been reports of such waves breaking people's knees while sweeping by unsuspecting populations.

How much does the **ground move** in earthquakes?

During the 1964 Alaskan quake, the Pacific plate slid a good 60 feet laterally. The ground shifted vertically 10 to 20 feet in areas around Anchorage. Horizontal offsets of 10 to 20 feet were found after the 1906 San Francisco quake.

One effect of the 1994 Northridge, California, earthquake was that the town moved about three inches closer to downtown Los Angeles, which means the average commuter has about 180 fewer feet of driving each year.

What are some of the **secondary effects** of earthquakes?

Fire is an all-too-frequent companion to the earthquake. The 1906 San Francisco quake caused a half-billion dollars in direct damage, while the resulting fires caused $3 billion in losses (in today's dollars).

One of the aftershocks of the devastating 1995 earthquake in Kobe, Japan, was not seismic in nature, but rather an airborne hazard. Demolition crews, when they razed hundreds of quake-damaged buildings, released large amounts of carcinogenic asbestos fibers into the air. Citywide asbestos levels were six times higher than normal, and around some work sites the values soared to 300 times normal. Also, ozone-depleting Freon gas, released from thousands of destroyed refrigerators and air conditioning systems, was added to the global atmosphere.

After the 1994 Northridge earthquake, there was at least one delayed death, with a fungus, not structural damage, to blame. Soil disturbed by earthquakes could release the spores of a rare organism called *coccidioidomycosis,* which can cause fatal complications when inhaled by humans. Such spores are found in the soil in the southwestern United States.

How often is **Tokyo** struck by major earthquakes?

The historical record shows that metropolitan Tokyo is struck by a major temblor about every 70 years. The last great quake destroyed large parts of the city in the 1920s.

How far can the **shaking of earthquakes** be felt?

One of the more powerful earthquakes ever recorded occurred on 9 June 1994 and was centered 360 miles below a sparsely populated region of Bolivia. At least a magnitude 8.3, it was the largest deep quake ever recorded, centered some 390 miles below the surface. It was felt in Peru, Argentina, Brazil—and Minneapolis, Seattle, Chicago, and Toronto. Indoor plant leaves waved as if in the wind, rocking chairs rocked by themselves, and buildings swayed for several seconds in these North American cities.

What are the **chances of a big earthquake** outside of California?

Earthquakes are not just a Los Angeles or San Francisco problem. They can occur along the entire length of the Pacific coastline. The earthquake hazard in the Seattle area is just now beginning to be widely recognized. A monstrous quake was believed to have shaken the Pacific Northwest around the year 1680. Further south, a 1985 quake registering 8.1 rumbled under the Pacific seafloor; the seismic waves were amplified in the sediment-filled valley of Mexico City, causing widespread devastation and taking 9,500 lives.

The area east of the Rockies is not completely safe from seismic risk. In the United States, 39 states have a chance of experiencing a strong quake in the foreseeable future. Seismologists compute a three-in-five chance that a damaging quake will strike east of the Rockies sometime in the next 20 years.

Has **New York** ever experienced an earthquake?

We know earthquakes are not confined to the West Coast, but the northeast is not considered earthquake country by most people. A moderate earthquake did rattle New York City on 19 June 1871. A magnitude 5 quake struck in 1884. A major geological fault zone runs directly under downtown Boston. In 1755 a magnitude 6 temblor occurred in the Atlantic Ocean off the coast of Massachusetts, toppling chimneys and walls in Boston. The tremor was felt from Chesapeake Bay to Nova Scotia.

Has the **southeastern United States** ever had a serious earthquake?

A severe earthquake devastated portions of coastal South Carolina during the last century. In the summer of 1886, a series of small tremors had rattled the area. But on 31 August, as repairs were underway to clean up from a recent hurricane, a devastating quake, estimated at magnitude 7.5, shook some 14,000 chimneys to the ground. Falling debris and collapsing buildings claimed 110 lives.

What is the **safest town** in the United States?

According to one survey that calculated the risk from tornadoes and earthquakes, a small town in New Mexico named Crossroads was estimated to be the "safest" in the nation.

What is **earthquake weather**?

Is there such a thing as earthquake weather? Some residents of the West Coast think that fair skies associated with a mild high-pressure system that is stalled over the western United States are typical of conditions during major quakes (as in the 1989 Loma Prieta quake). While some have theorized that the increased mass of the atmosphere over the region due to the higher atmospheric pressure might play some role in a quake, this is a question that will require many years of research in order to be

resolved.

What role did volcanoes play in **forming the Earth's atmosphere?**

The prevailing wisdom is that much of the Earth's atmosphere is the result of out-gassing from volcanoes. The most common volcanic emissions are carbon dioxide and water vapor, along with lesser amounts of nitrogen, methane, and argon. The planet's atmosphere was dominated by carbon dioxide until about 3.5 billion years ago, when the advent of photosynthesis began increasing the amount of oxygen. The volcanic theory of atmospheric origin has many supporters.

How well can volcanologists **predict eruptions?**

There are very encouraging signs that major volcanic eruptions can be predicted in advance, thus reducing the potentially vast loss of life. The 15 June 1991 eruption of Mount Pinatubo in the Philippines was presaged by numerous small earthquakes that were monitored by teams of Filipino and American seismologists. Evacuation orders had removed over 85,000 people from the path of destruction. So, when the mountain finally blew after 500 years of slumber, the death toll was relatively small (about 300 people in the eruption and ensuing mud slides).

But there have been failures. In 1985, a Colombian volcano named Nevado del Ruiz was coming to life. The town of Armero, with 29,000 people, was nestled below its summit. The town lay in the path of previous mud and ash flows that had descended rapidly from the volcanic peak. In fact, 1,000 people had died in Armero in 1845 from such a mudflow. While the scientists accurately predicted the impending disaster, local officials did not come up with an effective warning program. When the volcano finally erupted, it melted vast amounts of snow near the summit. A river of mud one-fifth the size of the Amazon swept down the mountainside. At least 22,000 people were instantly buried alive.

Mount St. Helens erupting in 1980.

In 1993, seismologists were attending a conference on the dangers of volcanoes. It was held near the slope of the Galeras volcano in Colombia, an active volcano that threatens almost a half-million people. Several of the scientists ventured onto the mountain to collect samples of gases and lava. A small eruption occurred while they were there, killing six scientists and three tourists. Much work remains to be done in volcanology and emergency preparedness.

What is the **Ring of Fire**?

The belt of high volcanic and seismic activity that rings the Pacific Ocean is often called the Ring of Fire. It marks the boundaries between the plate underlying the Pacific Ocean and the surrounding plates. It runs up the West Coast of the Americas from Chile to Alaska, down the Aleutian Islands and into the East Coast of Asia from Siberia to New Zealand. Over 75 percent of the world's active volcanoes are located along this zone, which is part of a 40,000-mile-long zone that winds around the Earth like the seams of a baseball.

Do volcanoes affect **climate**?

The residence time of particles of air pollution emitted into the lower atmosphere can usually be measured in terms of hours to days. But particles injected into the lower stratosphere by erupting volcanoes (or nuclear blasts) can remain suspended for months or years. Sulfur dioxide and sulfuric acid droplets are especially long-lived in the stratosphere. The effect of the increased atmospheric pollution is to block the sun's incoming radiation, thus cooling the planet. Such climate changes as occurred in the "Year without a Summer" (1816) have been attributed to major volcanic eruptions.

The tens of millions of tons of sulfur dioxide injected into the atmosphere by the massive 1991 eruption of Mount Pinatubo in the Philippines promptly reversed the trend of global warming considered a possible sign of the greenhouse effect. The eruption caused the probability of warmer than normal temperatures to go back to the levels of the early 1960s. Since 1895, when records of average global air temperatures began, the four warmest years have been 1990, 1991, 1994, and 1995. But temperatures were down in 1992 and 1993 due to the effect of the Mount Pinatubo eruption. The summer of 1992 in the United States was noticeably cooler than normal.

Who first proposed a link between **climate and volcanoes**?

Benjamin Franklin was a remarkable man, contributing to a wide range of human endeavors. He may also have been the first scientist to suggest a link between volcanic activity and fluctuations in the Earth's climate. He noted that a dry "fog" persisted over Europe for much of 1783 and 1784 and was accompanied by exceptionally cool weather. He speculated that it may be linked to the huge eruption of the Laki volcano in Iceland.

What is the largest volcanic eruption in recent centuries?

The amount of material blasted into the stratosphere in 1815 by Indonesia's Mount Tambora was 100 times greater than that of Mount St. Helens. It was the greatest amount of ejecta released into the atmosphere in more than 3,000 years. The cloud of dust and sulfuric acid hurled into the stratosphere caused the infamous "Year without a Summer" in the United States in 1816. And its "bang" wasn't to be ignored. The noise was heard to a distance of 970 miles. At least 10,000 people were killed instantly, with another 80,000 lost in the ensuing epidemics and famine.

What was the **loudest noise** in recorded history?

When the Indonesian volcanic island of Krakatau blew its top in 1883, it may have produced the loudest noise in recorded history. The noise was reported in New Guinea, 1,800 miles from the island, and in Alice Springs, Australia, some 2,000 miles distant. On the island of Rodriguez in the Indian Ocean, the chief of police reported hearing sounds from the East that he likened to the roar of distant guns. Thus, the explosion of Krakatau was heard nearly 3,000 miles away. The volcanic debris caused fiery sunsets around the world through 1890.

Did a **volcanic eruption** almost destroy humanity?

It is entirely possible that the human race came very close to extinction some 74,000 years ago with the massive eruption of Mount Toba on the Indonesian island of Sumbawa. It was probably the most violent eruption in the last million years. Coming at a period when glaciation was already cooling the planet, the added impact of "volcanic winter" from its massive dust shroud caused bitter cold, loss of vegetation, and possibly the extinction of some precursors to the human race.

How many **kinds of volcanoes** are there?

Volcanoes are usually cone-shaped mountains surrounding a vent connecting them to reservoirs of molten rock (magma) far below the surface. When the magma is forced upwards by gas or other pressures it can either break through in lava flows or shoot high into the air as columns of lava fragments, ash, dust, and gases. The accumulation of debris from eruptions causes the volcano to grow in size. There are four kinds of volcanoes:

Cinder cones. Built of lava fragments, they have a slope of 30 to 40 degrees and seldom exceed 1,500 feet in height. Examples are Sunset Crater in Arizona and Paricutin in Mexico.

Composite cones. Made of alternate layers of lava and ash, they are characterized by a slope of up to thirty degrees at the summit, tapering off to five degrees at the base. Examples are Mt. Fuji in Japan and Mount St. Helens in Washington.

Shield volcanoes. Built primarily from lava flows, their slopes are seldom more than 10 degrees at the summit and two degrees at the base. The Hawaiian Islands are clusters of shield volcanoes.

Lava domes. These volcanoes are made of viscous, pasty lava squeezed like toothpaste from a tube. Examples are Lassen Peak and Mono Dome in California.

How many **active volcanoes** are there?

Between 850 and 1,500 volcanoes worldwide have erupted during the past 10,000 years, and are therefore considered active. At any one time there are typically 8 to 12 volcanoes erupting worldwide. The United States has over 65 active or potentially active volcanoes, more than any other country except Indonesia and Japan.

What is the **world's largest** active volcano?

Hawaii's Mauna Loa is the world's largest active volcano. It rises to more than 13,650 feet above sea level. If one considers its base resting on the floor of the Pacific Ocean, the volcano approaches the vertical dimensions of Mt. Everest.

Has anyone ever witnessed the **birth of a volcano**?

At least twice in this century, people have watched the very earliest moments of a volcano's life. At around 4 P.M. on 20 February 1943, farmer Dioniso Pulido was working in his field in the village of Paricutin in central Mexico. He saw a long fissure suddenly cutting across the land. Moments later the ground thundered and ash shot up from the crack. A smoky sulfurous cloud roared out of the hole, and trees began bursting into flame. The fiery eruption continued through the night, and by the next day a cinder cone some 160 feet high was planted smack in the middle of his cornfield. By 1952, when the volcano went silent, it had grown to a height of 1,390 feet. No one was directly killed by the Paricutin volcano, but its ash cloud sparked frequent lightning bolts, which did kill three people. The fine volcanic ash also suffocated some 4,500 cattle and 500 horses.

On 14 November 1963, the crew of the fishing boat *Isleifur II* awoke early for a days' work fishing near the Westman Islands of Iceland. At first several crew members noted an odd smell. Then the man on watch felt a strange motion of the boat as if it were in some unseen whirlpool. Then, in the distance, through the morning haze, he saw what looked like a rock, but there were no rocks in that part of the ocean. He realized it was smoke. Before long the crew members watched dark columns of ash and smoke rising from the sea surface in the birth of the volcanic island of Surtsey. By late afternoon the smoke column towered to over 20,000 feet and was visible to the residents of the capital Reykjavik 70 miles away. After four days the newly born island was 200 feet high and 2,000 feet long. Scientists intensely studied this event, documenting, among many things, the production of lightning as seawater quenched the molten lava and imparted massive electrical charges to the volcanic cloud.

Why is Iceland indebted to volcanoes?

Iceland, which is covered far more with black lava than white ice, lies half on the North American plate and half on the Eurasian plate. It is composed entirely of volcanic material from 200 active volcanoes. As the two plates spread apart, more and more lava upwells in eruptions, adding to the area of the islands. The spreading process adds about 1.5 acres to the island nation each year. Residents can bathe in hot springs warmed by geothermal energy and grow bananas in greenhouses kept warm in part by the heat from the interior of the Earth.

How were the **Hawaiian Islands** formed?

The Hawaiian Islands have all been formed by volcanic activity. There is a hot spot deep within the Earth from which magma is continuously being extruded. As the Pacific plate drifts over the hot spot, successive volcanic islands have been formed. A new island, called Loihi, is currently in the making. It is forming 20 miles southeast of the Big Island. The seamount rises some 17,000 feet above the ocean floor and is only 3,000 feet below the surface. It will become available for occupancy in several thousand years.

The underground rivers of lava draining down the slopes of Mount Kilauea, which have continued now for 13 years, travel as much as nine miles before they splash into the ocean, causing gigantic clouds of steam. The continued growth of the Big Island will eventually result in gravity causing a large chunk of the island to split off and fall into the sea, causing monstrous tsunamis.

What are **pyroclastic flows**?

These avalanches of fire, gas, ashes, and lava are the volcano's most potent weapon. They occur when the rising ash cloud becomes so heavy that it collapses back upon itself and then rushes down the volcanic slopes at a speed of hundreds of miles per hour. In 1902, on the Caribbean island of Martinique, the Mt. Pelee volcano had been rumbling for several days. The local governor did not want to cancel an upcoming election, so he declined to evacuate the city of St. Pierre. On the morning of 8 May, two days before the election, a pyroclastic flow with temperatures greater than 1,000°F engulfed the town in a matter of seconds. Of St. Pierre's 29,000 residents, only two survived. One was a prisoner serving time in a windowless dungeon in the city jail.

What happened at **Pompeii**?

Pyroclastic flows were the cause of the total loss of the cities of Pompeii, Herculaneum, and Stabiae on the flanks of Italy's Mt. Vesuvius in the year 79. A brief period of ash fall prompted some people to flee, but the pyroclastic flow buried thousands of people in their tracks. Volcanic debris later reached 65 feet in depth in some places. Roman historian Pliny the Elder was one of the casualties of Vesuvius.

What happened at **Mount St. Helens**?

On 18 May 1980, the largest volcanic eruption in the 48 contiguous United States in modern history killed at least 61 people. The volcano, located in southwestern Washington State in the Cascades Mountain Range, is one of many non-dormant volcanoes in the western United States. The property damage was estimated at $1.5 billion.

After a 123-year slumber, small quakes began on 20 March 1980, totaling over 10,000 before a 5.1 magnitude earthquake at 8:32 A.M. destabilized the northern flank of the mountain. A wall of volcanic debris, a half cubic mile in volume, some 300 feet high, and the largest avalanche ever witnessed, slid down the mountainside. The mountain then simply exploded, with the blast wave moving outwards at 670 mph. The blast removed the top 1,300 feet of the mountain and blasted 275,000,000 tons of Earth into the atmosphere.

Mount St. Helens erupting.

Other volcanoes in the region are also potential candidates for eruption, including Mount Rainier, which towers above the Seattle-Tacoma region at a distance of only 20 miles.

What are the details of the **Mt. Pinatubo eruption**?

The 130-mile-wide ash clouds from the largest eruptions of Mount Pinatubo in the Philippines on 15 June 1991 were measured by satellite to reach heights approaching 40,000 meters (125,000 feet)—well into the stratosphere. North of the former summit, the explosions left a crater 2,500 meters across. If all the ash and rock ejected from Mount Pinatubo were collected and dumped onto Manhattan Island, it would be 1,000 feet thick. An effective early warning system limited the loss of life to only 300 people.

The eruption of Mount Pinatubo ejected some 15–20 million tons of sulfur dioxide gas into the stratosphere. There they formed tiny droplets of sulfuric acid. This resulted in brilliant sunsets as well as the reflection of some of the Earth's heat back into space, inducing a short-term global cooling. Mean global temperatures were depressed by 0.5 to 1.0°C for the next several years. The dust and sulfuric acid cloud took three weeks to make the west-to-east trip around the world, and then lingered for several years. Space shuttle astronauts who had flown before and after the eruption said they had never seen the Earth look as hazy as it did afterward.

What are the **deadliest volcanic disasters** known?

The deadliest volcanic disasters since the eighteenth century are:

Volcano	Date of eruption	Number of casualties
Mt. Tambora, Indonesia	5 April 1815	92,000
Karkatoa, Indonesia	26 August 1883	36,417
Mt. Pelee, Martinique	30 August 1902	29,025
Nevada del Ruiz, Colombia	13 November 1985	23,000
Unzen, Japan	1792	14,300

Are the **death tolls** from volcanic activity rising?

Population growth is making humanity ever more vulnerable to volcanic disasters. Between 1600 and 1900, it is estimated that the death toll was about 315 people per year. Since the start of this century the number has increased to 845 each year.

What are some of the **largest known eruptions** in history?

Some of the largest known volcanic eruptions are ranked according to the amount of material (in cubic kilometers) that was ejected into the atmosphere during the eruptions. The Laki eruption also produced the largest known lava flow in recorded history.

Name	Year	Country	Ejecta Volume (in cubic kilometers)
Yellowstone	600,000 B.C.E.	U.S.A.	10,000
Toba	74,000 B.C.E.	Indonesia	1,000
Tambora	1815	Indonesia	100
Santorini	1470 B.C.E.	Greece	10
Laki	1783	Iceland	10
Krakatau	1883	Indonesia	10
St. Helens	1980	U.S.A.	1

What is the **largest known volcanic eruption** in North America?

Going back in time about 600,000 years, an eruption of unimaginable proportions occurred in the western United States in none other than Yellowstone National Park. Today there can be found a sunken crater some 45 by 30 miles. The Wyoming blast was probably ten times bigger than that of Tambora. Thick dust fell over most of the area that is now the western United States. And there is no guarantee that it couldn't happen again. The 200 hot springs and geysers that attract tourists to the region are a testament to the vast pool of 2,000°C molten lava that bubbles just below the surface. Estimates say the Yellowstone volcano, which is the size of Rhode Island, erupts every 600,000 years or so.

What are the **top five volcanic eruptions** of the twentieth century?

1. Santa Maria (1902)
2. Katmai (1912)
3. Cerro Azul/Quizapu (1932)
4. Mount Pinatubo (1991)
5. El Chinchon (1982)

Is there any basis for the **legend of Atlantis**?

Researchers are gradually concluding that the legendary undersea island of Atlantis never existed, but they have uncovered the source of the myth. The volcano Santorini on the island of Thera, near Crete, was part of the thriving Minoan civilization that existed around the year 1470 B.C.E. The island eventually blew up, leaving only the present-day crescent that is formed from the volcano's caldera (a volcanic crater). There may have been some advance warning, since no skeletons have been found buried in the ruins. The island's dramatic disappearance seems to fit some of the legends related by Plato many centuries later from which the Atlantis myth seems to have emerged.

Are volcanoes a **threat to aircraft**?

Flying bombs of lava are obviously to be avoided by aircraft, but the real volcanic threat to aviation is a bit more subtle. Airliners can develop severe engine problems if they accidentally ingest the coarse, grit-like material found in the plumes of volcanic eruptions. The 1989–90 eruption of the Redoubt Volcano in Alaska seriously disrupted transpolar jet traffic. One jumbo jet ingested so much volcanic ash that the engines temporarily failed, endangering the lives of 244 passengers. Meteorologists now track volcanic dust clouds so jetliners can steer clear of this hazard. In September 1992, commercial jets as far away as New York were shifting and juggling routes and altitudes to avoid the dust cloud from Alaska's Mt. Spurr volcano.

The accurate re-routing of aircraft around volcanic plumes is one of the unexpected benefits of the nation's weather satellite program. By tracking clouds of ash **265** from volcanic eruptions, satellites played a key role in helping jets over the Pacific avoid the huge clouds from Mount Pinatubo.

What is a **paleoclimatologist**?

A paleoclimatologist is a scientist whose specialty is the reconstruction of past climates from a variety of hints and clues. These might include tree ring histories, polar cap ice cores, pollen layered in lake sediments, and fossil marine organisms.

Are there volcanoes elsewhere in the **solar system**?

Scientists have observed a cloud of sulfur being ejected into the atmosphere from the volcano Pele, which is located on a moon of the planet Jupiter. The eruptions were first discovered when the Voyager I space probe passed the moon Io. It was recently observed spouting off again using Earth-based telescopes. Volcanoes have also greatly altered the surface of Mars and Venus.

AIR POLLUTION & THE ENVIRONMENT

What is **air pollution**?

Air is a complex mixture of gases and aerosols (minute particles). Pollutants are defined as chemicals or substances that are present in the atmosphere in sufficiently high concentrations to be in some way harmful to humans, other species, or ecosystems as a whole. Pollutants can arise both from human activities (called anthropogenic sources) and from natural sources (dust storms, volcanoes).

What are the **major classifications** of air pollutants?

The two basic types of pollutants are gases and aerosols, with the latter consisting mainly of solid materials (particulates) as well as droplets of liquid such as sulfuric acid. Most polluting gases are invisible to the naked eye, with the exception of some oxides of nitrogen. Even air that appears to be very clean contains myriad minute solid particles. These aerosols can number in the millions per cubic foot of air. Polluted cities can have hundreds to thousands of times more particles than rural regions.

What are **primary and secondary** air pollutants?

Primary air pollutants are those that are directly emitted from a source, such as carbon monoxide from an automobile's tail pipe. Secondary pollutants form as a result of chemical reactions in the atmosphere among multiple pollutants. Ozone, a key constituent of smog, results in part from automotive emissions, but you could not find any by sampling your vehicle's exhaust; auto emissions form only part of ozone. Similarly, sulfur dioxide, a primary pollutant, can be converted into various secondary sulfate species once in the atmosphere.

What are TSPs?

TSP stands for total suspended particulate. These airborne contaminants come in a variety of sizes; most particulates are smaller than 10 microns in diameter, and some can even be smaller than 1 micron. By comparison, the width of a human hair is 100 microns or more. TSPs come directly from smoke stacks, auto exhaust, or dust kicked up by wind or human activities. They can also form as secondary pollutants. For instance, sulfur dioxide from a power plant reacting with ammonia gas from a feedlot produces the particulate ammonium sulfate, which can reduce atmospheric visibility.

Not all TSPs are respirable, that is, able to be inhaled into the human respiratory system. The larger TSPs are often screened out by the several defense mechanisms of the human body.

What are some **sources of particulates** in the atmosphere?

Both human activities and natural processes release vast amounts of particulate matter into the atmosphere. Some rough estimates, in millions of tons per year, include:

Natural sources

Sea salt	1,000
Soil, dust	200
Volcanoes	4 (highly variable)
Forest fires	3

Human activities

Coal combustion	36
Agriculture	10
Iron/steel manufacture	9
Combustion of wood	8
Cement manufacture	7
Incineration	4
Combustion of oil	2

What is **sulfur dioxide**?

Sulfur dioxide was perhaps the first industrial pollutant to be recognized as such. The chemical formula is SO_2, and it arises from the burning of sulfur in oxygen. One of the major sources of sulfur dioxide air pollution is the sulfur found in almost all coal. The sulfur content of coal is quite variable, but in the extreme, one ton of sulfur dioxide

A power plant ironically places a large No Smoking sign on a smokestack.

gas can result from the burning of ten tons of high sulfur (bituminous) coal. Techniques are now available to partly remove the sulfur from coal or remove the SO_2 gas before it goes out the smoke stack. Sulfur dioxide is a potent irritant to the human respiratory system.

What is **carbon monoxide**?

Carbon monoxide, chemical formula CO, is produced by combustion of materials containing carbon in the presence of limited oxygen. Carbon monoxide is colorless, odorless, tasteless—and deadly. Even small amounts of carbon monoxide can severely disrupt the blood's oxygen-carrying capacity. In effect, by breathing enough carbon monoxide, one begins to suffocate. The symptoms initially include drowsiness, headaches, and disorientation.

Automobile exhaust contains significant quantities of carbon monoxide. In traffic-clogged urban streets or poorly ventilated garages, values can reach 100 parts per million (ppm). This can be a definite health hazard. Humans should not be exposed to concentrations over 8 ppm for any extended period of time.

Carbon monoxide can also be found within the home, resulting from malfunctioning gas appliances, furnaces, or blocked chimneys. CO detectors for the home have been selling very well over the last few years due to a number of well-publicized deaths from carbon monoxide. But the home detectors are not without their flaws. They tend to have a high rate of false alarms. During 1995 in Chicago, a strong temperature inversion allowed carbon monoxide from the city's automotive exhausts to accumulate in the atmosphere and finally seep inside homes. Chicago firefighters received over 10,000 calls in a 48-hour period from startled homeowners.

What is ozone?

Ozone is a colorless form of oxygen that is extremely chemically active. It was discovered in 1840, in Paris, by C. F. Shoenbein. Its chemical composition (one molecule has three atoms of oxygen, as opposed to the usual two atoms of "normal" oxygen gas) was identified a few years later by J. L. Soret. There is too much ozone in the air of our polluted cities like Los Angeles and Chicago, and too little in the stratosphere, resulting in what's described as the "ozone hole." Ozone is extremely rough on the human respiratory system, damages many types of vegetation, especially leafy vegetables, and causes the rapid deterioration of many widely used materials, such as rubber.

What are the **oxides of nitrogen**?

Almost all combustion of fuel results in production of various oxides of nitrogen (including NO, N_2O, NO_2). While power plants and automotive vehicles are the principal sources of the oxides of nitrogen, there are also significant natural sources of oxides of nitrogen. In and of themselves they are somewhat less harmful than pollutants such as SO_2 and CO, but they play a key role in generating ozone through photochemical reactions in the air.

Mountaintop view of the sea of smog covering the Los Angeles basin.

What are **VOCs**?

VOCs are volatile organic compounds. There are literally hundreds of chemicals, collectively called VOCs, that are emitted from industrial activities and are key components of photochemical smog. There are also many natural sources of VOCs.

What is **smog**?

The word smog comes from a combination of smoke and fog and may have been coined in 1905 by a British physician named Harold Des Voeux, who later went on to found the Coal Smoke Abatement Society in England. But smog, as the term is used today, is in fact neither smoke nor fog. Los Angeles–type photochemical smog is a chemical witches' brew that forms from sunlight-driven reactions of exhausts from automobiles, power plants, industry, and individual homeowners. The principal ingredients required to create this type of smog are oxides of nitrogen and VOCs. Among the many components comprising smog are ozone and minute solid particulates (which make it visible). A chemical called PAN is what causes the eye irritation so often reported in the Los Angeles region during severe smog episodes.

As the 1980s came to an end, and after two decades of control efforts aimed at limiting Los Angeles–type smog, almost 100 U.S. cities and regions were still in violation of federal clean air standards for ozone, the primary contributor to smog.

271

Los Angeles–type air pollution is spreading around the planet as industrial development continues. And surface ozone levels are threatening to increase to the levels where they endanger over half the world's food crops. Estimates are that long term exposure to U.S. levels of ozone could reduce agricultural production in many of the world's farm belts by up to 5 percent.

There have been some victories in the battle against smog. In a recent period, total emissions of gaseous nitrogen oxides from automobiles in Michigan dropped almost 50 percent. This reduction in auto emissions, one of the major building blocks of this type of smog, was due both to retiring older cars as well as making changes in the chemical makeup of gasoline.

Is air pollution a problem of the **modern age**?

Air pollution is anything but a modern problem. The earliest cave dwellers undoubtedly had severe air quality problems once they discovered fire. The Hopi Indians in Arizona discovered the use of coal in making their pottery, but legend has it that it smelled so awful (sulfur dioxide) that tribal leaders ordered a kachina—a masked figure representing a spirit being—to go to all the homes and forbid the use of coal indoors. In 1257, Queen Eleanor, wife of King Henry III of England, refused to stay in Nottingham Castle because of the choking smoke arising from the coal fires in the town below. During the late eighteenth century, the air of London was so fouled by the burning of high sulfur–content coal that infant mortality reached horrifying levels. It was estimated that half of the children born in the city died by their second birthday—with air pollution as a major cause.

What was the first **anti–air pollution law**?

The first known anti–air pollution law was enacted in England in 1306. The burning of high-sulfur coal in England during the fourteenth century began to cause severe environmental problems, so much so that King Edward I proclaimed what might have been the first environmental legislation. Coal burning was prohibited during sessions of Parliament, at penalty of death. And one man indeed was caught with the hot coals of evidence and was summarily dispatched.

What where the **London killer fogs**?

London is famous for its "pea soup" fogs as popularized in the Sherlock Holmes novels. But before recent air pollution controls, it wasn't just a naturally occurring fog. London suffered many severe air pollution episodes during the 1950s. Sulfur dioxide from coal burning, combined with thick fog, produced a deadly mixture of soot and sulfuric acid droplets. In 1952, stagnant air over a four-day period allowed sulfur dioxide and particulate concentrations to reach unprecedented levels. The black and yellow "smog" was so thick that people literally couldn't see their hands on their out-stretched arms at times. Traffic came to a standstill. One driver found himself in a cemetery without a clue how he got there. Only the blind were able to navigate around with some security. When the air cleared, 4,000 were dead. As many as 100,000 were believed to have become sick. Deaths from bronchitis increased ten times, influenza by seven fold, and pneumonia by five times. Parliament finally enacted the British Clean Air Bill in 1956 and the days of the pea soup fogs were numbered as soft (bitu-minous) coal use was largely banned. The human race is burning hydrocarbon-based fuels at an alarming rate; we are using these resources at one million times the rate at which they were produced over the last 500 million years.

In 1966 a similar tragedy took place, though on a somewhat smaller scale, in New York City. The build-up of pollutants from the city power plants and industries during November of that year resulted in over 600 "excess deaths" on one weekend alone. These events were among the many that propelled the Clean Air Act through the U.S. Congress.

What are the **National Ambient Air Quality Standards**?

The major air quality protection legislation is called the Clean Air Act, which was passed by Congress in 1967 and amended in 1970 and subsequent years. It established standards for air quality that must be met by the combined actions of industry and individuals in each region of the country. The National Ambient Air Quality Standards (NAAQS) are the values that air pollution measurements cannot exceed without the threat of governmental penalties.

These standards are as follows:

Particulate matter	50 micrograms per cubic meter	annual average
	150 micrograms per cubic meter	24-hour average
Sulfur Dioxide	80 micrograms per cubic meter	annual average
	365 micrograms per cubic meter	24-hour average
NO_2	100 micrograms per cubic meter	annual average
Carbon monoxide	9 parts per million	8-hour average
	35 parts per million	1-hour average
Ozone	120 parts per billion	hourly average not to be exceeded once per year over three years
Lead	1.5 micrograms per cubic meter	calendar quarter

What happened in **Donora, Pennsylvania**?

It was not until the environmental activism of the 1960s that air pollution became a national concern. But several disasters began to focus attention on the rapidly declin-ing state of the atmosphere in the United States. In October 1948, in the valley town of **273**

Donora, Pennsylvania, about 30 miles south of Pittsburgh, a stagnant air mass allowed pollution levels to build. Half the population of 14,000 became ill, and 20 people died. A similar disaster struck the Meuse River Valley in Belgium in 1930 when thirty people died and thousands more were made ill.

What is **long range transport**?

Not so many years ago, it was assumed that air pollutants emitted from smoke stacks became too dilute to be of any consequence after traveling perhaps only a few miles. The diluting capacity of the atmosphere has proven not to be infinite, however. Meteorological research has demonstrated that even single smoke plumes can stay together as cohesive and measurable entities for hundreds of miles or more. Dust storms have been tracked across entire continents and oceans. Pesticides sprayed in Louisiana have been found depositing in the Great Lakes. In addition, the combined affect of hundreds or thousands of smaller sources can ultimately pollute entire air masses. This should not have been a real surprise, since radioactive debris from atmospheric atomic tests in the 1940s and 1950s was routinely tracked for several circuits around the planet. The pollutants associated with acid rain are known to travel for thousands of kilometers from their sources.

And not only air pollutants can travel long distances. Spores of a fungus causing a disease in rice, which had been ruining crops in west central Africa, were found to have been carried by the wind across the Atlantic to infect crops in the Dominican Republic. There was considerable puzzlement on a small Caribbean island when, one day in 1988, a rain of small red grasshoppers occurred. Meteorologists studied wind patterns aloft and concluded that they originated in a massive infestation in northwestern Africa and were literally blown over the Atlantic.

How much **pollution** is emitted by the United States each year?

A source of pollutants is considered "significant" by the U.S. Environmental Protection Agency if it emits more than 15 to 25 tons per year of particulates, 40 tons per year of SO_2 and NO_2, 100 tons per year of CO and 0.6 tons per year of lead. And all those tons add up. The estimated annual U.S. emissions of sulfur dioxide are 23 million tons. There are also 21 million tons of oxides of nitrogen and 22 million tons of volatile organic compounds.

How much do **old cars** contribute to air pollution?

In many cities, some 50 percent of the automotive-related air pollution comes from just ten percent of the cars. A disproportionately large amount of smog-forming pollutants come from pre-1980 cars, and it has been considered cost-effective to pay people cash to get their clunkers off the highway. Some European countries have already begun experimenting with this idea.

How far did the **radioactive cloud** from Chernobyl travel?

On 26 April 1986, the worst nuclear accident in history occurred at the Chernobyl plant in Ukraine in what was then the Soviet Union. An explosion literally blew off the thousand-ton lid of the reactor vessel and released 3.5 to 5 percent of the reactor's fuel (7 tons) and 10 percent of the graphite reactor itself high into the atmosphere. This represented some 50 to 100 million curies of radiation. There were 31 immediate casualties, while 240 people sustained severe radiation sickness and 150,000 people living in the region were evacuated, many permanently. The long-term effects on 600,000 persons are still being evaluated.

The radiation was detected far beyond the immediate vicinity of Chernobyl. High rates of fallout were detected 1,300 miles away, including some over parts of Scandinavia. The reindeer herd, one of the prime food sources in northern Scandinavia, became contaminated, due to the animals ingesting radioactive lichen. Swedish scientists thought they found a region of reduced lightning activity in the part of Sweden that had the highest fallout. And scientists from the United States tracked the cloud of radioactive debris for several circuits around the globe before it became too dispersed to be measured.

What is the **cleanest** major energy source?

Natural gas is considered to be the "clean fossil fuel" of the next century. What nation has the largest reserves? The former Soviet Union has more than ten times that of the United States.

What is **radon**?

Radon is a radioactive gas that, when inhaled into the lungs, can produce lung cancer. It is estimated that as many as 20,000 deaths from lung cancer occur in the United States each year due to long-term exposure to radon. A naturally occurring gas, radon recurs from the natural radioactive decay of uranium and radium in the soil. It is generally found at very low concentrations in the atmosphere. But it can seep into the basements and lower floors of buildings and accumulate to dangerously high concentrations. The federal government says long-term averages of radon radiation should not exceed 4 picocuries per liter. To convert that into more understandable terms, that amount of radon has the equivalent lung cancer–causing potential of smoking half a pack of cigarettes a day. Homeowners should seriously consider testing for radon. More often than not problems that might be found can be fixed relatively simply and inexpensively. The solutions can often be as simple as sealing basement cracks or adjusting ventilation devices.

Can **carbon dioxide** be dangerous?

While carbon monoxide is one of the deadliest of common air pollutants, CO_2 is generally considered harmless. After all, it exists in large quantities in the natural atmosphere and is used to make the bubbles in soda pop and beer. But in extremely high concentrations, problems can arise. In 1990, stands of trees began dying for no apparent reason in the Inyo National Forest in California. At the same time, tourists using the park's cabins began to experience bouts of dizziness and nausea. The mysteries were both explained as seepage of carbon dioxide gas from buried volcanic faults. In the small quantities normally found in the air, CO_2 is harmless. But the volcanic gas displaced the oxygen from the soil and in some cabins CO_2 concentrations reached above 25 percent, sufficient to cause harm and even death. Improved ventilation has now made the enclosed structures in the park safe.

Are **ships** a source of pollution?

In the 1960s, when scientists first started examining weather satellite photos, they were surprised to find long streaks of brighter clouds embedded within marine cloud decks. These streaks, some of which extended for over 1,000 miles, were found to result from pollutants emitted from the stacks of passing steamships.

Are **airplanes** a source of pollution?

There are now over 10,000 commercial jet aircraft plying the Earth's skies. That number is expected to double by the year 2010. And each plane has several engines that are burning fossil fuel and emitting exhausts. Much of it is composed of water vapor, which can often be seen rapidly turning to ice crystals in the contrail behind the high-flying plane. In some parts of the United States it is believed that artificial cirrus clouds produced by the spreading of jet contrails have actually begun to affect cloud cover statistics. Aircraft currently account for three percent of the carbon dioxide emitted planetwide by human activities. Aircraft engines also emit carbon particles (soot), sulfur dioxides, and oxides of nitrogen. The amounts of these emissions are not known: since aircraft performance changes with pressure and temperature, emission measurements can't be easily made on the ground. So NASA scientists are spending time flying in small jet planes right behind larger jet plane engines, way up near the tropopause, taking measurements of the exhaust. Ideally the measurements will allow scientists to better understand the contributions of the jet passenger fleet to atmospheric pollution.

Have **smog levels** in the United States gotten better or worse?

As a result of environmental controls, the air in the United States is cleaner than it was several decades ago—but 133 million people still live in areas failing to meet the

Federal Air Quality Standards for ground level ozone, the major component of smog. Another major pollutant, carbon monoxide, adversely affects up to 78 million citizens.

The improvements are largely the result of the Clean Air Acts, which were passed by Congress in 1967 and 1970. Air pollution emission controls are working in some areas. By 1992 cities such as San Francisco and Kansas City had controlled their smog sources well enough that they were no longer in violation of federal ozone standards. Los Angeles was also better, but still had a long way to go.

While air quality in general has improved in the United States over the past 20 years, especially for sulfur dioxide and particulates, ozone (smog) has remained difficult to control. While Los Angeles peak ozone values have fallen some 55 percent, the air is still often unhealthy.

Is your lawnmower an air polluter?

Off-road internal combustion engines—like lawnmowers, chain saws, weed trimmers, and leaf blowers—produce large amounts of carbon monoxide pollution. And one thousand chainsaw motors produce the same amount of smog-forming chemicals as 620 cars. The gasoline engines powering the millions of backyard tools may contribute 5 to 10 percent of the nation's smog-forming pollutants. With a gasoline-powered lawnmower, the pollution output is equivalent to driving your car several hundred miles. In 1996, however, much "cleaner" garden gadgets started coming on the market.

How **deadly** is air pollution?

Particulate air pollution in the atmosphere (smoke, dust, soot, etc.) is estimated to cause between 300,000 and 700,000 premature deaths each year around the world. In the United States, 66 percent of Americans live in areas that violate federal air quality standards for one or more pollutants. Lung disease is the third leading cause of death in the United States.

The health hazards of smog, a complex chemical soup rich in ozone, includes eye irritation. Serious smog attacks can irritate eyes and mucous membranes and cause respiratory system problems. Long-term effects of ozone can cause impaired breathing, increased susceptibility to lung disease, and permanent lung and cardiac problems.

277

What U.S. cities have the **worst smog**?

Los Angeles still holds the dubious distinction of being number one in this category. But not so far behind in the air pollution derby are the metropolitan areas of Baltimore, Chicago, Milwaukee, Muskegon (Michigan), New York, Philadelphia, Houston, and San Diego.

How bad is **air pollution** in other nations?

Air pollution is a problem in the United States, but our air is pristine compared to many underdeveloped nations. In Mexico City, there were only 31 days in 1993 when the air was considered fit to breathe. Breathing the air in Bombay, India, is now the equivalent of smoking ten packs of cigarettes a day.

Air pollution in cities with ancient roots such as Athens, Greece, is causing extreme problems for many classical architecture and art treasures. By one estimate, more damage has been done by air pollution to the Acropolis in the last four decades than all of the assaults of the past 24 centuries.

Nations experiencing rapid industrialization, such as China, are also developing serious air quality problems as their pollutant emissions are still largely uncontrolled in the name of economic growth.

What is the **most polluted city** on Earth?

Mexico City. Twenty million people can make a lot of pollution, and when they all live in one high valley surrounded by a ring of mountains, it is a recipe for disaster. As many as 5,000 people die each year from diseases related to inhaling the air. Among the cleanest of the 20 largest cities on the planet in terms of air quality are New York, London, and Tokyo.

Is **asthma** becoming more of a health problem?

Yes. Asthma is the only pediatric disease in the United States, aside from AIDS, in which the death rate has been rising. The trend is especially noticeable among children raised in urban and inner city areas. It is thought that long-term exposure to a combination of high outdoor and indoor pollution may be a major contributing factor.

What are pollution **"hot spots"**?

Pollution levels can vary dramatically throughout a large metropolitan area; hot spots are the areas where pollution is particularly highly concentrated. Carbon monoxide levels adjacent to a busy freeway or highway intersection can be many times higher than the city-wide average. The highest values may often be found in large indoor parking ramps and tunnels where poor ventilation allows auto exhaust to build to extreme levels. Plumes from tall smokestacks may drift above most residents, but they then fumigate those living on hillsides downwind of the factory.

What is the **Brown Cloud**?

Los Angeles has its famous smog. Mexico City is world famous for its blanket of polluted air. And Athens' air pollution is eating away at historical structures. Denver, not to be outdone, has its own "Brown Cloud." As more and more people move into Colorado's Front Range, the cloud is getting browner and bigger, sometimes stretching north to the Wyoming border or south towards Colorado Springs. It is most pronounced in winter, sometimes blocking out the views of the snowcapped Rockies. The sources are highly varied, ranging from automobiles to power plants and industries, as well as road dust and agricultural burning.

What is **pollution prevention**?

Rather than trying to deal with the impact of pollutants after they are used in industrial processes or released into the atmosphere, more and more industries are trying to simply not use toxic or polluting materials at all. Pollution prevention works. Switching to low-sulfur coal, or even much cleaner natural gas has helped cut sulfur dioxide levels in many areas. Many chlorofluorocarbons (CFCs) used in high-tech industries have been replaced by more "environmentally friendly" materials. Cutting back on excessive packaging helps reduce the pressure on landfills. Getting the lead out of gasoline, paints, and many other substances used in daily living has dramatically cut lead levels in our air and environment—and in the human blood stream. Between 1976 and 1991, lead levels in U.S. blood dropped some 78 percent. Lead contamination from the environment has many serious effects, including lowering the IQs of children exposed to the toxin.

What are **air toxins**?

Many chemicals are emitted into the environment that are in some way harmful. There are more chemicals in the modern American home today than would be found in a well-equipped laboratory of the nineteenth century. According to one EPA estimate, 2.7 billion pounds of toxic air pollutants are emitted into the U.S. atmosphere every year. From 1,500 to 3,000 cancer deaths are believed to result from these toxins. Common household items can be toxic to sensitive individuals. Some permanent press clothes contain formaldehyde, which gives off fumes. Some air fresheners contain xylene, napthalene, and the like. Mothballs are pure paradichlorobenzene.

Has a **baseball game** ever been called off due to air pollution?

Baseball games have been called off for all sorts of reasons: rain, snow, cold, fog, darkness. Now you can add toxic fumes. In May 1993, a game between the Iowa Cubs and the Buffalo Bisons was canceled because the area around the stadium was evacuated due to exploding barrels of toxic chemicals in a nearby building fire.

How can you help **save the environment** by turning on a light?

Lighting consumes 25 percent of all U.S. electricity, with Americans buying three billion light bulbs each year. If all U.S. citizens changed to energy-efficient fixtures, we could reduce CO_2 emissions by 202,000,000 tons! One company estimated for every high-efficiency light fixture it installed it saved $30 per year and reduced emissions of CO_2 by 387 pounds, sulfur dioxide by 2.8 pounds, and oxides of nitrogen by 1.3 pounds. Next time you go shopping for illumination, think compact fluorescent bulbs. They do cost more to buy, but over the bulb's lifetime you'll save $50 in electricity. Moreover, for each such bulb you use you'll prevent a half-ton of carbon dioxide from being emitted into the atmosphere from power plants.

Can we afford to **clean up our air**?

The price of clean air is going up. Since 1970, as a result of the Clean Air Act and its amendments, it is estimated that the United States has spent over $200 billion to reduce emissions of such pollutants as sulfur dioxide, carbon monoxide, and particulate matter. It's costly to clean the air, yet we can't afford not to. If the Los Angeles basin met federal clean air standards, it is estimated that medical costs alone would decrease by nearly $10 billion annually. Some cost-benefit studies suggest that for every dollar spent on environmental improvements, the benefits reaped are several times that amount.

Why is **lead** an environmental hazard?

Lead has been found to be one of the most toxic substances in our environment. Ingestion of lead by infants and children has been linked to sharp declines in mental functions, including IQs. Lead has also been recently linked to hypertension in adults. From the 1940s through the 1970s, automobile exhaust emitted vast quantities of lead as a result of performance additives in the fuel. Other sources of lead include lead plumbing and lead additives in paint, the use of which has now ended. Great strides have been made to limit lead emissions into the air, especially from gasoline. Yet it is estimated that in the Los Angeles basin, almost five tons of lead still enter the air each day, though less than 20 percent is from automotive exhaust. Over many decades, lead has also accumulated to dangerous levels in the soil of some urban neighborhoods. The EPA now severely restricts the release of lead into the environment. Worldwide, some 3.5 million metric tons of lead enter the environment each year.

Is **lead pollution** a modern problem?

While emissions of toxic lead into the environment have soared on a global scale, especially since the 1940s, our ancestors did a fair job of soiling their own nests. During the Roman Empire some 100,000 metric tons of lead was mined yearly. The uncontrolled smelting of lead released huge quantities into the atmosphere. The record of this activity is preserved in the Greenland glacier, where the lead particles were trapped with each year's accumulation of snow and ice. Lead values in the atmosphere were way above natural amounts from 500 B.C.E. to 300 C.E. They dropped significantly with the decline of the Roman Empire, only to begin increasing rapidly in the seventeenth and eighteenth centuries. There has been another sharp drop since the 1970s, when intense efforts began to reduce lead emission into the world's atmosphere.

What is **acid rain**?

The massive quantities (40 million tons in the United States alone) of chemicals such as sulfur dioxide and oxides of nitrogen that are emitted into the atmosphere from industrial sources and vehicles don't just disappear. They undergo complex chemical changes and become distributed throughout the ecosystem. These gases have a significant impact upon the chemistry of precipitation as raindrops become carriers of weak sulfuric and nitric acids. The term acid rain may have been first used in 1872 by British chemist Robert Angus Smith. The problem rose to prominence in Scandinavia in the 1960s when lakes became acidic and fish populations died off. The cause was traced to acid rain (and snow) caused by emissions of pollutants from western and central Europe. In North America, acid rain was found to be a potential problem in the Northeast as well as portions of eastern Canada.

Extensive studies of the problem during the last decade, done as part of the National Acid Precipitation Assessment Project (NAPAP), showed that there was some damage already occurring and the potential for more, but that it was generally not as extreme as had been feared earlier. However, the 1990 Clean Air Act Amendments have mandated substantial reductions in both sulfur and nitrogen acid-forming compounds during the next half decade.

How is the **acidity of rain** measured?

Acidity—or at the other end of the scale, alkalinity—is measured by a scale known as the pH (potential for hydrogen) scale. It runs from 0 to 14. Zero is extremely acid, whereas 14 is totally alkaline. Neutral is a 7. The scale is logarithmic, so a change of one unit actually represents a tenfold change. Thus a pH of 3 is ten times more acidic **281** than a pH of 4. Any rain (or snow, cloud, fog) below 5.0 is generally considered acidic. Rainfall average pH values in eastern Europe and Scandinavia have often been in the 4.3 to 4.5 range. In the eastern United States and Canada many averages are in the 4.2 to 4.6 range. Some comparison values of pH are listed below:

Concentrated sulfuric acid	1.0
Lemon juice	2.3
Vinegar	3.3
Acid rain	4.3
Normal rain	5.0 to 5.6
Normal lakes and rivers	5.6 to 8.0
Distilled water	7.0
Human blood	7.35 to 7.45
Sea water	7.6 to 8.4

Are there **natural sources** of acid rain?

Rainfall is naturally slightly acidic. Natural (distilled) water has a pH of 7.0. Yet, due to dissolved carbon dioxide (like weak soda), rain water is actually a weak solution of carbonic acid (like salt-free club soda) with a pH of about 5.6. Dissolved salts and other chemicals can also alter the pH of natural rain water. When minerals lofted by a dust storm become entrained in rainwater, it is even possible to have some alkaline rain at times.

Humans are not the only source of acid rain. Volcanoes also can be prodigious sources of "pollutants" such as sulfur dioxide. The Laki volcano in Iceland erupted in June 1783 and resulted in 100 million tons of sulfuric acid rain falling over Europe, an amount equaling today's global annual acid rainfall. The volcanic acid rain ruined most of Iceland's crops, and spread a blue haze over much of Europe. Oxides of nitrogen are released from microbial action in the soil, as well as from lightning discharges.

What is **acid fog**?

Everyone has heard of acid rain. But acid fog? Turns out in many cities, and in mountainous regions, which are often wrapped in fog, the pH of the fog droplets can be several tens of times more acidic than the rain. Fog with acidity equal to vinegar has been found in major cities. Acid fog (and clouds) might be a contributor to forest decline in mountainous regions.

Has the United States decreased its dependence on foreign energy?

From 1958 forward, the United States has consumed more energy than it produced, meeting the difference with imported energy. During previous cold winters and OPEC

Scraping away the surface reveals clean white snow beneath the soot-covered snow soiled by airborne pollutants entering Scandinavia from western Europe.

oil embargoes, there was concern that heating fuel supplies could be in dangerously short supply. So how are we as a nation doing on energy conservation? As evidenced by improved gas mileage in cars, we have made many efforts to conserve energy. The nation is producing 2.4 million barrels of oil per year, less than in 1973. The United States now consumes 24.6 percent of the world's energy, and as a result we are now importing 13 percent more foreign oil than before the first of the big energy crises. Per capita energy consumption in the United States is 2.6 times higher than in France or Japan.

Is **solar energy** practical?

Between 1980 and 1992 the cost of solar-generated electricity dropped some 1,000 percent. It has continued to drop rapidly since that time. One industry group suggests that solar power (photovoltaic cells, for example) will be cost-competitive with conventional electric utilities by the turn of the century. More and more rural homes and industrial sites, such as radio and microwave repeaters, are relying on solar energy.

When is **solar heating** a bad idea?

Solar heating sounds like one of the most environmentally responsible concepts there is, unless of course you are overheating your home or office during summer because your lack of blinds, shades, or glass coatings allows too much solar heat inside. This **283** overheating keeps your air conditioner working overtime. It is estimated that up to 3 percent of U.S. electricity consumption is used for driving air conditioning units trying to get rid of excess solar heat in structures.

Is **wind energy** practical?

Yes. Wind energy is becoming ever more practical and economical. The tax incentives in past decades for wind energy resulted in many companies being created to sell wind generators, more for quick profit than out of a desire to produce reliable, cost-effective equipment. As a result many early wind energy pioneers purchased devices that had high maintenance costs and frequent failures. Over the last decade, new designs in turbines and blades have become vastly more efficient, reliable, and inexpensive. Many parts of the world are increasingly turning to wind energy as a practical means to generate electricity or supplement supplies from other sources. The practical use of wind energy to generate electricity has been steadily gaining ground. There are now well over 20,000 wind turbines in use worldwide, though this still represents only about 0.1 percent of the world's use of electricity.

How much wind energy could the **United States generate**?

The wind contains energy. The total wind power of the atmosphere has been estimated at about 3.6 billion kilowatts. On the whole, the United States is a windy country, and, if it so desired, it could extract some 370,000 percent more electrical energy from the wind than it does now.

Windmills were first used extensively in the Middle East in the eleventh century. Two hundred years later they became established in Europe. In the 1400s the Dutch were using wind power to drain swamps and marshes, and before the start of this century, the Danes were using wind to generate some 40 megawatts of electric power. Small windmills were widely used in the rural parts of the United States before inexpensive electricity became available during the 1930s. Wind energy is an old idea, but it's one whose time may be coming again. California currently leads the nation in wind energy, and plans are being made to "farm" up to 10,000 megawatts of wind energy in the north central United States before the end of the decade.

How much of the national energy supply is from renewable resources?

According to a Department of Energy estimate, of all the nation's potential energy resources, 92 percent are renewable categories such as wind, solar, hydropower, biomass, geothermal, and alcohol fuels. Therefore, renewable energy sources are far more plentiful than coal, oil, gas, and uranium. Despite this potential, renewable resources are currently only contributing 12 percent of the nation's energy needs, with hydropower being by far the largest contributor.

What is wind prospecting?

Wind energy requires wind. And just like in real estate, the best sites are determined by location. Average annual wind speeds must be at least 10–12 mph in order for most wind generators to produce sufficient electricity to pay for themselves. So finding the windiest site in a region becomes very important, especially when it was realized that the amount of electricity increases with the cube of the wind speed. In other words, a site with a 15 mph average wind can produce 3.38 times more energy than one with a 10 mph average speed. Variations in land use, tree cover, and elevation over just one mile can easily cause these types of microclimate variations in mean wind speed. Thus arose the art of wind prospecting. Trained meteorologists can use their experience, combined with measurements, to find those local sites with just a little more wind, but a lot more energy.

Can you see air pollution from **space**?

When the first of the high resolution Earth resources mapping satellites were launched around 1970, scientists were stunned to see plumes of smoke from individual industrial complexes traveling for hundreds of miles downwind. By the mid-1970s, scientists using the new geosynchronous satellites began to notice large hazy areas covering extensive regions of the eastern United States. These were soon found to be huge clouds of air pollution, rich in sulfate aerosols and ozone. These "smog blobs," as they were called by some, could cover a dozen states or more at a time and persist for a week or more until the blob was swept out into the Atlantic by a Canadian cold front ushering in some cleaner air.

Analysis of images taken by NASA astronauts over the years have found many phenomena—including fires. Over 1,400 cases of huge plumes of smoke, many from massive fires used to clear land for farming, have been logged. The smokiest part of the world appears to be the Amazon area and southern Africa.

What is the **dustiest major city**?

Las Vegas ranks in the top five regions in the nation for airborne particulates, thanks not only to industry and vehicular traffic, but also to its surrounding dry deserts that kick up big dust clouds whenever the wind blows, which is often and at high speeds. **285**

On the whole, the United States is far less "dusty" than it was at the start of the century. Eastern cities of a century ago had copious quantities of dust, produced from the burning of coal and wood and other industrial practices. These sources were largely eliminated by efficient air pollution controls. While today particulate pollution is measured in millionths of a gram per cubic meter, in decades past, dust measurements were expressed in tons per square mile!

But the dust problem is not totally under control. You're driving down a country dirt road on a beautiful spring day, soaking up nature, feeling at one with the environment. Perhaps you feel a little guilty because you are polluting the pristine rural environment with the foul exhaust from your internal combustion engine. It turns out that's only half the story. Your car exhaust is adding pollutants to the air, but during dry weather, your exhaust is crystal clear compared to the trail of road dust left behind. Some two to three pounds of dust can be kicked into the atmosphere per vehicle mile traveled on a country road. Road dust is a major contributor to such pollution problems as Denver's infamous Brown Cloud.

What are **natural** air pollution controls?

Two-thirds of all the wood cut on this planet is used for fuel. Over half the world's population still relies on wood as its primary source of heat and energy. Yet instead of using trees to pollute, people should be preserving them because trees are one of nature's best air pollution control devices. One acre of trees can filter some 13 tons of dust and gaseous pollutants from the air each year. The roots of mangroves in swampy areas are very efficient at removing waterborne pollutants.

What is **Arctic haze?**

You might think the North Pole would be a good candidate for having the cleanest air in the planet. During winter and early spring, however, the Arctic has been found to be wrapped in haze believed to come from many distant pollution sources. Concern in the international community prompted action and recently the Arctic atmosphere began getting a bit cleaner. The dense Arctic haze was traced to industrial pollution from Europe and the former Soviet Union. Reductions in European pollution emissions and a gradual shift from coal and oil towards the more environmentally friendly natural gas in Russia has helped.

The South Pole is still pretty clean, but there is growing concern about management of waste from the numerous research stations that dot the southern continent.

Are **computers** environmentally friendly?

The Environmental Protection Agency projects that by the end of the decade computers could consume almost 10 percent of all electrical power in the United States. New energy-efficient chips being designed have the potential to eliminate the need for 10 power plants and reduce the CO_2 emissions into the atmosphere by the equivalent of 5 million cars.

To inform consumers about which appliances and computers use the more energy efficient components, the EPA has instituted its Energy Star program.

Are some **air pollutants** beneficial?

One of the major sources of soil nutrients in the Amazon basin is the Saharan desert. Tens of millions of tons of soil and trace elements drift as much as 5,000 miles across the Atlantic each year and settle upon the jungle, supplying essential fertilizer that is unavailable from the poor soil of the Amazon.

The same African dust that feeds the Amazonian ecosystem may also be a prime source of fertilizer for the plankton of the Atlantic. These plankton are the first link in the complex ocean food chain. Thus dust storms over Mauritania may help feed fish in the waters around the Azores.

The increasing amounts of carbon dioxide in the atmosphere have triggered much concern about possible global change including global warming. But experiments have also shown that the increased CO_2 levels could increase the growth of many plants.

Does air pollution **damage crops**?

One impact of urban smog is that it is becoming more and more difficult to grow certain vegetables. Sensitive species include spinach, which is becoming harder to grow downwind of major cities such as Los Angeles, Chicago, and the East Coast megalopolis. Crop losses due to air pollution in the state of Maryland alone are estimated to reach $40 million each year. Of that total, over $2 million is from human air pollution impact upon tobacco.

Ozone air pollution in the eastern United States is estimated to cause reduction in agricultural yields of at least $3 billion annually. Forest growth is also being severely retarded in some areas of the United States and Europe as a result of the combined effects of ozone and acid rain.

What is **light pollution**?

When you hear the word pollution, you think of smoke or smog in the air, but you probably don't think of light. Yet for astronomers, "light pollution" is becoming a major problem as cities and their millions of light bulbs edge ever closer to once remote observatories. At some telescopes, the impact has become severe. Astronomers **287** are being forced to establish new facilities in extremely remote areas of the globe in order to escape both light and air pollution.

Are **dust storms** polluters?

In late February 1977, strong winds over the high plains of Colorado and Texas scoured the bare fields ready for planting, sending a massive cloud of dust eastward. Satellites tracked the dust cloud well into the Atlantic. Particulate air pollution levels in Oklahoma were 20 times higher than allowable from man-made pollutants. Up to three million tons of dust settled out onto Oklahoma during a dust storm. Some of the highest levels ever recorded for total suspended particulates were recorded when the dust cloud swept by Mississippi and Alabama.

Can **weather** affect air pollution levels?

Atmospheric conditions have a pronounced effect on air pollution. Many industrial installations release pollutants out of their stacks at a relatively constant rate. Yet the pollution concentrations experienced at the ground can vary by many orders of magnitude. The reason for the variation lies in the vertical changes of wind and especially temperature. During inversion conditions, plumes can be trapped close to the ground. If the temperature falls with altitude at a rate of about 5.4°F per 1000 feet and does so for several thousand feet, then a deep "mixed layer" is present, which vertically dilutes the pollution and lowers the concentration. Readings of carbon monoxide near a heavily traveled freeway will be dramatically higher than during the afternoon on nights when a "nocturnal inversion" is present that traps the CO close to the ground.

Warm humid air masses also favor the development of regional smog (ozone) episodes. A side benefit of the cool summer of 1992 in the eastern United States was that the photochemical smog (specifically ozone) had few chances to cook up to significant levels. For instance, the Louisville, Kentucky, area typically has 8 days per summer with ozone levels in the unhealthy range, but recorded none during that entire summer. By contrast, during the extremely hot summers of 1987 and 1988 in the eastern United States, many urban areas set all-time records for ozone pollution.

The field of air pollution meteorology is very active and has created many tools, including complex computer models, which link the emissions of pollution to atmospheric conditions in order to predict the details of pollution concentrations that affect people and crops.

What **major air pollutant** has decreased dramatically?

With humankind continuing to pollute its atmosphere at alarming rates, it is consoling to know at least one form of air pollution has largely been eliminated—radiation from atmospheric testing of nuclear weapons. Between 1945 and 1968, 324 atomic and hydrogen bombs were detonated, with a combined force of 511 million tons of TNT. Included in the fallout were huge quantities of iodine-131, strontium-90, cesium-137, and carbon-14, all of which are highly radioactive.

During the days of the Cold War, when all the nuclear powers were testing thermonuclear weapons in the open air, there were usually unknown or unexpected results. In late May 1953, the U.S. government tested an atomic bomb in the atmosphere over Frenchman's Flats, Nevada. Two days later, radioactive hailstones, some the size of tennis balls, pelted Washington, D.C.

How much pollution did the Kuwaiti oil fires release?

At the end of the 1991 Persian Gulf War, retreating Iraqis ignited over 500 oil wells, resulting in the burning of 5 million barrels of oil and 70 million cubic meters of gas per day. This in turn released 500,000 tons of CO_2 and 40,000 tons of SO_2 per day into the atmosphere. The latter figure equals the combined SO_2 and CO_2 pollution and emissions of France, Great Britain, and Germany. At their worst, the Kuwaiti oil well fires were estimated to be releasing over 18 trillion grams of fine particulate matter into the atmosphere each day, an amount roughly equivalent to the combined particulate pollution exhaust of all the world's motor vehicles.

What was the atmospheric impact of the Kuwaiti oil well fires?

When Kuwaiti oil wells were set afire during the Gulf War, there was much concern over the possible climatic impacts of smoke from such massive fires. Computer programs used to simulate the climate impacts of a major superpower nuclear exchange were pressed into service. The computer models concluded that while severe regional pollution problems might result, the impact on the global scale was expected to be small.

The dense clouds of smoke from the oil well fires of Kuwait in the spring of 1991 were so thick that daytime temperatures in their path were sometimes 10°F to 20°F cooler than in surrounding areas. Black rain fell over parts of the Middle East as smoke from the hundreds of oil well fires became involved in rain clouds and colored the precipitation.

Not so long ago many thought "the solution to pollution is dilution"; in other words, dump it into the air and soon it will be at such low concentrations no one will ever know. The Kuwaiti oil field fires proved that that notion is a bit of an oversimplification. Two months after the start of the fires, observations of atmospheric soot content over places as diverse as Hawaii and Wyoming were reading 5 to 10 times greater than normal.

What is yellow rain?

In 1930, a yellow rain fell over much of southern England. Red rains and brown snow have also been observed. Actually, colored precipitation is not that unusual. It is generally associated with storms that ingest large amounts of dust from distant desert storms. In Europe, the cause is usually dust from the Sahara and Morocco being swept northward into a storm system. Red snow has been seen in Minnesota, the result of dust storms in the southwest United States. The term yellow rain was also used in describing chemical warfare used in southeastern Asia.

What is **odor pollution**?

Bad odors can be one of the more evident aspects of air pollution. Humans can detect odors at surprisingly small concentrations. The concentration at which 50 percent of the population can detect an odor is called the olfactory threshold, and it can be as small as one part per million, billion, or trillion, depending on the chemical species.

High winds fanned a fire at the Central Storage and Warehouse in Madison, Wisconsin, on 3 May 1991. The building housed huge stores of cheese and meat products. Even through a driving rainstorm, the grease fire continued to burn for three days. Then the hot weather that followed promoted quick rotting of the food residue, resulting in major odor pollution.

How pure is **rain water**?

Sometimes people talk about "pure rain water." Rain water is very clean, but not entirely. Every raindrop forms by condensation of water vapor onto something, often a speck of salt or dirt, which remains with the rain water until it reaches Earth. When the drop dries up, it leaves its nucleus behind.

WEATHER AND THE HUMAN BODY

What is **bioclimatology**?

Bioclimatology is the science that investigates the effects of the atmosphere upon living organisms, including humans, animals and plants. This field is more advanced in Europe, where daily forecasts of the anticipated effects of the daily weather upon health (including such conditions as arthritis and angina) are issued by some national weather services.

What is **carbon monoxide** poisoning?

Late fall and early winter often bring reports of preventable deaths due to carbon monoxide poisoning in the home. Each year as many as 1,500 Americans die needlessly because of carbon monoxide (CO) poisoning, more than from tornadoes, hurricanes, and lightning combined. Any combustion without enough air in an enclosed space can produce the deadly gas; faulty and improperly maintained home heating systems are one possible cause. Inhaling large quantities of the colorless, odorless, tasteless gas has one result—death. If a gas flame is burning yellow and not blue, that is a sign of incomplete combustion and possible CO production. Have your heating system serviced prior to the start of the winter. Have your fireplace chimney cleaned (birds may have built nests in it, inhibiting ventilation, or creosote may have built up to fire-hazard levels) and follow your gas or oil furnace manufacturer's maintenance guidelines. Carbon monoxide alert units (similar to smoke detectors) are available, but only three percent of U.S. homes currently use them.

For a free safety checklist, send a stamped, self-addressed envelope to the National Consumers League, 1701 K. Street NW, Suite 1200, Washington, D.C. 20006.

What is **frostbite**?

Frostbite results from the body's survival mechanisms kicking in during extremely cold weather. The body's first imperative is to protect the vital inner organs, which it does by cutting back on circulation to your extremities: feet, hands, nose, etc. If these parts are exposed to the cold and receive less warming blood flow, they eventually freeze.

One way to avoid frostbite is to avoid going outside during severe cold, especially if the wind chill is -50°F or below. If you must go, be sure to protect the exposed parts of your body, such as ears, nose, toes, and fingers. Mittens are more effective than gloves for warming your hands. Keep your skin dry. Stay out of the wind when possible. Drink plenty of fluids since hydration increases the blood's volume, which helps prevent frostbite. Avoid caffeinated beverages, however, as they constrict blood vessels and prevent the warming of your extremities. Alcohol should also be avoided since it reduces shivering, which is one of your body's ways of keeping warm. And be especially wary of smoking cigarettes in extremely cold temperatures. According to one physician, when you smoke, the blood flow to your hands practically shuts off.

What should you do if you think you have **frostbite**?

Victims often are unaware that they are getting frostbite due to the numbness from the cold. When frostbitten areas begin to thaw, the pain starts. The folk wisdom about rubbing the frostbitten area with snow is false. In fact, you really should not rub the area with anything. Try to avoid using hands or walking on toes that have recently thawed. Thaw in the tub, not by the fireplace. A warm water bath in the 100°F to 110°F range is ideal.

What are the **different degrees** of frostbite?

There are four degrees of frostbite. Ice crystals forming on your skin is a sign of first degree frostbite. In the second degree of frostbite your skin begins to feel warm, even though it is not yet defrosted. If your skin turns red, pale, or white, that indicates third degree frostbite. In the fourth stage, the pain lasts for more than a few hours, and you may see dark blue or black areas under the skin. See a doctor immediately if these symptoms arise. Gangrene is a real threat.

What are **chilblains**?

Prolonged exposure to cold, damp weather can result in a condition known as chilblains. The symptoms on hands and feet include redness, burning, itching, and chapping.

In May of 1996, Dr. Beck Weathers suffered from severe frostbite while climbing Mount Everest. The black splotches on his face are scabs caused by frostbite; his hands were also frozen and doctors thought he might lose most of his right hand.

Does weather affect **arthritis**?

While conclusive proof has not been obtained, many arthritis sufferers swear they can "predict" the weather by the pain in their joints, as in the old adage, "A coming storm our shooting corns presage, our aches will throb, our hollow tooth will rage." Research at the University of Pennsylvania sealed some patients with rheumatoid arthritis in a climate-controlled chamber. They found that with rising humidities and falling pressure (the conditions that frequently occur before a winter storm), arthritic symptoms did indeed increase.

Can exposure to cold induce a **heart attack**?

Bioclimatological studies have shown that in northern European countries the mortality rate of coronary heart diseases is generally much higher than average in January through February, and lowest in mid-summer. This may be partly due to the increased clotting tendency of blood during cold weather.

Why does your **nose run** in cold weather?

It is not necessarily because you have a cold. If very cold air is suddenly inhaled, the mucous membranes inside your nostrils first constrict, then shortly thereafter rapidly **293** dilate as a reflex reaction. This permits an excess of mucous to form, resulting in a runny nose or the "sniffles." With continued breathing of cold air, the nose begins to adjust back to a more normal condition.

Does cold weather cause colds?

Probably not directly. While it seems there is more flu and cold during the winter months, their basic causes are not cold, damp air but the dozens of different micro-organisms that cause colds. The fact that people spend much more time together inside buildings during the winter months probably greatly increases the ability of the germs to spread from one host to another. It is also known that at the start of the indoor heating season, hospital admissions rise in many states. Part of the reason is apparently the many carbon monoxide poisonings resulting from malfunctioning heating systems that have been turned back on without repairs or proper maintenance. In milder cases, carbon monoxide poisoning may be mistaken as a bad case of the flu.

What is **heat exhaustion**?

Heat-related illness stems from too much exertion and/or lack of fluid intake during high heat stress periods. And all other factors being equal, the severity of heat trauma increases with age. Heat cramps in a 17-year-old might be heat exhaustion in someone aged 40, and heat stroke in a person over 60 years of age. Heat exhaustion results in sweating, weakness, paleness, cold and clammy skin, irregular pulse, fainting, and vomiting. Heat exhaustion is one of the often unrecognized hazards of summer. On 21 July 1991, during the height of a searing heat wave, over 100 people attending the Dayton (Ohio) Air Show were felled by heat exhaustion. Over 40 of them had conditions serious enough to require being taken to hospitals. When the heat and humidity go way up, people should slow way down.

Heat waves can kill. Even in this age of air conditioning, don't overdo it. Wear loose, lightweight, light-colored clothing. Slow down. Don't over-stress your body and drink plenty of fluids. Keep in the shade when possible. If you tend to sweat a lot, consider taking salt tablets (unless you are on a low-salt diet). Be especially careful on the first few hot days of summer.

What is **heat stroke**?

Heat stroke (also called sunstroke) is a severe medical emergency. Heat stroke is accompanied by high body temperature (106°F or above), hot dry skin, rapid and strong pulse, and possible unconsciousness. Summon emergency medical assistance or get the victim to a hospital immediately. Delays can be fatal. Move the victim to a cooler environment. Reduce body temperature with a cold bath or sponging. Loosen clothing, use fans and air conditioners to start cooling the body. Do not give fluid.

Heat waves can be killers. It is estimated that some 15,000 Americans would perish from heat-related health problems during a hot summer without air conditioning.

Can you survive a direct hit by a **lightning bolt**?

Yes. In fact most people struck by lightning actually do live to tell about the experience. However, the survival rate would be even higher if those who appear "dead" from a lightning strike were immediately given CPR to jump start their temporarily paralyzed heart. If you are with a group of people who are struck by lightning, help the "dead" first. Those showing no motion or other signs of life can very often be revived by immediate application of CPR.

Do people develop **phobias** about weather?

Psychologists have found that many people develop phobias related to weather. *Ombrophobia* is a fear of rain while *chionophobia* is a similar fear of snow. *Anemophobia* is the fear of wind while *keraunophobia* is the fear of lightning and its mate is *tonitrophobia,* a terror of thunder (not uncommon in dogs). *Homichiophobia* is a fear of fog.

What is **SAD**?

It may not be the cold that depresses many people in the winter, but the dark. It is estimated that some 35 million Americans suffer to some degree from SAD (Seasonal Affective Disorder), or the "winter blues," caused by reduced sunlight. It is suspected that there is actually a change in the brain's chemistry in response to the diminished sunlight. SAD may also be genetic as it has been found to run in families. Remedies include light therapy and trips to bright, warm, tropical places.

Can the tilt of the **Earth's axis of rotation** make you gain weight?

The orientation of the Earth's axis of rotation can have an effect on your weight. Since the planet's axis is tilted 23.5°, we have seasons, and nights grow longer as winter approaches. In winter, many people, especially women in their 30s or older, develop SAD (Seasonal Affective Disorder). This "dark season" depression often results in carbohydrate craving, weight gain, and oversleeping. And during the winter, people are genetically programmed to eat more, especially foods with higher fat content.

How does your body respond to day and night?

Rather like clockwork. Humans have a built-in biological clock producing our circadian rhythms, which make us tired at night and awake during the day—no matter what shift we happen to be working. The human organism wants to be asleep at 4 A.M., and is full of vim and vigor at 10 A.M.—taking its cues from the daily cycle of night and day. Studies have shown night shift workers have more accidents at work as well as when driving home. They also have a higher incidence of cardiovascular problems. Several famous disasters, including the Chernobyl and Three Mile Island nuclear mishaps, occurred during the graveyard shift. Jet lag, the curse of pilots and airline cabin attendants, is the result of an unsuccessful fight against circadian rhythms. It has even been found that West Coast baseball teams, when travelling to the East Coast, score fewer runs and commit more errors.

Does weather affect **appetite**?

If you live in a cold climate and spend a lot of time outdoors, your body may require 1,000 to 2,000 extra calories a day to keep warm. Persons living in an Arctic climate do indeed consume more food calories than a tropical dweller. On the extreme, daily caloric requirements can increase by 2,000 during a prolonged trip in cold weather. There is some evidence that exposure to ongoing cold increases the body's demand for dietary fat.

Why do some people **sneeze** when they look into the sun?

An unusual allergy is the one that causes perfectly healthy people to suddenly sneeze when they leave a building during daytime or look up at the sun. Called allergic conjunctivitis, which is an inflammation of the lower eyelid, it is a reaction to the ultraviolet portion of the sun's spectrum. Sunglasses and antihistamines can help.

How much **pollen** is released into the atmosphere?

Each year American vegetation spews forth some two billion pounds of pollen into the air you breathe. That's roughly ten pounds per sneezing, wheezing, sniffling human. For the tens of millions of hay fever sufferers it hardly seems like a fair fight.

What is **ultraviolet light**?

Ultraviolet light (or UV) is among the smaller wavelengths in the solar spectrum, with wavelengths just shorter than those causing violet light. They have profound effects on biological activities. UV is often used to sterilize objects and is now widely used in water purification. This powerful radiation is not so kind to larger living organisms, like people. Though UV rays cannot be seen with the naked eye, you can certainly sense their presence by the sunburns developed while lying on the beach. There are two types of UV rays: Ultraviolet-A and Ultraviolet-B. The UV-B rays are primarily responsible for burning skin and are associated with skin cancers. They are strongest in summer. The UV-A rays can penetrate the skin even deeper and cause cancer as well as other skin damage. The UV-A rays are prevalent year round.

The amount of ultraviolet-shielding ozone in the stratosphere isn't the same worldwide. The ozone shield is much thinner at altitudes closer to the equator. Thus, with equal exposure to sunshine, the chance for skin cancer is higher in the southern United States than in central Canada.

What are some **health effects** of ultraviolet light?

Ultraviolet rays from the sun can cause a number of health problems. In the United States, it is believed that over 600,000 skin cancer cases occur each year, 90 percent of these due to overexposure to the sun's rays. Melanoma is the deadliest of all skin cancers. Its incidence has been steadily rising at a rate of 4 percent per year in the United States since 1973, and 1,200 percent since 1930. Exposure to too much sun is the culprit. About 10,000 Americans die each year from sun exposure–induced skin cancer. Arizona and Colorado are the states with the highest incidence of skin cancer.

The Center for Disease Control notes the typical skin cancer patient is a fair-skinned male, 65 and older, who doesn't tan easily, spends a lot of time outdoors exposed to the sun, was sunburned a good deal before age 18, and has a large number of moles. As few as three severe sunburns during childhood appear to raise the risk of melanoma later in life. One severe blistering in childhood can double the chances of developing melanoma as an adult.

Early detection and removal of a melanoma is critical to survival. A pigmented lesion of the skin that is larger than the diameter of a pencil eraser, varied in coloration, and unevenly shaped with an irregular border should be seen by a doctor immediately.

The increase in human skin cancer due to additional harmful ultraviolet rays reaching the Earth's surface as the ozone layer becomes depleted is well documented. Other, more subtle impacts may also have disastrous consequences. Some estimates suggest that productivity of phytoplankton in the Southern Hemisphere oceans may have dropped 6 to 12 percent. This could potentially disrupt the ocean's food chain.

Excessive UV rays can also play a role in cataracts and can sometimes damage the immune system according to the Center for Disease Control.

What is the **sunburn index**?

Sunburn and summer are synonymous for many people, but they need not be. The Environmental Protection Agency (EPA) and the National Weather Service (NWS) have teamed up to provide a new sunburn potential forecast based upon the amount of the sun's skin-burning ultraviolet rays that are expected to reach the surface on a given day. The index ranges from 0 to 15, and it indicates how quickly a person's unprotected skin could begin to suffer damage. A forecast value in the 7 to 9 range means people with fair skin could start to burn in 7 to 8.5 minutes without proper protection. The index value for light-skinned people is:

0–2	Minimal	30 min.
3–4	Low	15–20 min.
5–6	Moderate	10–12 min.
7–9	High	7–8.5 min.
10+	Very high	Less than 6 min.

Persons with darker skin can tolerate exposures four or more times longer, but should still exercise reasonable care.

Does the **risk of sunburn** increase with altitude?

The Earth's atmosphere blocks out most of the harmful ultraviolet rays that burn your skin. In Denver, where there is less atmosphere above you, there are 10 to 20 percent more UV rays than at sea level. If you are mountain climbing at 14,000 feet, there are 30 percent more UV rays. And if there is snow cover, there is a greater impact since up to 90 percent of the UV rays reflect back off the white snowy surface onto your skin.

Does too much exposure to the sun cause any other **skin problems**?

Not long ago, "getting some rays" was the epitome of a healthy lifestyle. Then came the realization of the aging affects of the sun on the skin and the possibility of skin cancer. A more recently discovered and milder effect is known as Polymorphous Light Eruption, or PMLE. Dermatologists report that up to 10 percent of the population can get a periodic itchy, red, bumpy rash from sun exposure, especially after the first prolonged dose of springtime rays.

In Farmington, New Mexico, two older girls apply mud from the Animas River on a younger friend in an attempt to prevent sunburn, and to help keep her cool in the August heat.

What is the **SPF system**?

Sunscreens are rated by the sun protection factor (SPF) number. The SPF number tells you how much longer you can stay out in the sun than you would have been able to without sunscreen protection. To find out how long you would be able to stay in the sun without burning, multiply the SPF number on the label by the amount of time you ordinarily spend in the sun (without sunscreen) before you begin to burn. If you usually start to "pink" in 20 minutes, a product with an SPF 15 would allow you to stay in the sun for 300 minutes, or 5 hours.

When preparing to go out in the sun, apply the sunscreen some 20 to 30 minutes beforehand. Also, use hats and protective clothing when feasible.

299

Does weather affect **hay fever**?

Airborne pollen and mold levels can vary greatly from year to year depending on the weather. Spring 1995 was warm and sunny in Georgia, conditions that sound wonderful to most people. The early warmth and lack of rain, however, caused values of pollen measured in Atlanta's air to be *25 times higher* than the value considered "extremely high."

Heat and drought made the summer of 1991 tough in Illinois, not just for farmers, but also for allergy sufferers. Chicago is located near the "ragweed capital" of the United States. May 1994 in the Chicago area was one of the driest on record. As a result, mosquitoes were hard to find. But the lack of rain allowed huge clouds of pollen to fly through the air, tormenting allergy sufferers.

The lazy, hazy days of summer are often preceded by the wheezy, sneezy days of spring allergy season. A tip: pollen counts usually peak in the outside air between 5 A.M. and 8 A.M., when the winds are generally at their lightest.

How do the **birds and bees** migrate?

With magnets. It has been shown that certain species of migratory birds, fish, and insects can sense the Earth's magnetic field and use magnetosome cells in their brains to orient themselves and navigate during migration.

Is human breath destroying the **pyramids**?

In Egypt, up to two million visitors per year are breathing inside the great pyramid of Chephren at Giza, contributing to the destruction of the structure. Each visitor is estimated to exhale some 0.7 ounces of water vapor, which raises the humidity inside the structure and causes damage to the limestone blocks of the pyramid. A new ventilation system was recently installed to combat the problem.

Can you be **allergic** to cold weather?

Some people are. They might walk down the driveway on a cold day to get the newspaper, only to break out in hives two minutes later. A test to see if you are cold sensitive: place a sandwich bag filled with ice on your arm for two minutes. If an itchy welt forms, talk to your doctor. An antihistamine prescription might help control the problem.

What is **indoor relative humidity**?

During the cold winter months, without additional humidification, the heated air inside some homes can have relative humidities well under five percent. The amount of water vapor that air can hold doubles for every 20°F it is warmed. Air that is saturated outside at -10°F (foggy with 100 percent relative humidity), when it comes indoors and is heated to 70°F, will have a relative humidity in the six percent range.

What should the **relative humidity** be inside your home?

The ideal average humidity inside a building ranges between 30 and 60 percent. Air that is too dry causes cracking of furniture and respiratory discomfort. In many northern portions of the country, during the winter season, relative humidities inside homes without any additional source of moisture can drop to under three percent—drier than the Sahara Desert during the heat of a summer day.

If indoor relative humidity consistently averages *over* 60 percent, problems can arise. High indoor humidity is possible in today's highly insulated homes, in which a lot of water enters the atmosphere from building materials as well as that used for activities like clothes and dish washing. Without precautions, it is relatively easy to develop a severe mold and mildew problem, especially in closed-in areas. Many people are allergic to the spores from these organisms. Condensation on cold wall and window frame surfaces can also lead to structural damage to your home and will be quite costly to repair.

What are **indoor thunderstorms**?

They may make a crackle rather than a rumble, but those electrical discharges jumping from your fingertips on a winter's day inside your home are actually small-scale lightning discharges. They are a sign of very dry air. A sure sign that the indoor relative humidity is too low—those discharges of static electricity after you cross the carpet. For each inch of electrical discharge there are 40,000 volts of static charge build-up.

A few pints of water evaporated each day into the home or building's ventilation system will end the indoor thunderstorms.

What is **hypothermia**?

Hypothermia is a severe medical condition due to exposure to cold air or water. The warning signs include uncontrollable shivering, memory loss, disorientation, incoherence, slurred speech, drowsiness, and apparent exhaustion. A body temperature below 95°F requires immediate medical attention. Warm the body's trunk first, not its extremities. Provide warm broth rather than very hot beverages.

In hypothermia, the body's core temperature has dropped to dangerously low levels. Even a 3- or 4-degree fall in your body's temperature can produce paled skin, violent shivering, and impaired speech. During spring, when deep lakes have not yet warmed, careless boaters falling overboard are at great risk. Falling overboard in 40°F water can be more than a damp experience, it can be life threatening. People can lose consciousness after 30 minutes, and death from hypothermia becomes increasingly likely within two hours.

Is **snow shoveling** hard on the heart?

Snow shoveling places extreme strains on the heart. If you are middle aged or older, unless you are in very good condition, don't be a hero. Hire a young neighbor to do the shoveling. Heart attacks are a major cause of death during and after major snow storms. Men constitute 75 percent of all winter fatalities due to cold weather. Half of all such deaths are people more than 60 years old.

What is the **temperature of your skin**?

It is not 98.6°F. With an air temperature around 72°F, your skin would typically be about 86°F. As the outside temperature increases, your skin temperature goes up too, reaching about 95°F, after which it stays relatively constant.

What is the **hottest air temperature** the body can survive?

One California university professor conducted tests in which he escaped without serious injury a temperature of 250°F for almost 15 minutes. One British scientist claimed to have survived a temperature of 364°F for 60 seconds.

What wind is called "**the Doctor**"?

In many tropical regions, the oppressive daily heat is often relieved by the arrival of the cooler sea breeze. This refreshing breeze is termed the Doctor in many parts of the world, including South Africa, Australia, Jamaica, and the West Indies. The term is also applied to other cooling winds, such as the dry harmattan wind of the west coast of Africa.

What is **Legionnaire's Disease**?

Legionnella is a bacterium that thrives in hot-water systems, whirlpools, condensers, and commercial-scale air conditioning systems. Considered an environmental disease, this bacterium each year causes a pneumonia-like disease in somewhere between 10,000 and 100,000 Americans, with one in six cases proving fatal. The disease was first identified from the illness of 221 American Legion members who attended a conference at Philadelphia's old Bellevue Stratford hotel in 1976. Thirty-four Legionnaires died of the ailment, which was traced to the hotel's air conditioning system.

Fortunately, home air-conditioning systems generally are not particularly hospitable to this deadly micro-organism.

Does the moon affect human behavior?

Many people claim that the phase of the moon affects human behavior. The term lunatic, in fact, derives from the notion that people tend to get a little crazy during the full moon. Many police departments swear they get more calls for erratic behavior when the moon is near full. But a recent review of 37 scientific papers that attempted to relate the phase of the moon to violent crime, suicide, crisis center hotline calls, psychiatric disorders, and mental hospital admissions, found that there is absolutely no correlation between lunar illumination and mental states.

What is **humidifier fever**?

While using humidifiers during winter is a good idea, be sure to follow the manufacturer's instructions. If not properly cleaned, bacteria can grow within some systems and result in "humidifier fever." Modern sonic humidifiers usually require the use of distilled water. Otherwise, the units can spray a fine white dust all over one's house from the minerals found in most tap water. Home air-conditioning units, while not a host to Legionnella bacteria, can play host to a variety of other organisms, including mold. Air conditioners should be cleaned according to the manufacturer's directions.

How much heat does the **human body** give off?

Ever notice how warm a room crowded with people can get? The reason is pretty simple. The typical human body, which is akin to an internal combustion engine, gives off about the same amount of heat as a 100-watt light bulb. We also exhale considerable amounts of carbon dioxide gas, the product of our body converting food into energy.

Between 7 and 55 percent of body heat is lost through the head. Without a hat during cold weather, you can lose as much as half of your body's heat energy.

How sensitive is the **human nose** to odor?

The human nose is pretty effective. It can detect the smell of fuel oil, or of roses, at molecular concentrations of about one part per billion. To give perspective: one in a **303** billion is equivalent to one person out of the entire population of China, or one second in 32 years' time.

How strong is a **strong draft**?

When inside a building, a little air circulation is a good thing, but too much "indoor wind" becomes an annoying draft. Generally humans cannot notice air motion of less than a half-mile per hour. Most are aware of drafts at 1 to 2 mph, and anything over about 3 to 3.5 mph would be considered a nuisance of a draft.

Are **Christmas trees** a health hazard?

An allergist found that up to 20 percent of the population suffering from airborne allergies may react negatively to the various hydrocarbon gases given off by Christmas trees. The symptoms are the usual coughing, rashes, congestion, and wheezing.

Christmas trees also present a far more serious health hazard: fire. It is essential, if you have a live Christmas tree, that you keep it well watered, slowing down the drying-out processes in the desert-like indoor air of winter. The various oils and hydrocarbons in the needles are like fire bombs waiting to go off. Once you have seen a dry tree burst into flame, you will appreciate why caution is an absolute must.

What is **"mountain sickness"**?

Mountain sickness, also called high altitude sickness or hypoxia, is a series of physiological responses, such as headaches, lassitude, dizziness, shortness of breath, and palpitations, caused by the decreased oxygen intake at higher altitudes. Some individuals report symptoms at 6,000 feet, though it usually takes an elevation of 10,000 feet or more for the symptoms to become pronounced. They usually fade after a week or so at high altitude, but some individuals never really acclimate.

Aviators know that being at 10,000 feet in an unpressurized airplane cabin (or on a mountain highway) means you are getting 30 percent less oxygen than at sea level. Hypoxia's first signs can be drowsiness and slowed reflexes. In order to function for extended periods above this altitude, supplemental oxygen becomes essential. One's judgment soon becomes impaired as the result of hypoxia.

It is possible for humans to become acclimatized to higher elevations, which are accompanied by lower air pressures. Many civilizations have thrived above 10,000 feet in the Himalayas and the Andes. The mining town of Cerro de Pasco is nestled in the Peruvian Andes above 14,300 feet (about the altitude of the highest of the Rocky Mountains). People also live and work in sulfur mines atop Mount Aucanquilcha in Chile, some 18,000 feet above sea level.

How high can you go without **oxygen**?

Sustained life is nearly impossible above 20,000 feet without supplemental oxygen. A pilot who suddenly loses oxygen at 25,000 feet may lose consciousness in just a few minutes. At 30,000 feet, unconsciousness will occur in about one minute for those not acclimated to the altitude. If somehow you found yourself at 50,000 feet without an oxygen mask, you would have between 11 and 18 seconds to enjoy the view before passing out.

Do **alcohol and winter** mix?

Alcohol and winter do not mix, at least when vehicles are involved. In 1993–94, 21 snowmobilers died in Wisconsin alone. One-third of the fatalities were alcohol-related. So have your hot toddy *after* you go on the trails or ski slopes. Also, it is a myth that a swig of alcohol "warms" you up. Aside from the apparent warming sensation, alcohol actually makes your body more sensitive to cold (as well as impairing your judgment).

How did **panic about the weather** result in a tragedy?

In August 1994, a thunderstorm killed 143 people in Brazzaville, Congo. The tragedy was not caused by flood, lightning, or a tornado, but rather by panic. A large crowd attending a church service panicked at the onset of a cloudburst. In the resultant stampede the victims were trampled to death.

This tragic story highlights the need for families, companies, and schools to have a disaster plan in place for tornadoes, floods, hurricanes, lightning, and other serious weather conditions. Fear of the unknown causes panic. Knowing what to do when danger threatens can go a long way to preventing disasters.

What are some **weather safety fallacies**?

The Disaster Services unit of the American Red Cross and the National Weather Service have published a list of weather safety myths that seem to linger on, including:

Windows on the "lee" side of a hurricane should be opened. Not true. During a hurricane, winds and wind-borne debris can be blown from all directions. The best thing to do is securely cover your windows.

Windows should be opened in a tornado or a hurricane to equalize barometric pressure. Absolutely untrue. If a tornado threatens, forget the windows and seek shelter immediately. If a hurricane warning is issued, cover up all the windows before the storm strikes.

Drive at right angles to a tornado. No way. While this may be preferable to driving right at a twister, if you're close to a tornado, you shouldn't be driving at all.

Many people have been killed in cars trying to escape. Seek interior shelter immediately or get into a low-lying ditch to avoid blowing debris.

Put tape on windows to avoid hurricane wind damage. Tape may cause glass to form larger pieces when it breaks, but otherwise it is nearly useless. Instead, cover the window with sturdy plywood or storm shutters.

Include candles in your supplies for use after a storm. In fact, you should avoid using candles or any open flame under such conditions. In some cases more people have been killed by fires after the disaster than by the incident itself. Replace candles with more batteries for flashlights.

You are safe from a lightning strike if you are wearing rubbers or rubber-soled shoes. Not true. A lightning bolt that has just traveled five miles through the atmosphere is not going to be stopped by a thin layer of rubber.

Who is most likely to be killed by severe weather-related events?

Men aged 30–39 are the largest weather-related casualty group. In fact, men are the victims of almost 70 percent of all weather-related fatalities (excluding traffic accidents). And when it comes to lightning, men constitute 85 percent of the fatalities.

Is the air cleaner inside your home or outside?

Usually the air is cleaner outside. Many air pollutants are found at concentrations many times higher inside homes and offices than in the outside air. The Environmental Protection Agency (EPA) found that of 11 common pollutants, concentrations were typically two to five times higher inside than outside, with some being 20 to 70 times more concentrated inside. One exception is ozone, a key constituent of Los Angeles–type smog, which is generally at 10 to 50 percent less inside than outside (unless some electrical device is sparking and creating indoor ozone).

Vast sums of money are spent to clean up the outdoor air, though Americans spend between 80 and 90 percent of their time indoors. Thus, over a lifetime, more than 80 percent of the breaths we take are of indoor air. The EPA has declared indoor air pollution one of the nation's top five environmental problems.

What are the most common symptoms of poor indoor air quality?

Eye irritation, dry throat, headache, fatigue, sinus congestion, skin irritation, shortness of breath, coughing, dizziness, nausea, sneezing, and nose irritation are the most
common symptoms. Of course, many viruses and bacteria produce the same symp-

toms. But if these symptoms repeatedly occur when inside a building and abate after one leaves, that is a sign of potential bad indoor air quality.

What causes indoor air pollution and "sick buildings"?

When does a building become "sick" due to bad indoor air quality? By definition a building is considered sick if after spending several hours in the building, more than one in five of the occupants report various symptoms including headaches, nausea, dizziness, sneezing, fatigue, or respiratory problems, which are relieved only when people go into the clean outside air.

Government studies have determined that there are various sources of sick buildings (which include homes, factories, and offices). Over 50 percent of the problems arise from inadequate ventilation, while 17 percent are a result of dust and vapors released inside the building. Outside pollutants trapped inside account for 10 percent, while bacteria molds and fungi produce 5 percent of the problems. Some 3 percent come from chemical fumes from furnishings, building materials, and chemical products.

Many newer buildings, in the effort to be energy efficient, have insulated to the point where insufficient fresh air enters. Thus, whatever pollutants are being emitted in the structure are not diluted by fresh air, but accumulate to ever-higher levels.

The EPA spends $500 million per year on studying outdoor air pollution and $13 million on indoor air pollution. Yet one in six of the four million commercial buildings in the nation have such poor air quality that 20 percent or more of the workers in them suffer persistent health problems. Ironically, several years ago the EPA was found to be headquartered in a sick building.

What is the most common particulate in indoor air?

The most common particulate floating around in the air of a typical home is not dust or cigarette smoke or cooking grease—it is dandruff from humans, dogs, and cats.

What are some sources of indoor air pollution?

There are literally hundreds of potential sources of pollutants in the indoor environment. Many of them come from the materials used in building construction. Formaldehyde has been widely used in many products such as plywood. It emits gasses over time and even in minute concentrations can make some sensitive individuals ill. Carbon monoxide, from cigarette smoke and poorly maintained combustion appliances and fireplaces, is a hazard. Auto exhaust from improperly vented attached garages can seep into living spaces. Dandruff from people and animals abounds. Fumes from poorly maintained gas appliances can be a problem for some. Mold and mildew in chronically wet spots in the house or basement can release spores that make some people ill. Even pollen from some house plants can be an issue. Fumes from various household chemicals can cause some chemically sensitive people to react strongly. And without proper ventilation, concentrations can build up to unacceptable levels. At the minimum, people should remember to use ventilation devices over such appliances as stoves and clothes dryers.

What is **radon**?

Radon is a radioactive gas that, when inhaled, can cause lung cancer. It is estimated that as many as 20,000 deaths from lung cancer occur in the United States each year due to long-term exposure to radon. But before you blame your local nuclear power plant for radiation pollution, note that radon is a naturally occurring gas. It recurs from the natural radioactive decay of uranium and radium in the soil. It is generally found at very low concentrations in the atmosphere, but it can seep into the basements and lower floors of buildings and accumulate to dangerously high levels.

Should you have your **home tested for radon**?

Yes, especially if you have a basement, homeowners should seriously consider testing for radon. More often than not problems that might be found can be fixed relatively simply and inexpensively. The solution can often be as simple as sealing basement cracks or adjusting ventilation devices. For those with more serious levels there are sub-slab ventilation devices that can draw radon-rich air from beneath the basement slab and vent it directly to outside air, preventing its infiltration into the house. Testing generally costs less than $60 and is well worth the peace of mind if the report comes back negative, or the avoidance of potential harm if there is a problem and it is found early. Do not assume that because your neighbor's radon report was negative, yours would be to. The values of radon can vary substantially from house to house on the same block.

Schools, just like other buildings, can collect radioactive radon gas in quantities thought to promote lung cancer after extended periods of exposure. One school district in Colorado found that over 50 percent of its schools had one or more rooms in which radon level exceeded EPA guidelines.

Are you safe from radon in **high rise buildings**?

Radon is usually thought of as a problem in basements or the lowest floors of buildings, due to its origin in the soil below the buildings. Yet high values of radon have been found on the upper floors of 25-story high-rise towers. Why? Because in many such buildings the heating, ventilating, and air-conditioning systems are usually located in the basement, where they ingest the radon and circulate it throughout the building, which often is nearly hermetically sealed.

What do **dust mites** have to do with indoor air pollution?

Everything. The dust mite, not even discovered until about 30 years ago, lives just about anywhere that humans do because people are their food source. More specifically, the hundreds of thousands of minute skin flakes that we each shed daily sift down into our bedding, wall-to-wall carpeting and sofa upholstery—where myriad hungry dust mites live and feast. In a typical home, it would not be unusual to find one or two million dust mites. The dust mites themselves are not a problem for humans, but their excretions are. Twenty times a day they excrete, and that adds up to about 1.2 billion dust mite waste particles going into the home environment each month. Too small to be seen, they are highly respirable, meaning they get into your nose and lungs. Allergists suspect that perhaps as many as half of all asthma attacks are allergic reactions to dust mite waste.

What can you do about **dust mite waste** in the air you breathe?

There are three things one must do to improve indoor air quality: control the source, ventilate, and filter the air. This holds true for dust mite waste products and other undesirable airborne contaminants.

For dust mite source control, it helps to wash your linens in hot (more than 130°F) water. Cold-water washing is just like a day at the beach for dust mites. Using plastic pillow and mattress covers helps keep them down. Removing wall-to-wall carpeting in favor of stone or wood eliminates one of their favorite nesting places. But you can't get rid of them all. Ventilation can be accomplished by opening windows, when weather permits, or installing one of the new class of energy efficient ventilation systems. Filtration of airborne particulate is best accomplished by using electronic air cleaners. These can be installed within ducts for whole house air cleaning. If that is not an option, individual rooms can be spot cleaned using portable units.

What is the **Pigpen effect**?

Indoor air pollution is a major health problem. And now we have the "Pigpen effect"—named after the character in the *Peanuts* cartoon. Researchers have shown that people actually stir up their own personal clouds of dust and particles wherever they go, as they shuffle across carpets, plop into overstuffed chairs, and bounce into their beds.

Do plants help alleviate indoor air pollution?

Particulate pollutants can be controlled by air filters, but gaseous pollutants, which include formaldehyde, benzene, and tricholoroethylene, have been tougher to control. Now NASA research suggests many common house plants are quite effective at cleaning noxious gases out of the air. Included in the list are philodendron, spider plant, golden pothos, several types of dracaena, chrysanthemums, ficus, and English ivy. Of course, some house plants also produce pollens to which some may be allergic. The pots in which the plants live can also be a source of mold and spores. Since many people tend to overwater their plants, mold is frequently found lining the sides of clay pots. Switching to plastic or glass containers, or watering more moderately, can help eliminate this problem.

Are **parakeets** a source of indoor air pollution?

Indoor air pollution is a growing problem. Increasingly air-tight homes are trapping pollutants at ever-higher levels. The sources are as varied as cigarette smoke, dust mites, and even parakeets. One recent study suggests that having a pet bird for more than 10 years may double or triple your risk of lung cancer. Airborne yeast bacteria that breaks down bird waste and particles of bird feathers can cause lung irritation, rather like asbestos. Keeping your bird's cage clean and airing out the room can help.

How much **fresh air** should you bring into your home?

Every building, except maybe the Biosphere II in the Arizona desert, exchanges air with the outside. During both the heating and air-conditioning seasons, when windows are typically closed tight, there should ideally be one exchange of fresh air about every three hours to avoid a build-up of pollutants. Many older, highly "leaky" homes from the era before high-tech insulation and double-paned windows might have one exchange of fresh air every hour—along with huge energy bills. Some modern "energy efficient" buildings, by contrast, certainly minimize energy loss by having perhaps one exchange of fresh air in ten hours. But they are also trapping any pollutants that have been emitted into the structure. All other factors being equal, the modern, almost hermetically sealed home or office would thus have ten times the indoor pollutant levels of an older structure.

How can you **insulate and ventilate** your home at the same time?

It's no wonder consumers are getting a little puzzled. First, in the decades after the several energy crises of the 1970s, energy conservation and insulation were the bywords. And builders and remodelers responded with dramatic improvements in energy efficiency. Now we are told that we need to get more fresh air back into our buildings.

A new technology does exist to allow you to have it both ways. Devices variously called air-to-air heat exchangers or energy recovery ventilators now allow building and homeowners to bring in plenty of fresh air, while capturing most of the energy dollars from the building's exhaust air. The heated (or cooled) and polluted indoor air is vented through a rotating disk constructed of a "modern, space-age material," which captures up to 85 percent of the energy and transfers it to the incoming clean air stream from the outdoors. Using the device allows one to have the equivalent of an open window without throwing energy dollars out of it, and at the same time preventing bugs, burglars, and street noise from entering.

Does **high atmospheric pressure** have a positive effect on humans?

Some studies suggest people tend to work better, eat more, and sleep more soundly when high atmospheric pressure is indicated on the barometer.

How deadly is a **parked car**?

Temperatures inside parked cars left in the sun with their windows up can peak at over 140°F. While it may seem hard to believe, every year many pets and even some small children die as a result of being left in such cars. During an early season heat wave in southern Wisconsin in May 1991, a 15-month-old baby was locked in a parked car for several hours and died from heat exhaustion. It's never a good idea to leave your family pet, much less a child, in a parked car on a sunny day with the windows rolled up, even just to briefly run inside a store.

How dangerous is **secondhand cigarette smoke**?

It is now fairly certain that between 10,000 and 40,000 Americans die every year from cardiac problems due to the impact of breathing secondhand smoke. Smoking is more than just an individual decision. Los Angeles smog is better for you than being in a room filled with smokers.

Inhaling cigarette smoke indoors, whether primary or secondhand, can magnify the unhealthy effects of indoor air pollutants. New research suggests that breathing **311** tobacco smoke may pose an even greater risk for sudden infant death syndrome (SIDS) than previously thought. Keeping tobacco smoke away from babies could reduce the death rate from SIDS by up to two-thirds, according to British medical researchers.

What is the "indoor greenhouse effect"?

Scientists are concerned with rising levels of carbon dioxide (CO_2) outside. But air measurements made inside rooms crowded with people in modern office buildings often show CO_2 levels that are 300 to 800 percent higher than in the air outside. The reason for the high levels inside: people exhaling CO_2 in a building in which there is inadequate ventilation.

Can **climate change** impact disease?

There have been many concerns raised about the impact of so-called global warming due to the greenhouse effect. One potential series of problems arising from such a climate change is now being taken more seriously, namely the effect upon disease. Change in both short- and long-term weather patterns can allow many opportunistic pests to thrive, ranging from rodents (such as those carrying the deadly hanta virus) to bacteria, protozoa, viruses, and insects. In 1994, after a long Indian monsoon, 90 consecutive days of 100°F heat drove rats into cities. In Surat they carried the bubonic plague, which killed 63 people, caused a major panic, and cost the nation some $2 billion.

Many diseases and their transmission are very sensitive to climate change. A 1991 cholera epidemic in Peru took 5,000 lives and is suspected to have been triggered by the unusually warm ocean waters associated with El Niño. Warmer water temperatures are also linked to cholera outbreaks in Bangladesh. The extremely warm summer of 1988 allowed the mosquito species carrying dengue fever to migrate up mountain slopes in Mexico from 3,000 feet to 5,500 feet, therefore exposing far more people. The outbreak spread as far north as Texas. Overall more than 140,000 people in the Americas were infected and 4,000 died. Malaria rarely thrives in regions with an annual temperature of less than 60°F. Today about 45 percent of the Earth has malaria-conducive temperatures. If the most dramatic predictions of global warming were to materialize, up to 60 percent of the planet would be subject to malaria by the twenty-second century.

How can you **heat your house** with a few glasses of water?

During the winter heating season, a house heated to 80°F with a 20 percent humidity feels comfortable to many. Yet by increasing the indoor humidity to just 40 percent, the same house feels equally comfortable at 72°F. A few pints of water per day in a humidifier can lower your energy bills. Your wood furniture is also likely to last a lot longer as it will not dry out so rapidly.

What do **lake breezes** have to do with spiders?

In the summer of 1991, Chicago lake shore high-rise residents had to contend with an invasion of spiders. These arachnids, or *araneus serricatus,* may actually "fly" in across Lake Michigan. They are believed to spin little parachutes and waft across the lake on brisk easterly lake breezes, coming to rest on the tall buildings. The wind also blows other smaller insects into the spiders' new webs, providing a steady diet that enables some spiders to grow to be an inch across.

How can you escape an **undertow and rip current**?

A rip current, sometimes also caused a riptide, is a strong current of surface water of short duration flowing seaward from the shore. It marks the point of return to the sea of the water piled up by the actions of wind and waves. It often appears (from above) as a narrow band of agitated water. The strong forces of the flowing water can suck a swimmer out to sea. When caught in such a strong riptide undertow, swimmers feel helpless and often panic as they struggle to get back to shore. The swimmers need to remember the current is only a few tens of feet across. Therefore don't try to swim inwards toward shore, against the current. Rather, swim parallel to the beach. Very soon you will be out of the swift current and able to swim normally. Each year many swimmers drown when caught in such undertows. The best protection is not to go swimming during rough seas when warnings for high surf have been posted.

How do **insects** respond to weather?

Before air conditioning, many people would travel to the cooler mountains to escape summer's heat. Some insects do the same thing. On the Colorado High Plains, each spring harmless but pesky miller moths—uncountable but seemingly trillions strong—drift from their hatching grounds in eastern Colorado and Kansas westward into the cooler Rockies. The annual migration in 1991 was so heavy that moths clogged mechanical equipment, knocked radio stations off the air, and made some public buildings unusable for several days. Three residents of Colorado had to have miller moths surgically removed from their ears.

Insects congregate in regions with rising air motions, such as sea breeze fronts. Birds also know that this is where the insects they dine on are located, so they tend to flock to the same region. Radar operators know this since both the bugs and birds provide large enough targets that precipitation-free fronts can be tracked using these biological radar reflectors.

WEATHER FORECASTING

What have great minds said about the prediction of atmospheric behavior?

Those who struggle with the mind-boggling complexity of the behavior of air and water in the atmosphere and the oceans—meteorologists, oceanographers, and hydrologists—can take solace from this insight of Galileo many centuries ago: "I can foretell the way of celestial bodies, but can know nothing about the movement of a small drop of water." One of the first problems Albert Einstein tackled after he received his doctorate degree was that of turbulence, which remains to this day one of the fundamental challenges of atmospheric sciences.

Does the **moon** affect the weather?

The phases of the moon do appear to have some impact on terrestrial weather. Statistical analyses have suggested that the moon's phase was slightly correlated to such conditions as the number of thunderstorms, air pressure changes, cloudiness, and even the concentration of ice nuclei. The reason for this connection is not at all clear. Recent work at Arizona State University has found the Earth's lower atmosphere is warmest during the full moon. The increase is only $0.02°C$, but this discovery may help to unravel the lunar influence puzzle.

How accurate is the *Farmer's Almanac*?

With respect to their weather forecasts, accuracy is not the almanacs' strong suit. **315**

However, the various almanacs published each year do have some good recipes for brownies and fudge. Their tide tables are dead on. The list of national holidays is handy. They also seem to have the astronomical predictions under control.

In 1504, Christopher Columbus was having trouble talking the natives of the island of Jamaica into providing his crew with food and supplies. So, after consulting his early "almanac," he told natives if they did not cooperate, he would take the moon away. The lunar eclipse that followed greatly impressed the natives.

What is **NOAA Weather Radio?**

Would you like to have a continuous source of local weather forecasts and data, along with instant notification of hazardous weather warnings? The National Weather Service's NOAA (National Oceanic and Atmospheric Administration) Weather Radio continuously broadcasts to most of the United States. Inexpensive weather radio receivers, many battery powered and with a "tone alert" feature, are widely available in consumer electronic stores. Whenever a watch or warning for severe weather is issued by your local National Weather Office, an audio alert will let you know there is severe weather brewing.

Where can you get **computer access** to weather data and forecasts?

If you would like access to real-time weather data, forecasts, or warnings using your home computer and modem, there are several cost-effective databases readily available from commercial weather companies, many through on-line services. Check the computer magazines or *Weatherwise* magazine for advertisements of these services.

What sources of weather data and forecasts are available through the **Internet?**

It has been estimated that there are already 1,600 World Wide Web sites dealing with weather and related issues. More and more forecasts and data, including NEXRAD radar, GOES satellite, and surface and upper analyses, are easily obtained via modem. You would be well served to have a fast modem since many of the images can otherwise take quite a while to download. The following list offers an introduction to weather sites on the Internet:

Beginner's sites

Weather Channel	www.weather.com
Weather Underground	cirrus.sprl.umich.edu/wxnet/

General sites

www.nws.noaa.gov	

National Hurricane Center	www.nhc.noaa.gov
National Aeronautics and	
Space Administration (NASA)	www.nasa.gov
	explorer.arc.nasa.gov/pub/weather/
Ilana Stern's FAQ, Part 2	www.scd.ucar.edu/dss/faq

Advanced user's sites

Global Atmospherics	www.gds.com
Storm Chaser's Homepage	taiga.geog.uiuc.edu/chaser/chaser2.html
Research Applications Program,	
National Center for Atmospheric Research	www.rap.ucar.edu
Intellicast	www.intellicast.com
WXP: The Weather Processor	wxp.atms.purdue.edu
Ohio State University	
Atmospheric Sciences Program Homepage	twister.sbs.ohio-state.edu

When did the **first attempts** at weather forecasting begin?

Probably ever since the first humans walked the Earth they have been trying to predict the weather. Most likely many shamans and witch doctors (who were often astute observers of the natural world) attempted foretelling the state of the sky. Scientific forecasting starts with observations. The Babylonians recorded wind directions on an eight-point compass as early as 900 B.C.E. Rainfall records were known to be kept by the Greeks in the fifth century B.C.E. and in India a century later. The earliest known systematic climatology was attempted by Oxford scholar William Merle, who kept weather records from 1337 to 1344. Some early, quasi-scientific attempts at forecasting were made in 1692 in the English weekly newspaper, *A Collection for the Improvement of Husbandry and Trade,* which provided readers with barometric readings and wind reports for that week in the previous year. In 1771, a journal called the *Monthly Weather Paper* was devoted to English weather prediction. With the arrival of modern telegraphy to collect and disseminate weather information, the British Meteorological Office began issuing daily weather predictions in 1861. The first known media weather forecast was on 3 January 1921 from the University of Wisconsin's experimental radio station in Madison, Wisconsin.

When was the **National Weather Service (NWS)** started?

In 1870, the U.S. Congress created a national weather service operated under the U.S. Army Signal Corps. In 1891, the organization was transferred to the Department of Agriculture and the name was changed to the U.S. Weather Bureau. The first "chief weatherman" was Cleveland Abbe of the Cincinnati Observatory. In 1940, the Weather Bureau became part of the Department of Commerce, where it remains today, though **317**

A space shuttle view of a major East Coast storm.

it is now officially called the National Weather Service. The grand old National Weather Service will soon mark its 106th birthday as a civilian agency.

Who issued the first **official storm warning** in the United States?

On 8 November 1870, the first official, public storm warning was issued in the United States for a storm on the Great Lakes. It was issued by the U.S. Signal Corps, specifically by Professor Increase Lapham of Milwaukee.

Where is research conducted on forecasting **severe storms and tornadoes**?

Oklahoma has a lot of extreme weather, with severe storms and tornadoes leading the list, so it's not by chance that Norman, Oklahoma, is the home of the U.S. government's National Severe Storms Laboratory. This group conducts pioneering research on tornadoes and other aspects of severe local storms, often in connection with the University of Oklahoma. The Storm Prediction Center, which issues tornado and severe thunderstorm watches for the nation, has recently moved to Norman from Kansas City (where it was called the National Severe Storms Forecast Center). Oklahoma has a lot of weather data: an automated weather station has been established in each county of the state, the first such statewide system in the nation. But nationwide, many universities and other government laboratories also conduct severe storm studies, as do private research companies under contract to government agencies, such as NASA and the National Science Foundation.

Who formulated the **polar front** theory?

During World War I, Norwegian meteorologists, working in a neutral country, were largely cut off from weather information due to restrictions imposed by the warring nations of Europe. Norway established a dense network of stations within their own country, and a dedicated group of scientists, now known as the Bergen School, studied these data intensely. From their analyses emerged the concept of air masses and fronts. These developments are particularly indebted to the father and son team of Vilhelm and Jacob Bjerknes. Given the times, the fronts were so named because they represented the battlegrounds between advancing and retreating armies of air (air masses). Instabilities on the polar front, the demarcation between polar and tropical air, were studied, and the basic theory of mid-latitude storms was developed. This theory became a cornerstone of modern meteorology and forecasting.

When did **cold and warm fronts** first appear on U.S. weather maps?

The now obvious concept of air masses and their boundaries (warm and cold fronts) took almost two decades to be accepted by American meteorologists after the discovery during World War I by Norwegian meteorologists. It was not until the mid 1930s that the red and blue barbed lines now so familiar to television weathercast viewers were first allowed to be drawn on official U.S. weather maps.

What is **numerical weather prediction**?

The discovery of air masses, fronts, and instabilities on the polar front (low-pressure **319** systems) was the result of analyses of weather data by the Bergen School scientists in Norway who, because their background was in fluid dynamics, understood that there was in effect an "equation" for weather. Air is a fluid and as such obeys the fundamental physical laws of fluids, called the hydrodynamical equations. These include the Newtonian equations of motion, the thermodynamic energy equation, the equation of mass conservation, the Boyle-Charles equation of state, and the equations of continuity. In theory, if one knows the current state of the atmosphere, then these equations can be used to predict its evolution over time. In other words, one can scientifically predict the future state of the atmosphere—forecast the weather.

Who first proposed **numerical weather prediction**?

The fundamental notions of numerical weather prediction were first stated by Vilhelm Bjerknes in 1904. In 1922, a rather eccentric British scientist named Lewis F. Richardson formally proposed in a book that weather could be predicted by solving the "equations of atmospheric motion," ideas he worked out while serving as an ambulance driver during World War I. He soon realized that the amount of calculation would be formidable. He proposed a weather prediction center in which a giant circular amphitheater would contain some 26,000 accountants equipped with calculators who would make their additions and subtractions as commanded by a sort of conductor. Richardson's first attempts at weather prediction bombed because the method predicted pressure changes far larger than any that had ever been observed. This was later found to be due to his failure to understand some basic problems in numerical methods. His idea—which was basically sound—was thus dismissed and forgotten for over 20 years.

When were the **first practical attempts** at numerical weather prediction?

The electronic computer was conceived in the 1940s, when mathematician John von Neumann developed the prototype of the stored program machine, the forerunner of today's modern machines. He turned his interests to numerical weather prediction. In 1946, at Princeton's Institute for Advanced Study, a meteorology project was formed. Meteorologist Jule Charney became interested in the problem of numerical weather prediction, and after discovering the reasons why Lewis Richardson's first attempts at numerical weather prediction 25 years earlier had failed so miserably, was able to formulate equations that could be solved on a modern digital computer. The first successful numerical prediction of weather was made in April 1950, using the ENIAC computer at Maryland's Aberdeen Proving Ground. Within several years research groups worldwide were experimenting with "weather by the numbers." The first operational weather predictions, using an IBM 701 computer, were begun in May 1955 in a joint Air Force, Navy, and Weather Bureau project.

When was the first official tornado warning issued?

Fearing panic, the government long decreed that it was illegal to forecast a tornado in the United States. This continued until after World War II, when damage to several military installations by devastating tornadoes in the early 1950s prompted a crash program in scientific severe storm and tornado prediction. The first public tornado predictions were issued by what was then called the U.S. Weather Bureau in 1953.

What role have **computers** played in numerical weather prediction?

Numerical weather prediction has advanced greatly in four decades, in part due to the spectacular growth in speed and memory of digital computers. Meteorologists have invariably required the biggest and fastest machines to do their job. One of the very first computers dedicated to atmospheric science research was the IBM 1620—it could perform about 1,000 additions per second. Today's larger machines can race along at billions of additions per second.

With the high speed number crunching of numerical weather prediction, atmospheric scientists use and generate a lot of data. And that data has to be stored someplace. The National Center for Atmospheric Research, located in Colorado, estimates that its scientists alone maintain computer files totaling 29 terrabytes—29,000,000,000,000 bytes—in size. That is roughly the capacity of a million or so PC hard drives.

Who was **C. G. Rossby**?

Carl-Gustaf Rossby was born in Sweden in 1898 and studied under Vilhelm Bjerknes at the Geophysical Institute in Bergen, Norway. After moving to the United States in the 1930s, he was instrumental in founding the Departments of Meteorology at the University of Chicago, the University of California at Los Angeles, and the Massachusetts Institute of Technology. He made fundamental contributions to understanding the giant waves in the middle atmosphere that mark the jet stream and control the propagation of surface high- and low-pressure systems. Understanding these Rossby waves was a major breakthrough in the scientific understanding of atmospheric motions. He may be the only meteorologist ever to have graced the cover of *Time* magazine.

What are **mesoscale and regional models**?

The atmosphere has many scales of motion, ranging from the planetary scale Rossby waves to individual puffy cumulus clouds. Atmospheric scientists have investigated these various phenomena to the extent that they now can be described using computer simulations. For day-to-day routine weather forecasting, the National Weather Service uses computer models that simulate features such as jets streams, fronts, cyclones, anticyclones, and their attendant weather. They simulate weather on a global scale. But computational limitations prevent them from forecasting many of the details, such as whether a sea breeze will form over a given part of the coastline or a thunderstorm may pop up over the Kennedy Space Center. For specialized smaller scale weather forecasts, many organizations run what are called regional or mesoscale models. While they may cover smaller geographical regions, their much finer resolution also requires substantial computation power. Powerful computer work stations are now being pressed into service to predict local air pollution, the path of potential toxic releases from chemical or nuclear plants, launch weather at the Kennedy Space Center, and severe storm formation, to name a few.

What is **NOAA**?

The National Weather Service is part of NOAA—the National Oceanic and Atmospheric Administration, which in turn is part of the Department of Commerce. NOAA also operates the nation's environmental satellite system and various research laboratories.

Are there **military weather services**?

Both the U.S. Air Force and the Navy operate weather services to meet their specific needs. The Air Force operations are largely centralized at the Global Weather Center at Offutt Air Force Base near Omaha, Nebraska. The Navy conducts much of its activities from the Fleet Numerical Weather Center in Monterey, California. The military services make their own surface and upper air weather observations and operate conventional and Doppler weather radars worldwide. There are also special military weather satellites.

What special forecasts were made for the **1996 Olympics**?

The city of Atlanta has suffered through some major weather disasters in recent years, with the three most costly weather-related disasters being the March 1993 blizzard, the 1994 floods, and Tropical Storm Opal in 1995. Hurricane Bertha in July 1996 briefly threatened Georgia and some of the Olympic venues. In response to the need for extremely accurate forecasts, especially of hazards such as lightning and wind storms, the National Weather Service established a state-of-the-art Olympic weather center. Some of the most advanced meteorological tools ever assembled supported the Centennial Olympiad.

What is St. Swithin's Day?

According to ancient English legend, if it rains on 15 July, St. Swithin's Day, it will rain for the next 40 days. The origin of the tale dates back to the rainy period that followed the removal of the medieval cleric's remains from his humble grave to a more sumptuous resting place after his canonization. It apparently took many tries before the weather let up enough to allow the remains to be reburied.

What effect does the Gulf Stream have on U.S. weather?

One of the fastest flowing ocean currents in the world is the Gulf Stream as it passes the eastern coast of Florida. It transports huge amounts of warm water in a narrow band up the eastern seaboard. At night, especially during winter, thunderstorms sometimes form over this warm band of water. In the winter, cold air streaming off the continent meets the Gulf Stream warm air, and things begin to happen. This is the breeding ground for Nor'Easters, which routinely pound the mid-Atlantic states and New England with winter snows. Using satellites and research aircraft, meteorologists have begun studying a class of storms that form initially over the warm North Atlantic waters in winter. Called "bombs," these storms behave in some ways like hurricanes. Their development can be explosive, with central pressures dropping an incredible 1.80 inches in less than 24 hours.

What is an upper air chart?

The jet stream is of great importance to weather forecasting. The "polar jet" actually separates the polar and tropical air masses of the planet. Mid-latitude storms form along this boundary and generally travel in the direction of the jet stream winds aloft. Upper air charts help define the global wind and temperature patterns at various pressure levels in the atmosphere. The most widely used chart is the 500-millibar chart. It defines the height of a constant 500-millibar pressure surface, which is located around 18,000 feet. Other widely used upper air charts are the 850, 700-, 300-, 250-, and 200-millibar charts.

What is a sounding?

Each day at hundreds of sites worldwide, weather services launch rawinsonde balloons to profile the winds, temperature, humidity, and pressure well into the stratosphere.

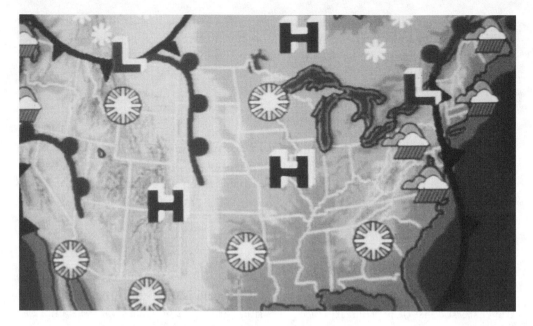

The television weather map, now computer-generated and available 24 hours a day.

These data, called soundings, are used to construct upper air maps and are also plotted on special charts called thermodynamic diagrams. Various techniques have been developed to help ascertain the stability of the atmosphere. These help in the prediction of the type and severity of thunderstorms.

What can you tell from the **television weather map**?

More information can be gleaned from a television weather map than you might expect. While graphical styles vary widely from station to station, there are some general similarities. On the television weather map (in the Northern Hemisphere) the wind circulates clockwise around the H (high-pressure system) and counterclockwise around the L (low-pressure system). The warm and cold fronts separate air masses with different temperature and humidity characteristics. The triangles and bumps drawn on the fronts are there for a reason. The triangles indicate a *cold* front. The semicircles indicate a *warm* front. They both point in the direction toward which the new air mass is moving. Alternating triangles and semicircles indicate a stationary front where the cold and warm air masses are in balance and thus barely move over the course of a day or so.

If isobars, the lines connecting points with the same constant barometer reading (as reduced to sea level), are used on your local television weather map, you can also deduce wind speed and direction. The surface layer wind directions are roughly parallel to the isobars and the wind speed increases as the spacing between the isobars becomes smaller. This is especially true when large-scale weather systems (major winter storms and strong high-pressure systems) dominate your region's weather.

Who prepares the weather forecasts you see on television?

Most of the basic weather forecasting support in the United States originates with the National Weather Service (NWS). Many stations simply use the information provided by the local NWS office. Many stations, however, hire professional meteorologists or use private forecasting companies to interpret the basic weather data and present the information to viewers. Though starting with the same observations and computer forecast model results, government and private forecasters may come up with a somewhat different interpretation of their meaning and thus the forecasts can vary somewhat from forecaster to forecaster. It is rather like several doctors examining an X-ray or lab report and making slightly different diagnoses.

Who may issue **severe weather watches and warnings** in the United States?

By mutual agreement, the National Weather Service has sole responsibility for issuing public severe weather and hurricane watches and warnings. Responsible private meteorologists will not attempt to countermand official watches and warnings. While a television or radio forecaster may augment the information or perhaps explain in depth why a warning was issued, it is almost universally agreed that there be only one voice in formulating weather watches and warnings to the public, and that is the National Weather Service.

What is the **AMS Seal of Approval**?

Do the television weathercasters on your nightly news really know what they are talking about? Well, meteorology is an imperfect science, but 85 percent of forecasts prepared by professional meteorologists are essentially correct. And if the weathercaster has been awarded the "Seal of Approval" by the American Meteorological Society (AMS), you can rest assured that he or she is well equipped to provide you with the best information available to the science today.

Weather forecasters take a lot of heat when they miss a prediction. Forecasting has improved considerably over time, though. The ancient Greek philosopher Theophrastus wrote a book that included many rules for forecasting weather, for example, the howling of a wolf indicates a storm within three days.

What is the **World Meteorological Organization**?

The World Meteorological Organization (WMO) coordinates the meteorological activities of the world's nations. The WMO is a United Nations agency that is headquartered in Switzerland and provides guidance to the Earth's various national weather services in terms of how to take weather observations, how to communicate them, and how to disseminate weather warnings and reports worldwide.

Who was **James P. Espy**?

The first "official" government meteorologist in the United States was James P. Espy, who received his appointment from Congress in 1842. His most famous contribution is his 1841 publication *The Philosophy of Storms,* in which he describes the basic characteristics of low-pressure systems (cyclones). While he did not understand the role played by fronts and air masses in mid-latitude storms, his ideas were vital to obtaining a clearer understanding of hurricanes.

When was the **anticyclone** discovered?

The concept of high-pressure cells, the opposite of cyclones and thus called anticyclones, was first proposed by Sir Francis Galton in England in the 1860s. These high-pressure systems do much to control the movement of large air masses across the continents.

What is the **Coriolis effect**?

One of the keys to understanding weather maps is to note that winds blow approximately parallel to isobars, the lines of equal air pressure around low- and high-pressure centers. But why doesn't the air just flow directly from the high to the low pressure? The rotation of the Earth causes air flow (and artillery shells and anything else that moves) to always veer to the right in the Northern Hemisphere (and to the left in the Southern Hemisphere). This effect was first described by Gaspard Coriolis in France around 1835. American scientist William Ferrel elaborated on the theory in the 1850s, showing the effect of the Earth's rotation on winds and ocean currents.

What are the names of some **local wind systems** from around the world?

The winds of the Earth have been given many names by many cultures in many places. Some are fairly well known. The *sirocco* is a model of an automobile as well as a hot, dry, and often dust-laden Mediterranean wind from the south or southeast that streams out of the deserts of North Africa or Saudi Arabia. But many names are a bit more obscure to most Americans.

If you are ever vacationing on Lake Toba in northern Sumatra, you could really impress the locals by displaying your knowledge of their names for some of the regional wind patterns, which include: *bolon, aloegoe, dahatoe, loehis, nirta, saoet, si giring giring, siroeang,* and the *tamboen.*

What do the following have in common: *lavanter, vendaval, ghibli, khamsin, estesians, vendarro, bora, tramontana, ponente, libecio, maestrale, marin, mistral, and graecale*? This is not the roster of a European soccer team, but the names of some of the well-known wind systems that occur in the Mediterranean region.

Some other names for local wind systems around the planet include: *austru, bentu, bhoot, bohorok, brubu, chubasco, datoo, elephanta, gharbi, ghibli, haboob, imbat, knik, oe, papagayo, sharki, sno, taku, williwaw,* and *zonda.* In Tunisia, a forecast of "hot today and *chili*" would make perfect sense to the locals. *Chili* is the name for a hot wind blowing from the North African desert. The forecast for Sydney, Australia, might just read: "sunny and hot today, with a brick fielder." The latter term is a local name for a dusty dry wind that blows from the south across the old brick fields outside the city. In Scotland, a waff is a slight puff of air or a gentle breeze. In Mexico you might encounter the *Tehuantepecer,* which is the name for a tropical wind that gusts offshore from the western Mexico and Guatemala coastal mountains, often reaching high speeds and extending a hundred miles out to sea.

If you were standing near the Pacific coastline of South America and were hit by a *puelche*, it would mean you were in an easterly breeze that had crossed the Andes mountain ranges. The term is also sometimes used for a land breeze. The opposite west wind, the sea breeze, is sometimes termed the *virazon.*

Haboob is a name given in parts of Africa to a strong wind causing a dust storm or a sandstorm. The wall of advancing grit can sometimes reach over 3,000 feet high and have wind gusts 30 to 60 mph.

A "cat's paw" is a light breeze affecting a small area, as would cause a patch of ripples on the surface of a still water surface. This is not to be confused with a "cat's nose," a term used in England for a cool, northwest wind.

On Ellesmere Island in Canada they sometimes experience "cow storms," gales of wind that are so strong they are reputed to "blow the horns off the cows."

What is Groundhog's Day?

All over the country on 2 February, groundhogs (woodchucks) are dragged from deep hibernation in their dens so it can be seen if they cast a shadow. The origins of this holiday are uncertain, but according to custom, in eleventh-century Scotland, on Candlemas Day (40 days after Christmas), if "the weather was fair and clear, there would be two winters in the year." How the groundhog got into the picture is unclear.

Punxsutawney, Pennsylvania, is the nation's Groundhog Day capital. In 1995, Punxsutawney's very own celebrity groundhog (named Punxsutawney Phil) "predicted" winter was on the wane. Within 48 hours his burrow and much of the Northeast was hit with one to two feet of snow and sub-zero wind chills.

What are some of the **local winds** in the United States?

The United States has its own collection of local winds with interesting names. For instance, the *knik* is a blustery wind that roars out of the southwest in the area around Palmer, Alaska. The *kona* is a warm and muggy southwesterly wind that sometimes interrupts the normal northeast trade winds in Hawaii. When in Hawaii be sure to experience the *mauka*. It's not a type of hula dance, or a tasty local fish, but a cool and refreshing breeze that descends from the mountain slopes during nighttime hours. Northeasters, also called Nor'Easters are strong winds associated with major winter coastal storms that usually bring heavy rains and sometimes blizzards to the mid-Atlantic and New England coastline. The *Santa Ana,* named after the city in California, is a hot, extremely dry wind that descends from the inland high deserts, roars through the mountain passes, and then spreads across the southern California coastal plain. It is often a major ingredient in the region's wildfire disasters. The *blue norther* marks the passage of an Arctic cold front into Texas.

When is **National Weatherperson's Day**?

When the hubbub about prognosticating rodents dies down after Groundhog's Day, the nation wildly celebrates its human forecasters. National Weatherperson's Day is 5 February.

Are **locust plagues** still a problem?

Locust plagues are not just a story from the Bible. Today, up to 20 percent of the Earth's land surface is subject to periodic invasions of the plant- and crop-destroying pests. The problem is especially acute from Africa to India and in parts of Australia. To monitor conditions suitable for locust outbreaks, weather satellites are now being routinely used with increasing success to spot the trouble before it gets out of hand.

How well was the **1993 Storm of the Century** forecasted?

The "storm of the century" in the winter of 1993 left 243 people dead from Cuba to Canada. The $1.6 billion in property losses were exceeded only by Hurricanes Hugo and Andrew. Snow fell as far south as the Florida panhandle and a record 13 inches buried Birmingham, Alabama. The enormous size of the Great Blizzard of March 1993 can be illustrated by noting the location of three of the higher recorded wind gusts: Mt. Washington, New Hampshire (144 mph), Flattop Mountain, North Carolina (101 mph), and Dry Tortugas, Florida (109 mph). This encompasses virtually the entire length of the eastern seaboard.

The good news is that the storm was extremely well forecasted up to several days in advance. Street crews were on early red alert. Airlines diverted airplanes so they would not be trapped in snowbound airports, saving millions and restoring service much earlier. Businesses were able to stock up on staples or sell perishable inventory as appropriate. The cost in terms of loss of life and property would have been far greater without the excellent forecasts from the National Weather Service.

How accurate are **seasonal hurricane forecasts**?

Colorado State University scientist Bill Gray has, for more than ten years, predicted the number of hurricane days and named tropical storms in the Atlantic using global indicators. These indicators include rainfall in western Africa, wind shear high above the Atlantic, and water temperature patterns such as those found with El Niño. And he's getting good at it. As an example, in the 1991 hurricane season, his group predicted 10 hurricane days (8.25 occurred) and a "destruction potential" of 25 units (versus 23 observed). More important, when Professor Gray predicted a near-record 1995 hurricane season, even he was reluctant to believe the numbers. But in the final tally, the 1995 Atlantic tropical storms were the second most numerous this century.

How accurate are **long range weather forecasts**?

Thirty-day forecasts of large-scale trends in temperature and precipitation are routinely prepared by the Climate Analysis Center of the National Weather Service. After thirty **329** years, these monthly outlooks are beginning to show some significant skill. They are far from perfect, but they are clearly better than relying on simple climatology (which would say that next January's weather will be just like the average of the last 100 Januarys). Seasonal (90-day) forecasts are less reliable but have also recently begun to exhibit improvements over pure climatology. The key to improving weather forecasting a season ahead may lie in understanding the oceans and their interactions with the atmosphere. It is becoming increasingly clear that changes in average water surface temperatures over large areas of the ocean can dramatically affect regional and hemispheric weather patterns. In the Pacific, El Niño is an especially dramatic example of this.

What is the **chaos theory**?

The chaos theory is now widely applied throughout science to explain erratic and unpredictable behavior in many natural and mathematical systems. Its principles were first uncovered in the 1960s by a meteorologist, Professor Edward Lorenz of the Massachusetts Institute of Technology. One of the limits on longer range weather forecasting may be the "butterfly effect." Lorenz is reputed to have asked: Could the flap of a butterfly's wings in Brazil spawn a tornado in Texas? Very small perturbations in the atmosphere can result over time in a storm developing differently. Currently the practical foreseeable limit to detailed numerical forecasting of daily weather is thought to be about two weeks.

What are **atmospheric teleconnections**?

Atmospheric teleconnections is a fancy term meaning that weather or ocean conditions in one part of the planet can influence events thousands of miles away. Starvation in Africa may be related to ocean water temperatures in the western Pacific. Scientists are not sure of the details, but statistics clearly show that periods of warm water—called El Niño—were strongly correlated to rainfall in Zimbabwe that resulted in an adequate corn harvest. But cold water temperatures can mean little rain in Africa and failure of the harvest. Many such atmospheric "teleconnections" appear to exist.

The 1982–83 El Niño event caused floods in Bolivia, Cuba, and the United States, hurricanes in Hawaii and Tahiti, and droughts and fires in Australia, India, Indonesia, and Mexico. The total economic impact was at least $8 billion.

How do **satellites** help industry?

Satellites save money. According to the National Oceanic and Atmospheric Administration (NOAA), the use of information supplied by weather satellites saves key industries such as agriculture, construction, shipping, fishing, utilities, and aviation more than $5 billion per year. Satellites can do more than just locate clouds. By tracking the motion of clouds, wind speeds can be inferred. New systems are being tested that can take vertical temperature and moisture soundings. Infrared sensors can routinely monitor sea surface temperatures, a development that has been a tremendous aid to fisherpersons who use these reports to help locate favorable harvesting areas.

What is "nowcasting"?

Nowcasting refers to forecasting over very short intervals of time, typically two hours or less. It involves detailed use of "real-time" data systems such as radar, satellites, and automated surface weather observation systems. With detailed analyses, forecasters can predict sudden and important changes in weather with considerable skill.

On the afternoon of 3 July 1993, the temperature was 58°F in Ft. Collins, Colorado, while 100 or so miles to the south it was 98°F. This looked like a cold front, and one that was screaming south driven by wind gusts up to 94 mph (straight line, non-thunderstorm winds). The wind blast was so intense that garden vegetables were literally seared to a dry crisp and killed. A forecaster performing analyses of conditions every 15 minutes (aided by a computer, of course) would have been able to tell the residents of Colorado Springs that they were about to get a rude shock that afternoon.

What is the **dry line**?

You have heard of cold fronts and warm fronts. But the "dry line?" During spring and summer in west Texas and Oklahoma, the dry line separates moist tropical air streaming out of the Gulf of Mexico from the bone-dry air of the desert Southwest. And since dry air is heavier than moist air of the same temperature, the dry line acts as a cold front—and triggers some of the most violent thunderstorms in the nation.

What is a **sea breeze**?

Sailors can often make a very good forecast by taking note of the daily sea breeze patterns in their area. A sea breeze is a local wind system that is caused by differences in heating between the land and water. The sun's energy heats up the land surface to very high temperatures, and the land in turn heats the overlying air. By contrast, the temperature of the water increases only very slightly due to its much greater heat capacity. The warm air over land rises and cooler air moves onshore to replace it. The sea breeze can often have a marked front, just like a miniature cold front. Sea breezes can push inland from less than a mile to over 100 miles, depending on the larger scale wind patterns. Especially along the Gulf and Florida coastlines, thunderstorms triggered by the sea breeze front are a major source of summertime precipitation. At night, the reverse process occurs, called the land breeze. The land areas become cooler than the warm ocean and the cool air drains offshore. Land breezes are much shallower and weaker than sea breezes. Lakes and broad rivers as small as 10 miles across can generate their own lake breezes on warm summer days when the prevailing winds are very light. **331**

What is **POP**?

POP is weatherperson lingo for probability of precipitation. Most forecasts are best expressed in terms of probability, and this is especially true of precipitation, for several reasons. First, there is always some degree of uncertainty in any forecast. Rare are the days when a forecaster can say with 100 percent certainty that it will or will not rain. Second, there is the problem that precipitation is highly variable in space. A forecast may be issued for a region of 25,000 square miles, but it would be unusual for a precipitation system to uniformly cover such large areas.

If a POP is given at 50 percent, this doesn't mean the forecaster is flipping a coin. First of all, if the climatological average of rain is 1 day in 4 (25 percent), the forecast means that rain is twice as likely as normal. Furthermore, it means that 5 days out of 10, precipitation (0.01 inches or more anytime during the 12-hour forecast period) will occur. If a 90 percent POP is issued for ten days, then the forecaster would show skill if on one of the ten days, no precipitation fell. Generally, precipitation is not mentioned in public weather forecasts if the POP is 20 percent or less.

How can **climatology** help plan a vacation?

Going to Disney World? During the summer season (June–September), almost two-thirds of the rain falls in afternoon thundershowers between 2 and 7 P.M. So plan to go on the rides in the morning and late evening if you want to keep dry. If you are an eclipse fan, use climatology to identify the eclipse's location along the path of totality with the highest chance of clear skies.

How accurate are **daily weather forecasts**?

While an unexpected rain shower soaking your picnic may make you question the skills of the local meteorologist, it must be acknowledged that weather forecasts are becoming more accurate. Today, predictions of general weather patterns some 5.5 days in the future are as accurate as those for 3.5 days made in 1977. The increasing sophistication of computerized weather prediction systems is a major reason for the improvement. And the public must learn how to make better use of forecast information. A 20 percent chance of showers doesn't mean a guarantee that it will not rain.

How accurate are **weather proverbs**?

If March comes in like a lion, it is supposed to go out like a lamb. That notion has about the same credibility as a theory that if the stock market comes in like a bull it will go out like a bear.

One of the more enduring legends is that the upcoming winter can be predicted by the width of the stripes on a woolly-bear caterpillar. If the brown bands are wide, says the superstition, a mild winter is coming up. Narrow bands mean cold. The American Museum of Natural History conducted a test of the woolly-bear's prognostic skills and it failed completely.

On the other hand, some weather proverbs do distill some weather wisdom. The adage that a ring around the sun or moon foretells rain is reasonable. Such optical phenomena are caused by high, icy cirriform clouds, which often precede precipitation-bearing low-pressure centers, especially during the winter months. Then there is the sailor's rhyme:

Red sky in the morning,
Is a sailor's sure warning;
Red sky at night
Is the sailor's delight.

Generally a storm advancing from the west can create a very red sunrise, while clearing evening skies as storms pull east can make for colorful sunsets. But if you were to base all your decisions regarding weather on this poem you would be right sometimes, but more often wrong.

What are the **Ice Saints**?

In parts of Europe, Saints Mamertus, Pancras, Savertius, and Gervais, whose feast-days fall between 11 and 13 May, are so named because those days became associated with spring frosts in certain regions.

What is the **world's toughest weather forecast**?

Some say that the Air Force team that makes forecasts in support of the U.S. space shuttle launches at the Kennedy Space Center may have the toughest forecasting job. Over a two-year period, 12 out of 16 launches were canceled or delayed because of thunderstorms, surface or upper air winds, and potential electrified cloud layers. The space shuttles are uniquely vulnerable to high-level winds and wind shears, clouds, precipitation, and atmospheric electrical effects. Apollo 12 was struck by lightning during launch, and the exhaust from the AC67 rocket carrying a $250 million satellite triggered its own lightning bolt. Though the ascending rocket survived the initial jolt, the electrical surge essentially reprogrammed the guidance computer, making the rocket want to try to fly sideways, something it could not do. Since these events, even greater care has been taken to avoid repeat accidents.

East central Florida has one of the highest thunderstorm frequencies in the nation. In the summer of 1991, scientists launched a major research effort to better understand the local thunderstorm problem, and have developed computerized systems to aid the local Air Force forecasters in their prognostications.

333

The sky was clear for this space shuttle launch, but thunderstorms frequent the Kennedy Space Center and make for many difficult weather forecasts for Air Force meteorologists.

Why do meteorologists need to know **geography**?

Most meteorological information is displayed in a variety of map projections and scales. It is essential to the forecasting process to be able to assimilate radar, satellite, and computer predictions, all of which may be displayed on different types of maps, and then integrate these images into a coherent picture in one's mind.

What are **marine wind warnings**?

Boaters should know the meanings of the various warnings used in marine weather forecasts. Small craft warnings are generally issued by the National Weather Service when impending winds up to 32 mph and related hazards will threaten boaters. Gale warnings are issued by the NWS when impending winds could reach between 32 and 54 mph. Storm and hurricane warnings are for winds above 54 mph and 74 mph, respectively.

What should be the **easiest city** to forecast in the weather?

Residents of many cities around the world claim "If you don't like our weather . . . wait five minutes." But this is not true for the residents of Quito, Ecuador, which may have the most constant climate of any major city. The mean temperature of each month varies little from 55°F, and precipitation is infrequent.

Is there is a difference between **partly sunny** and **partly cloudy**?

Yes. Partly sunny means it is mostly cloudy with a little sunshine and partly cloudy means it is mostly sunny with some clouds.

CLIMATE CHANGE

Is the **Earth's climate** changing?

Constantly. It was not so long ago that scientists believed the climate of our planet was more or less constant. The discovery of the ice ages confounded that notion. But even in the current interlude between ice ages, only in the last few decades has the evidence been accumulating to show that the Earth's climate is in a constant state of change. Today when the public hears about climate change, it is usually in the context of alterations due to human activities. The planet's atmosphere, however, has been fluctuating since long before Earth was populated by humans. One of the warmest eras ever on Earth occurred during the Cretaceous periods about 100 million years ago. Global temperatures averaged some 18°F warmer than they are today. Ocean levels rose and spilled over onto continents, splitting North America into two land masses.

What is **weather modification**?

Humans have long dreamed of being able to change the weather. The 1946 discovery that supercooled clouds could be seeded with dry ice or silver iodide and coerced into yielding the moisture in the form of precipitation ushered in an era of wild optimism about the possibilities of human "weather control." Ultimately the promise far outweighed the realities, but today there are ongoing projects that use seeding to augment winter snows in mountains, increase rainfall from convective storms, and dissipate supercooled fogs over airport runways. But to "dial up" the weather desired is totally beyond the realm of science and technology.

337

What is **inadvertent** weather modification?

While attempts at deliberately altering the weather have met with only modest success, humans have been changing local weather and perhaps global climate simply by going about their daily activities. Clouds form in the smoke plumes of factories and steel mills. Oceangoing ships leave trails of clouds extending 1,000 miles behind them as they pump billions of cloud nuclei from their smoke stacks into the marine air. Cities are warmer than their rural surroundings. There is some evidence that precipitation is higher downwind of some cities. Visibility has dropped dramatically in many parts of the industrialized world due to the vast amounts of smoke, soot, and sulfate particles belched into the atmosphere.

What is the **urban heat island**?

Most cities are warmer than surrounding rural areas, especially on calm, clear nights. Some inner cites might be 10 to 20°F warmer than low-lying rural areas. The urban warmth results from waste heat from energy use, changes in the heat capacity of the surface (concrete holds more heat than does grass), smoke acting as an insulating blanket, and reduced cooling due to less water evaporation. One small contribution to possible global warming might be urban sprawl. In the United States, over 750,000 acres of land are urbanized each year.

What is the **greenhouse effect**?

The greenhouse effect is a warming near the Earth's surface that results when the Earth's atmosphere traps the sun's heat. The atmosphere acts very much like the glass enclosure of a greenhouse. This effect was described by John Tyndall in 1861, though a similar notion had been mentioned by French mathematician Jean Fourier a century earlier. Without the natural greenhouse effect in which certain atmospheric gases trap heat within the atmosphere, our planet would be covered with ice. It is estimated that Earth would be some 60°F colder than present without the greenhouse effect. The real concern is that too much of a good thing might increase global temperatures beyond the optimal level. The thick blanket of clouds and carbon dioxide smothering the planet Venus produces a runaway greenhouse effect resulting in an average estimated surface temperature of about 600°F.

Carbon dioxide gas is a major contributor to the greenhouse effect. The levels of CO_2 in our atmosphere have varied considerably over geological time. But since the industrial revolution, humans have been dumping vast amounts of CO_2 into the atmosphere. The question thus was raised, can increases in atmospheric CO_2 levels caused by human activities in some way change global climate and, in particular, cause an artificial global warming?

Clouds of condensed water vapor rising from a power plant
mark a source of numerous emissions, including such greenhouse gases as carbon dioxide.

Who first proposed **major human impacts** on climate?

The idea of global warming from a runaway greenhouse effect due to the mass burning of fossil fuels is not a new one. It may have been first proposed at the end of the last century by Svante Arrhenius, a Swedish scientist. He estimated that global temperatures could possibly rise as much as 10°F. Today's sophisticated computer models continue to yield a similar broad range of potential impacts.

The human race continues to use massive quantities of hydrocarbons as fuel. Our present day oils, coals, and gases took approximately half a billion years to form. We are using them up, however, at one million times their formation rate. Combustion of fossil fuels and forest burning are causing a measurable increase in the amount of carbon dioxide in the atmosphere. Once emitted, a molecule of CO_2 may remain in the atmosphere for 50 to 200 years on the average. In the last three decades, the annual release of global carbon dioxide into the atmosphere has doubled. The concentrations remaining in the atmosphere are increasing at a rate of about a half percent per year.

How much warmer could the Earth become from global warming?

Some computer models of global warming estimate it is conceivable that the worldwide average temperature could rise by up to 9 to 11°F within a century or two. This **339** change is of the same magnitude, but in the opposite direction, as the cooling that, over thousands of years, produced the ice ages. Carbon dioxide emissions from human-caused combustion is a major cause of the possible "global warming" phenomenon. Methane gas (CH_4, produced by sources as diverse as rice paddies and termite colonies) is another contributing factor.

What are the major **greenhouse gases**?

There are five major greenhouse gases: water vapor, carbon dioxide, oxides of nitrogen, methane, and chlorofluorocarbons (CFCs). The last four have been significantly increased by human activities.

It's more than just the rising atmospheric carbon dioxide levels that have been implicated in any possible global warming due to the greenhouse effect. While CO_2 levels are increasing at the rate of about 0.5 percent per year, methane, a very potent greenhouse gas that is also known as swamp gas, is rising at 0.8 percent per year. Atmospheric concentrations of methane, currently about 1680 parts per billion (PPB), are almost three times the 600 PPB value in pre-industrial revolution air. Methane concentrations have risen in part from increased rice cultivation and livestock numbers.

One molecule of methane has 20 times the heat-holding power of a CO_2 molecule. And a molecule of CFC, which is produced only by humans, is 20,000 times more efficient. Due to population increase and changes in the planet's land use patterns, by the year 2000 atmospheric methane is expected to be six times higher than in the year 1800.

How much **carbon dioxide gas** have humans added to the atmosphere?

The Earth's atmosphere weighs 5.1 million billion tons. The burning of fossil fuels adds an additional 10 tons of CO_2 per year to the atmosphere—for *each person* in the developed countries. But how much carbon dioxide gas was in the Earth's atmosphere before the start of the industrial revolution and the widespread burning of fossil fuel? Chemists of the seventeenth century didn't even know about CO_2, much less how to measure it. But a clever late twentieth-century atmospheric chemist found a way to get the measurement. By drilling deep in glacial ice in Greenland, and by analyzing the air trapped within myriad small bubbles, CO_2 levels as far back as 1530 have been dated. The result showed that pre-industrial CO_2 was approximately 280 parts per million. Current values are well above 350 parts per million. During the last ice age, some 18,000 years ago, CO_2 levels were even lower, about 200 parts per million.

What is the **most important** greenhouse gas?

Contrary to popular belief, the most potent greenhouse agent in the atmosphere is not carbon dioxide, but water vapor. Water vapor, along with clouds, accounts for over 90 percent of the natural greenhouse impact of the Earth's atmosphere.

What are some sources of **methane gas**?

The activities of both termites and coal miners result in substantial inputs of methane gas into the atmosphere. Methane is expected to contribute up to 20 percent of any global warming that may occur. According to the Environmental Protection Agency (EPA), the physical act of mining coal (without having burned it) may contribute up to 12 percent of the methane entering the atmosphere. The billions of termites nibbling their way through the planet's wood also collectively contribute large quantities of methane. Livestock (more specifically, the gas emissions from livestock) are also prodigious producers of CH_4. The methane resulting from human consumption of beans and cabbage has not yet been computed.

What is some of the **evidence** presented to support the global warming theory?

Over the last few years, the press has been filled with stories purported to "prove" that human intervention with the greenhouse mechanism is producing the global warming that the computer models (which are far from perfect tools) suggest might happen. Some are listed below. But it should be remembered that climate varies naturally all the time. While global warming as a theory has been widely debated, actual measurements that support the theory are hard to come by.

May 1991 was hotter than usual in Kentucky. Louisville had the warmest May in 149 years of record. Lexington also had its warmest May in over a century. Curiously, no daily record highs were broken anywhere in the state during the month.

Some evidence seems to indicate that it's getting drier as well. Yearly precipitation over land areas of our planet declined in the 1980s following a three-decade trend towards wetter conditions. It's much too early to tell, however, if this is an effect of global warming or just part of the natural climate cycle.

One possible impact of global warming may be increased amounts of water flowing through U.S. streams during the cold season. This seems to contradict the notion that the planet is getting drier; such contradictions are typical of many of the pronouncements concerning the evidence for global change.

In the period 1941–88, increasing levels of water flowing through U.S. rivers and streams, especially in autumn and winter, may be one indication of some systematic climate change. Unfortunately, periodic summer droughts do not seem to have been canceled out.

The polar ocean ice sheets have long been suspected to be sensitive indicators of global temperature. Satellite observations did show a two percent shrinkage over an **341** eight-year period in the 1980s in the Arctic Ocean. No change was noted in South Polar regions, however.

Whether global warming due to human activities is just a theoretical prediction or something that is already occurring is a hotly debated topic in atmospheric sciences. Temperature data from surface stations around the world are beginning to show some apparent signs of the predicted warming. The period 1991–95 was the warmest half-decade on record, with 1995 setting a record high in the 140 years of reliable global temperature data. British researchers believe the planet is now warming at the rate of 0.2°C per decade. Globally, the years 1990 and 1991 were two of the warmest years since worldwide records began during the last century. This does not yet prove, however, that widespread global warming is underway, but it is one piece of evidence.

By most measures the winter of 1991–92 was the warmest on record in the 48 contiguous United States. January 1990 was noted for its warmth over much of the United States. For the first time in over a century some northern plains cities had no minimum temperatures of 0°F or less.

Has potential global warming been detected in the Earth's vegetation?

One possible indicator of a warming planet is the finding that forests and grasslands in the Northern Hemisphere are greening a full week earlier in spring than they did in the mid 1970s. This has been determined by very careful monitoring of planetary carbon dioxide levels. The normal spring decline in CO_2 is indicative of uptake by green plants awakening from the winter sleep. This effect has been noted earlier and earlier in the season, suggesting that milder winters are lengthening the growing seasons. If this trend continues it may be one of the better pieces of evidence available to confirm that a true global warming is taking place (though the cause is not necessarily pinned down).

Where might evidence of **global warming** first be detected?

Computer models suggest that one of the first signs of global warming may be a warming near the surface, especially in south polar regions, and, ironically, a cooling in the upper atmosphere, again particularly in the Southern Hemisphere. Recent analyses of 25 years of balloon soundings do indeed suggest a cooling stratosphere, especially in more southern latitudes. At the surface, Antarctica also appears to be getting warmer. Over the last 50 years the average temperature on parts of the ice continent has increased by 2.5°C. Some of the milder coastal regions are seeing 25-fold increases in the amount of vegetation. And in 1995, a gigantic iceberg some 23 by 48 miles in size broke loose and floated out to sea. The 600-foot-thick iceberg was the size of the country of Luxembourg. None of these findings are conclusive, but they do strengthen the case of those scientists who believe human activities are warming our planet. These effects are rather good evidence that at least some anthropogenic-caused global warming may be kicking in.

Is it global warming or "global less cooling"?

With all the debate about rising average temperatures and their possible link to CO_2 emissions it has been noted that much of the observed warming has been coming from rising minimum temperatures rather than any change in average high temperatures. In fact, in some areas, summer maximum temperatures are declining but winter minimum temperatures are sharply rising. These effects could easily be induced by changes in aerosols in the atmosphere or cloud cover due to air pollution from natural or human sources.

What evidence is there **against** global warming?

The newspapers have also been filled with many accounts of weather extremes that do not appear to support the global warming hypothesis, but rather satisfy the predictions of those claiming a new ice age is about to begin. A small sample:

A cold wave swept northern Europe in late 1995, bringing the coldest readings in a century. Some 750,000 people in Scotland were without water as pipes froze and burst. Similar problems arose in Norway, in part due to the lack of insulating snow cover.

Any trend towards global warming cooled down a bit in the north central United States during the summer of 1992. July was the coolest on record in many cities from Montana to New England. The temperature at Sault Ste. Marie, Michigan, failed to reach 80°F all summer.

The month of January 1990 was the warmest January ever in the United States. But by contrast, the previous month, December 1989, was the fourth coldest since such statistics began to be tallied in 1895.

Winter temperatures in Finland from 1961 to 1990 have averaged 2°C colder than the long-term average.

Recent winters in the eastern United States have included some of the coldest and snowiest in memory.

Are there any **good effects** of increased carbon dioxide?

Rising levels of CO_2 in our air might not be all bad. Researchers suggest that CO_2 not only stimulates plant growth, but also reduces plants' water consumption. One experiment found that as CO_2 was doubled, depending on the species, the amount of water required by the plant will drop between 17 and 27 percent and plant growth could increase by as much as 84 percent. If the various predictions of global warming were to come true, many barren northern regions would become suitable habitats for forests. But how fast can forests actually migrate? It is estimated that during the retreat of the last great continental ice sheets, the beech forests were able to follow only at the rate of 12 miles per century.

What **factors** can control global climate?

The public may think, from reading press accounts, that carbon dioxide emissions are the only thing that influences climate. The real issue is vastly more complex. Aside from the fact that other greenhouse gases are important, there are numerous factors that influence long-term climate trends, and many of them are natural processes that are quite oblivious to the presence of humans. Sea surface temperatures, sunspots, volcanic eruptions, and forest fires are just a few of the items that must be factored in.

Also, many human activities besides fossil fuel burning may play a role in changing the planet's climate. Albedo refers to the amount of the sun's light that is reflected back into space. The albedo of bare ground may be only 10–20 percent and, by contrast, 85 percent or more from fresh snow and 90 percent or more from dense clouds. The average albedo of the Earth as a whole (35–45 percent) is a critical term in understanding global climate. Large-scale changes in land use due to farming, grazing, urbanization, and deforestation affect not only albedo but soil moisture and cloud cover, which in turn affect global temperature.

As a measure of just how complicated the global change issue is, it appears that the simple practice of fertilizing fields may increase the potential for global warming. Along with tillage of grasslands, the result is that the land absorbs less methane and produces more nitrous oxide. Together these effects could account for 20 percent of the possible global warming that might occur. On the other hand, other agricultural practices might be able to mitigate global change processes.

The U.S. government has called for more research into global climate change before massive and expensive controls are placed on carbon dioxide sources. Recent federal budgets have earmarked over one billion dollars in research for global climate–related issues.

How much does the **global climate system** vary naturally?

Trying to separate out the human impact on climate from the natural chaotic signal is a major scientific challenge. The same effect (increasing temperatures) can arise from multiple causes. Therefore, jumping to conclusions in climate research can be a risky business. If one were to use temperature records from the city of London, one might conclude that the planet has indeed warmed—until it would be pointed out to you that your record merely reflected the growth of the London heat island, a local effect. It is known that climate has always fluctuated wildly. There is no reason to suspect it has suddenly stopped doing this as a result of natural processes. Some of the bits of evidence that scientists are sorting through include the following items.

By best estimates, the average temperature of the Earth is getting warmer. Scientists estimate that the planet's temperature has risen about 0.6°C during the past 100 years. But the cause is open to more debate since this change is well within the suspected normal fluctuations of the atmosphere's temperature. While the accumulation of greenhouse gases is certainly a favored theory, other natural factors such as changes in the output of the sun cannot yet be ruled out as at least a partial contributor. A period of increasing temperatures by itself cannot be taken as conclusive proof of human-induced global warming.

The period from 800 to 1250 in Europe is sometimes called the Little Climatic Optimum. The moderate weather was a main reason for the wave of Norse settlers heading for Iceland, Greenland, and even North America, as the mild climate allowed not only for livestock grazing, but also the growing of some crops. Vineyards even thrived in England.

A sudden planetary cooling has become known as Europe's Little Ice Age. It was most intense in the seventeenth century, during which the Thames River in London froze over solid at least ten times. The Viking expansion from Scandinavia into Iceland, Greenland, and Newfoundland during the early medieval period was abruptly terminated around 1400 at the start of the Little Ice Age, which forced the abandonment of many of these northerly outposts.

Attempting to assess whether or not human enhancement of the "greenhouse effect" is causing global warming is tricky business, if only because other influences are constantly acting upon the Earth's mean temperature. For instance, the El Niño of 1986–87 caused a global warming of as much as 0.6°C, while the eruption of Mt. Pinatubo caused a cooling, also of about 0.6°C.

The burning of fossil fuels has raised the carbon dioxide levels in the Earth's atmosphere to their highest levels in 160,000 years. But what caused, and later ended, an earlier, high-CO_2 event is unknown.

The planet's delicate thermal balance has been upset without the polluting intervention of humankind. The national Ocean Drilling Program has analyzed bottom sediment cores and found that around 57 million years ago, at the end of the Paleocene epoch, sea temperatures rose from 54°F to 64°F. This resulted in the disappearance of up to 50 percent of the species inhabiting the deep seas.

What is **paleoclimatology**?

The climate of the Earth is constantly changing, and it is the job of paleoclimatologists to use whatever shreds of evidence they can find to piece together the turbulent history of our planet's climatic past.

About 8,500 years ago, water temperatures around the coast of Greenland were about 10°F warmer than they are today. We know that because studies of fossil shells show certain species of warm water–loving mollusks thrived in the region during a "climatic optimum" period. It has long been thought by paleoclimatologists that during the ice ages the tropical regions of the Earth experienced very little cooling even while mid-latitude areas became icebound. More recent evidence from the fossils of temperature-sensitive organisms suggests that the equatorial regions did indeed cool down, perhaps by more than 5°F.

It was once much warmer in Oregon than it is today. In fact, tropical fruit once grew there. We know this because an Oregon teenage boy found a fossilized banana that has proven to be 43 million years old.

Scientists will use any clue they can uncover to try to reconstruct the past climate of the Earth. Now pine needles stowed away by pack rats some 30,000 years ago are yielding data. Examination of debris from pack rat middens in the western United States indicates changes in CO_2 levels in the atmosphere associated with the last glacial period.

Antarctica brings to mind scenes of endless icy wastelands and wind chills off the bottom of the scale, yet it once had a climate mild enough to support dinosaurs. Fossils of the 25-foot-long meat-eating *Cryolophosaurus ellito* have been exhumed only 400 miles from the South Pole. Thus, 200 million years ago, the climate was mild enough to support plants and animals necessary to feed the dinosaur.

Where are **climate data** archived?

The primary archive for weather data used in climatological and other studies is the National Climatic Data Center (NCDC), 151 Patton Avenue, Asheville, North Carolina, 28801. Their web page address is http://www.cdc.noaa.gov. NCDC archives a large fraction of the weather data taken by federal agencies and cooperative observers, with records going back well into the last century. Their daily update is rather impressive, including over 50,000 reports from surface stations, 2,500 rawinsonde balloon launches worldwide, 3,000 communications from ships and thousands more from aircraft reports. Weather satellites and radar add gigabytes more data each day. These data can be made available to anyone who is willing to pay the handling charges and service fees imposed by the government. Many other climate-related databases are found at institutes such as the National Center for Atmospheric Research (NCAR), in

NASA laboratories, the U.S. Geological Survey, the Department of Energy, and numerous national laboratories and university research centers.

NCDC recently moved into new facilities in Asheville. The need for upgraded space is evident when you note that NCDC stores some 320 million pages of paper, 2.5 million microfiche records, 403,000 tape cartridges and magnetic tapes, and satellite images dating back to 1960. The insurance and legal communities are among the largest users of these data, accounting for up to 30 percent of the 143,000 requests for data each year.

If life had not evolved on our planet, would our atmosphere be any different?

Yes. There would be less nitrogen, oxygen, and methane, but more carbon dioxide and carbon monoxide. The chemical composition of the Earth's atmosphere has changed dramatically over time as a result of living organisms. Today about 21 percent of the air we breathe is oxygen. But during the Carboniferous period some 300 million years ago, very high levels of plant activity shot oxygen levels up to 35 percent of the atmosphere.

Is the **sun's output** constant?

Meteorologists have long talked of the "solar constant" as the amount of energy emitted by the sun, and once considered it to be a rock-steady number. Not so. Solar output does indeed vary, on the order of 0.1 percent. Satellite observations show that the sun's energy output varies over the 11-year sunspot cycle, and probably on other time scales as well. Several billion years ago the sun was thought to shine with only about 75 percent of its current intensity. In the current era, while the solar energy output fluctuates slightly over time, astrophysicists expect it to keep on shining at roughly the current rate for another five billion years, which is pretty "constant" by most definitions.

Do **sun spots** affect the Earth's weather?

Yes, though scientists aren't exactly sure how. The energy emitted by the sun is not constant as once thought. How these slight fluctuations in the sun's output, associated with sunspots, perturb weather patterns is still an object of intense research. Much of the solar variation takes place in the shorter (ultraviolet) wavelengths to which upper atmospheric chemical processes are very sensitive.

What were the **ice ages**?

In the nineteenth century, Louis Agassiz helped compile the evidence that large parts of the Earth have been periodically covered with advancing and retreating ice sheets. The last ice age ended between 14,000 and 10,000 years ago. The climate of the most recent 10,000 years has been remarkably stable, at least by comparison to the wild climate shifts being deduced from analysis of the last 250,000 years of ice cores from the glaciers of Greenland. During that period there have been 35 major climate shifts that affected Greenland.

During the most recent ice age, when glaciers covered most of Wisconsin, Michigan, and New England, global mean temperatures were about 10°F cooler than at present.

During the Pleistocene era, extending back 600,000 years, there were four major glacial periods lasting 30,000 to 100,000 years. But in the interglacial periods, the Earth's temperature averaged 5°F warmer than today.

Scientists from the Lamont-Doherty Earth Observatory at Columbia University figure that the Earth is 8,000 years overdue for a major climate change. Instead of global warming, they are thinking of a possible ice age, and one that could start with little warning. The magnitude of such shifts in the past have been equivalent to changing the climate of Atlanta to that of Duluth, Minnesota.

While ice ages have played a dramatic role in the Earth's climate system over the past million or so years, they are relatively rare when the entire geological record is examined. Ice ages have accounted for less than one percent of the last 600 million years. The geological record suggests the first large-scale glacial epoch occurred about 2.3 billion years ago.

How much of the Earth was **covered by glaciers** during the ice ages?

Geologists suggest that up to 10 percent of the planet was buried in ice (though not necessarily simultaneously) during the maximum extent of the great ice ages.

How **quickly** can ice ages start?

Studies of ice cores from the world's thickest glaciers (some 10,000 feet tall) have suggested that past climatic regimes have changed much more frequently—and much more rapidly—than previously suspected. Some changes on the scale of ice ages appear to have occurred during less than the course of an average human life span and, in one case, in only 3 years. Sudden shifts in major ocean current patterns are a possible cause.

Is another ice age likely?

Yes. Unless human activities such as pollution and deforestation intervene, the most likely long-term changes in global climate will be a slide back into ice age conditions within several thousand years. Those with good memories might even remember that before the greenhouse effect/global warming flap began, the media frenzy about climate concerned the impending ice ages. The cold winters of the late 1970s prompted hundreds of magazine stories and television reports about how the glaciers would soon be advancing once again down Michigan Avenue in Chicago, and elsewhere.

How has global sea level fluctuated over time?

It is estimated that during the last ice age, 18,000 years ago, so much water was bound up in the polar ice caps that sea level was some 400 feet lower than it is today. We know that sea level changes significantly over long periods of time in response to climate fluctuations.

There is evidence that the global sea level has risen by as much as 20 cm (7.9 inches) during the last century. One possible sign of global warming is expected to be rising sea levels, both from melting ice caps and the expansion of the oceans as they warm. But what happens if the land itself is rising or falling? Many areas of Scandinavia are rising rapidly, rebounding from the massive weight of the last glacial epoch. Meanwhile, around Miami, the land areas are sinking at the rate of three inches a century. A new Worldwide Sea Level Observing System has been recently completed to try and sort out all these effects.

The Greenland ice cap today contains three million cubic kilometers of ice. If it were all to melt, the global sea level would rise more than 7.5 meters (24 feet). Topping that is the Antarctic ice cap, which contains 29 million cubic kilometers of ice. If it were to all melt, the global sea level would rise some 65 meters (210 feet). If all the ice sheets on the planet were to melt as a result of global warming and become part of the oceans, sea level would rise approximately 250 feet. Expansion of the sea water due to warmer temperatures would further add to the rise in sea level.

Who was Melutin Milankovitch?

Milankovitch, a Yugoslavian geophysicist, proposed in the 1920s that changes in the Earth's spatial relationship to the sun can significantly influence the amount and distribution of the sun's energy reaching our planet. In other words, the Earth's orbital changes can influence climate. There are many factors, including changes in the composition of the Earth's atmosphere, that can alter climate, but there are now well-known variations in the Earth's orbital mechanics that happen at intervals of 22,000, 40,000, and 100,000 years.

The orbit of the sun becomes more rather than less elliptical in 93,000-year cycles. The tilt of the planet's axis varies from 22 to 24 degrees in 41,000-year periods. The Earth also wobbles around its axis like a top every 25,800 years. While these theories were at first dismissed, evidence has steadily mounted that there are indeed cycles in global temperature that match these orbital gyrations. Deep ocean sediment cores have been valuable in showing there are multiple overlapping global temperature fluctuations similar to those predicted by Milankovitch. He was not actually the first to propose an orbital connection to the ice ages. Scottish geologist James Croll in 1864 amplified on a theory proposed in 1842 by Joseph Adhemar that ice ages occurred in a 22,000-year cycle caused by a corresponding cycle in the Earth's orbit around the sun. Since the initial theories did not match up well to the known dates of ice ages, the theory lost steam until resurrected and improved by Milankovitch.

What's your **personal contribution** to increasing the greenhouse effect?

The average American's use of energy in all its forms results in 20 tons of carbon dioxide emissions into the atmosphere each year. A citizen of Kansas, on the average, is responsible for 50 times the CO_2 emissions of a resident of Pakistan. If you are a lawyer, you might be doing even more than your share to promote global warming. Each lawyer is estimated to consume the equivalent of 17 trees annually for filing briefs, sending faxes, and making copies. Nationwide, that amounts to tens of millions of trees each year.

How can you **cut down** on your contributions to possible global warming?

According to the EPA's Green Lights Program, if everyone in the United States used energy-efficient lighting we would prevent emission of 232 million tons of carbon dioxide, 1.7 millions tons of sulfur dioxide, and 900,000 tons of nitrogen oxides. The Energy Star seal is now appearing on products from computers to thermostats to help the consumer identify energy-efficient devices. Be sure your next hot water heater and refrigerator have a high energy efficiency rating. Be sure good gas mileage is an "option" on your next new car. And keep your tires properly inflated—that alone can improve gas mileage. Such actions could reduce U.S. emissions of greenhouse gases by up to 13 percent.

U.S. emissions of carbon dioxide, the major "greenhouse gas" thought to contribute to the possible global warming scenario, have in fact begun to level off since 1988. Energy consumption accounts for 98 percent of U.S. man-made carbon dioxide releases. The use of cleaner burning fuels and the implementation of energy efficiency campaigns are beginning to pay off.

What is La Niña?

La Niña is the opposite of El Niño, referring to a period of cold surface waters in the Pacific. The two temperature states oscillate back and forth. The result is the Southern Oscillation, a term first used in 1932 by Sir Gilbert Walker in attempts to predict the year-to-year fluctuations in India's monsoon rainfall. He noted that when the pressure was high over the Pacific Ocean, it tended to be low in the Indian Ocean from Africa to Australia. This was the atmospheric response to the cyclical changes in sea surface temperatures, which in turn affected the sub-continent's rainfall. Sometimes scientists refer to ENSO—El Niño Southern Oscillation—to describe the entire phenomenon.

What is **El Niño**?

El Niño, a Spanish-language term referring to the Christ child, is also the name for an unusual warming of the surface waters of large parts of the tropical Pacific Ocean. Reappearing every several years, and peaking around Christmas time (thus the name), climatologists have found that it can have a profound impact on global weather patterns.

El Niño occurs rather erratically, every few years, and its cause is unclear. One scientist has speculated that it is an oceanic response to pollutants emitted by large volcanic eruptions, such as Mount Pinatubo. Over the last 45 years or so, El Niño events have started in 1952, 1958, 1964, 1966, 1969, 1972, 1976, 1982, 1987, 1991, and 1994. The most recent EL Niño episodes have been predicted with up to a year's notice by some of the complex global climate models being used to study the phenomenon.

The longest recorded period for El Niño, which typically lasts about 12 to 18 months, lasted from 1991 to 1995, a period marked by numerous weather disasters worldwide. Evidence suggests such long lasting El Niños should occur only every several thousand years.

What are some **impacts of El Niño**?

El Niño was once thought to be a purely local phenomenon occurring in the waters off Peru and Ecuador. Climate researchers have found the event covers a vast part of the world's largest ocean and in turn affects weather patterns globally. Just a few of the weather disturbances linked to El Niño are as follows:

The warm water associated with the 1972 El Niño along the coast of South America was accompanied by alarming declines in the anchovy population, which had major repercussions for the local fishing industry and the world animal feed commodity markets.

The 1976–77 El Niño resulted in drought in California and severe cold in the central and eastern United States.

The El Niño event of 1982–83 caused unusual and often catastrophic weather patterns around the world. It worsened the already devastating drought in Africa, along with the worst drought in 200 years in Australia, but aided in torrential rains in Peru and Ecuador. Intense winter storms smashed the U.S. Pacific coast, and the first typhoon in 75 years struck French Polynesia. All this because the water temperature of parts of the Pacific ocean were a few degrees warmer than normal. The death toll is estimated at 1,500 persons with $8 billion in property damage.

The 1991–95 El Niño events triggered a wave of diverse precipitation anomalies worldwide. Californians received an entire year's precipitation in January 1995 alone, while the worst floods in a century inundated Germany, Holland, Belgium, and France. In the meantime, drought seared Australia, Indonesia, and southern Africa.

Could **global warming** cause entire nations to disappear?

If some predictions of sea level rises associated with climate change were to come true, some island nations such as the Maldives and Marshall Islands, which are just barely above sea level now, could have some serious problems. Also, populated river deltas such as those of the Bengal, Nile, and Niger rivers could be threatened.

Have there been any changes noted in **ocean waves**?

Waves in the North Atlantic ocean seem to be getting higher. In the 1960s, the average reported height was about 7 feet. It has since increased to 9–10 feet. The climate of the Earth is constantly changing, and maybe that of the ocean is too. No reason is apparent for this increase, however, since storminess had not increased by that much during the same time period.

What are **CFCs**?

Manufactured chemicals called chlorofluorocarbons (CFCs) now supply about one-quarter of the current human augmentation of the Earth's greenhouse effect. Used in refrigeration, air conditioning, foam, and insulation, they tend both to warm the Earth's atmosphere and destroy the protective ozone layer in the upper atmosphere.

Atmospheric concentrations of CO_2 have increased from around 275 to 355 parts per

This satellite picture shows two types of clouds: a sharp cold front moving throught the eastern United States, and a cloud of dust kicked up from Oklahoma and Colorado. Dust clouds are natural sources of air pollution.

million during the industrial age. Methane has increased from 0.75 to 1.65 parts per million, mainly due to cattle and rice farming. CFCs have gone from 0 to 650 parts per trillion. The actions of the industrialized nations to severely limit production and use of CFCs is expected to stop the expansion of the stratospheric ozone hole by early in the next century and gradually (over many decades) allow the natural stratospheric chemistry to be restored.

What is the **ozone hole**?

Ozone, which shields the planet's surface from harmful ultraviolet rays, is concentrated in the stratosphere, particularly in polar regions. The ozone is formed by complex reactions of atmospheric gases under the influence of solar radiation. As in many chemical reactions, small changes in some chemicals involved in the process can have big results. CFCs injected into the system can destroy vast amounts of ozone and disrupt the natural formation and dissipation cycles. Satellite ozone measurements in the last decade began to detect an ever-deepening deficit of ozone (known as the ozone hole) in the Antarctic stratosphere, largely due to the influence of CFCs, which had been mixed into the global atmosphere after years of use by humans. NASA satellites monitor the Antarctic stratospheric ozone hole routinely. When it reached its largest size to date in September 1991, it was 15 percent larger than the previous year, and **353** covered almost 9 million square miles—or three times larger than the 48 contiguous United States.

The hole in the ozone is not just a problem for penguins and other polar region residents. While stratospheric ozone depletion is most pronounced in the South Polar regions, it is thinning out over much of the planet. A United Nations report suggests the thinner ozone will allow increased ultraviolet rays from the sun to strike people worldwide. A possible result? A projected 1.6 million new cases of cataracts and 300,000 new cases of skin cancer each year by the end of the century. Though not as dramatic as the decrease found over Antarctica, stratospheric ozone declines on the order of 5 percent have been noted over the Northern Hemisphere at latitudes as far south as Europe, the United States, and Canada. Satellites detected a hole of sorts in the stratospheric ozone layer over the southeastern United States in November 1994. These measurements were the lowest ozone values recorded in over 15 years of measurements, but it appears as if the low ozone levels were part of a natural process associated with the sub-tropical jet stream and not a man-made impact.

Are **frogs** potential victims of the ozone hole?

Very possibly. Scientists have been deeply puzzled by a rapid decline in frog populations almost on a worldwide basis. It seems that increased ultraviolet radiation reaching the surface due to the thinning of the stratospheric ozone layer may be killing the amphibian eggs before they can even hatch into tadpoles.

What nations are the most prolific producers of **greenhouse gases**?

The United States has the lead, with almost 20 percent of the global total. The area of the former Soviet Union contributes almost half that of the United States. Brazil, China, and India are next in line. On a per capita basis, however, citizens of oil-rich nations such as Qatar, the United Arab Emirates, and Bahrain, emit more greenhouse gases than do Americans. The only major nation to significantly reduce its output of carbon dioxide in recent times is France, due to its reliance on nuclear power.

Have atmospheric scientists ever won a **Nobel prize**?

The 1995 Nobel Prize for Chemistry went to three atmospheric chemists who unraveled the mystery of the Antarctic ozone hole. Drs. Mario Molina (Massachusetts Institute of Technology), F. Sherwood Rowland (University of California, Irvine), and Paul Crutzen (Max Plank Institute for Chemistry, Mainz, Germany) shared the award for demonstrating how man-made gases such as chlorofluorocarbons (once widely used in consumer products) could eat away at the planet's protective layer of stratospheric ozone.

Is there **malaria** in Switzerland?

Not normally, but recently a Swiss mail carrier became the first reported fatality from the tropical disease in this country nestled in the snow-capped Alps. The offending mosquito most likely was delivered via air mail, coming in on a jet and somehow managing to escape the insecticide sprayed in aircraft cabins on flights from the tropics. After escaping the airport, the mosquito was able to survive due to unusually warm temperatures in Switzerland during early July. It may be a stretch to say this is the result of global climate change, but it does serve as an example of how a complex world can become even more so if global warming scenarios prove to be true.

How might **land use changes** impact climate?

The Earth's surface has a direct impact on weather and climate. As a result of civilization, the human race has plowed up 15 million square kilometers for farmland, taken timber from 10 million square kilometers, and used 32 million square kilometers for livestock grazing. Much of the remaining 90 million square kilometers of the planet's land is unproductive, inhospitable, or inaccessible. Thomas Jefferson noted that afternoon sea breezes that once reached inland to only Williamsburg had begun, due to the clearing of Virginia's lands by settlers, to routinely penetrate beyond Richmond. In more recent times, severe overgrazing by cattle and sheep in many parts of the world have denuded the land of vegetation, and greatly affected evapotranspiration to the point where deserts became permanent. Changes in land use over southern Florida during this century as more and more room was needed for motels, hamburger stands, subdivisions, and the like have apparently affected the distribution of the afternoon sea breeze thunderstorms that provide much of the state's fresh water supply.

If the present rates of deforestation continue on a global basis, all tropical forests will be cleared in 177 years. The rate of deforestation is greatest in Brazil, Indonesia, and India. The Amazon rainforest, which originally covered an area of half the contiguous United States, is being reduced at many tens of thousands of square kilometers a year. Less than half the native forests of the Philippines remain. Large-scale deforestation is a concern to global climate researchers. While the current focus is on South America, much of North America and Europe were deforested during past centuries. The area covered by European forests has actually begun to expand in the last half of this century, although acid rain is now adversely affecting forest vigor.

Some scientists estimate that the worldwide destruction of forests has influenced global temperature patterns as much as the massive influx of greenhouse gases from human fossil fuel burning. Eight trees would have to be planted each year for each U.S. citizen in order to keep pace with forest cuttings and allow for a slow increase in our tree population. But it's worth it. Trees use up carbon dioxide in the atmosphere, help prevent soil erosion, and help modify extremes of temperature.

How much **topsoil** is being lost each year due to poor agricultural practices?

Globally, there are 3.7 billion acres of land that have soil and climate conditions suitable for growing crops. Soil erosion, however, is destroying at least 29 million of these acres per year. Worldwide some 75 billion tons of topsoil are blown away by the wind or eroded by water each year. This is ten times the rate of its natural replenishment.

While wind is a major factor in soil erosion, approximately 85 percent of the topsoil loss in the United States comes as a result of livestock production. It takes up to 1,000 years for an inch of topsoil to be created by natural processes. In the United States, it is estimated that a nationwide average of some nine tons per acre of topsoil are lost by wind and water erosion annually. Whether by wind or flood, America's topsoil is eroding, washing, or blowing into the oceans. The average depth was nine inches in 1776, but is now down to 5.9 inches.

What happens to **atmospheric carbon monoxide**?

With all the talk of increasing atmospheric carbon dioxide (CO_2), its relative, carbon monoxide (CO), is not usually considered. Carbon monoxide (a deadly component of auto exhaust) had been steadily increasing in the Earth's atmosphere, along with gases such as greenhouse culprits carbon dioxide and methane. Then, in the last few years, CO levels have unexpectedly begun to drop. One theory is that decreases in stratospheric ozone let more ultraviolet rays through to lower levels, thereby converting the CO to other gases.

What happens to the **carbon dioxide**?

Not all of the vast amounts of CO_2 pumped into the atmosphere by humans each year remain in the atmosphere. About 50 percent is thought to be absorbed by the oceans and vegetation. Yet the details of the fate of CO_2 remains a puzzle to climate change scientists. The oceans were long thought to soak up much of the excess CO_2, but it appears the Northern Hemisphere forests absorb some 3.5 billion tons each year, or twice that of the oceans. This gives Arbor Day a whole new significance, since the world's rain forests have shrunk to only about 50 percent of their original size and are being destroyed at the rate of 28 million acres per year.

What is the **average temperature** of the Earth?

The globally averaged surface temperature of the atmosphere is about 59°F. And if it were not for the natural greenhouse effect of water vapor and carbon dioxide, things would be much chillier. The global mean temperature would plunge to 4°F!

How large are the Earth's **artificial reservoirs**?

Humans have been conserving fresh water supplies for eons using reservoirs. But construction of reservoirs since the 1950s has been so extensive that artificial lakes now cover twice the area of the Laurentian Great Lakes (some 500,000 square kilometers). Moreover, if the water had been allowed passage to the oceans, global sea level today would be 3 cm higher than current levels.

What effect does **irrigation** have?

In order to grow crops, humans transport vast amounts of water from one place and apply them in another, often greatly affecting the vegetation in a region. Over large irrigated tracts of land, especially in the western United States, local wind patterns and thundershower development may be affected. In the 1970s, climatologists found rainfall increases in the irrigated portions of Kansas, Nebraska, and Texas during the summer months. This is a region that receives 14.5 million acre-feet of irrigation water each growing season. Unusual pockets of hail and tornadoes were found that may be associated with heavily irrigated regions.

About 12 percent of the 400 million acres of U.S. cropland are irrigated. Research has shown that daytime temperatures are typically a bit lower and humidities are higher over and downwind of large irrigated regions. These changes might even effect local cloud cover and thundershower development.

And it's not just the farmers affecting the atmosphere. Watering the lawn is a major contributor to growing water shortages. Fifty percent of water usage in Denver in summer is just for keeping lawns green. In Phoenix, when the numerous golf courses water their turf, there is a measurable uptick in the relative humidity. A half-acre of grass might use over a half million gallons per season. One way to conserve water would be to cut back a little on lawn sprinkling or, better yet, substitute rock gardens, native grasses, or landscape plants that use less water.

What **major energy source** does not contribute to global warming?

One partial solution to the continued buildup of CO_2 in our atmosphere is nuclear power. In fact, while no new nuclear plants were opened, nuclear power was recently the only major category to show an increase, up 6.2 percent over the previous year, to supply 21.7 percent of the U.S. total electrical power. Of course, the problem of nuclear waste must be solved if long-term use of the atom is to be viable. Fusion power (the same energy source as used by the sun) would be free of radioactive waste, but given the two- or three-decade time frame to develop it, the U.S. government has greatly cut back its research efforts on what could be the ultimate energy source. This **357** leaves wind and solar energy as the only other real alternatives, and government support for these environmentally friendly energy sources is also waning.

What is 1,200 miles long and the largest thing on Earth built by living beings?

The Great Barrier Reef, off the coast of Australia, is composed of coral. The marine micro-organisms that accomplished this marvelous feat of engineering are dying. No one is completely sure what the problem is. It may be man-made pollution and the impact of global climate change, or some aspect of unrecognized marine life cycles.

What role have meteors and asteroids played in climate change?

Scientists have long suspected giant meteorite impacts have had a significant impact on the Earth's climate and ecology. Currently they have identified about 110 preserved meteor impact craters on our planet's surface, some of which date back as far as 600 million years. Two very large craters located in Canada and South Africa may trace their origins back to impacts two billion years ago. Catastrophic changes in the Earth's climate system due to an asteroid impact in Mexico's Yucatan Peninsula 65 million years ago perhaps led to the extinction of the dinosaurs. Now research suggests that the constant pummeling of the planet from above during much earlier times may also have played a role in continental drift, giving the continental plates an extra shove, so to speak.

Analysis of dust in cores drilled in the sea floor bottom near the Azores suggests that our planet intercepts a blob of cosmic dust about every 100,000 years. Exactly how that might influence climate, however, is anything but clear. There may be a closer link between the stars and the human race than once thought, but not in an astrological sense. When stars explode in giant super novas, cosmic ray radiation causes the element beryllium-10 to be formed in our atmosphere. The traces of one such massive super nova about 35,000 years ago can be found as enhanced beryllium concentrations buried deep in the polar ice caps. Some speculate that the radiation bombardment could have sharply increased mutations in many life forms, speeding up the human evolutionary process.

What is nuclear winter?

Smoke injected high into the atmosphere cools the planet. This has been repeatedly demonstrated after many volcanic eruptions. The massive smoke palls generated by huge forest fires in British Columbia in 1982 cooled temperatures in the United States by 4°F to 7°F. In 1983, scientist Richard Turco investigated what the effect would be of massive nuclear exchange between the superpowers. By the late 1980s there were 50,000 nuclear weapons in the world. Their total explosive power was 15,000 megatons of TNT—or 3 tons of TNT for every person living on Earth. If a Hiroshima-sized bomb were dropped *every hour* from 1945 on, only one-third of the global nuclear arsenal would ever have been used!

If a full-scale nuclear war were to take place, with each nuclear detonation huge amounts of black sooty smoke would be lofted high into the stratosphere. Though much would be rained out over several months' time, it was estimated that the remaining smoke pall would cause a nuclear winter in which the planet's temperature would drop precipitously. Crop failures would cause worldwide starvation (of those not killed by the radiation). Many believe that the fear of nuclear winter, in which no nation would survive, finally hit home to the world's political leaders and helped the easing of the arms race in the 1980s. Yet the remaining stockpile of weapons is still sufficient to cause global nuclear winters 10 to 100 times over.

Did the **Kuwaiti oil well fires** contribute to global warming?

The massive plumes of smoke from over 600 oil wells set afire by retreating Iraqi troops in Kuwait presented a spectacular image of environmental destruction. Yet, when put in its global perspective, the 12,000 tons per day of smoke accounted for less than 10 percent of the emissions from burning biomass worldwide. The 1,800,000 tons per day of CO_2 was only about 2 percent of the worldwide emissions of CO_2 from fossil fuel and burning forests. The blocking of the sun by the dense smoke plumes did cause some short-term local cooling in parts of the Middle East and southern Asia. The nation of Bahrain, in the Persian Gulf, reported May 1991 to be the coolest in the last 35 years. It is unlikely that the fires, massive though they were, would have a detectable impact on global climate change mechanisms.

CAREERS IN METEOROLOGY

What is a **meteorologist**?

When the public hears the word "meteorologist," they often think of the person on television presenting the forecast. Many weathercasts are only weather reports, though increasingly many on-air weathercasters are highly trained scientific professionals. Meteorologists also work for the National Weather Service, preparing forecasts and warnings and making observations. But there are many other activities in which practitioners can be called meteorologists or alternately atmospheric scientists, atmospheric physicists, or atmospheric chemists. So what is a meteorologist? According to a guideline of the American Meteorological Society, it is a person with a specialized education, using scientific principles to explain, understand, observe, or forecast the Earth's atmospheric phenomena and/or how the atmosphere affects the Earth and life on the planet. This requires at least a four-year college degree in meteorology or related sciences. A masters or even doctoral degree is often required for many of the more advanced jobs. Many meteorologists also obtain degrees in fields such as chemistry, mathematics, electrical or computer engineering, or other branches of the physical sciences, and then become involved in studying the atmosphere.

Is meteorology a **good career**?

According to the most recent *Jobs Ranked Almanac,* the career of meteorologist now ranks 7th out of 250. This is a big move up from 38th place in the 1988 edition. The rankings are based on factors such as environment, employment outlook, stress, security, physical demands, and income. The really interesting part about the field is that **361** you can be involved at many levels ranging from taking observations to working on high-end theoretical problems on supercomputers. You can find employment within many federal government agencies, the military, state and local government, universities, broadcasting, utilities, private industry, or engineering consulting firms, or you can be a self-employed consultant.

How many **meteorologists** are there?

In the United States it is estimated that there are about 30,000 to 35,000 men and woman whose professional activities involve some aspect of the atmospheric sciences. Some of the professionals might call themselves atmospheric scientists, environmental engineers, or atmospheric physicists or chemists, but they all deal with the atmosphere in some way or another. Very closely allied to meteorology are the oceanographic, limnological (study of lakes), and hydrological fields. Some of the wide variety of jobs that involve some aspects of meteorology or another include:

Operational forecaster	Satellite meteorologist
Radar meteorologist	Agricultural forecaster
Climatologist	Commodities trader
Hydrological engineer	Aviation forecaster
Emergency planner	Instrument designer
Fire weather forecaster	Broadcaster
Flood forecaster	High school or university teacher
National laboratory researcher	Data communications engineer
Remote sensing specialist	Air quality forecaster
Air quality modeler	Hurricane researcher
Atmospheric chemist	Acid precipitation researcher
Atmospheric optics researcher	Radio propagation researcher
Severe storm forecaster	Numerical forecasting modeler
Air traffic control assistant	Computer visualization specialist
Bioclimatologist	Lightning researcher
Paleoclimatologist	Forensic specialist

Would **meteorology** be a good career for you?

Ultimately the answer to the above question is the same as for any career—if you would enjoy the work, then it's a good career. Here are some questions you may wish to ask yourself if you are considering a career in meteorology:

Am I curious about the physical world about me, and why it is the way it is? Have I always watched the sky, read books on science and weather, taken my own observations?

Would I like to work in a field of science that has many applications in human

Complex wind and power plant plume patterns in mountainous terrain are studied using computer models of atmospheric flow.

Am I intrigued by the concept of using mathematics as a language to describe things that happen in the natural world? Do I enjoy working with computers?

Do I have the ability to conceptualize three-dimensional physical phenomena?

Do I enjoy and do well in math, physics, chemistry, and computer courses?

Am I open to change, working in a field where developments occur at a break-neck pace?

For those interested in forecast and/or broadcasting: Am I willing to work shifts and be transferred to a number of job locations until becoming established?

Many meteorologists swear that you are born with the love of weather. Many were "weather freaks" as kids and thought nothing of staying up all night to watch a snowstorm. Many others, however, have entered the field from other disciplines as interests and opportunities availed themselves.

What are the **qualifications** to be a meteorologist?

According to *The Jobs Rated Almanac*, the number of new positions for trained meteorologists will continue to grow over the next decade. This would appear especially true for students with advanced degrees in the atmospheric sciences and who are highly trained in computer-related skills. While there are a few low-level jobs for those **363** with only high school education, the vast number of meteorologists and atmospheric scientists have four-year college bachelor degrees, and many have graduate level masters and doctoral degrees. The nature of most work in atmospheric sciences is such that a companion discipline besides classical meteorology is becoming almost essential. Meteorology students often take extra course work, double majors, or advanced degrees in areas such as physics, computer science, electrical engineering, physical chemistry, numerical methods, ecology, horticulture, or hydrology. In almost all cases a strong foundation in computer sciences and applications is essential.

Where do you get **training** in meteorology or atmospheric sciences?

The first university in the United States to have a formal meteorology degree program was the Massachusetts Institute of Technology. The department was founded in 1928 by Swedish scientist Carl-Gustaf Rossby. There are dozens of institutions of higher learning that offer formal degree programs. A complete listing of these universities, their faculties, facilities, and course offerings can be obtained from the American Meteorological Society, which periodically publishes its "Curricula in the Atmospheric, Oceanic, Hydrological and Related Sciences." A recent census showed that there were at least 77 universities offering Ph.D.-level programs, with an additional 9 only going to the masters degree level, and 18 providing only bachelor-level programs. The number of doctoral (Ph.D.) degrees currently granted by U.S. universities in meteorology and related fields is estimated at about 100 per year.

The military services will provide "in-house" training for the specific tasks they need to have accomplished. Many officers are also enrolled in university degree programs.

There are some correspondence courses available from various organizations. While useful, they may not always meet the requirements of prospective employers. Check out the suitability of such courses before investing too much time or money.

Do all meteorologists work for the **National Weather Service or television stations**?

Actually, these two groups, with whom the public has the most contact, comprise a minority of the nation's meteorological practitioners. About 1,000 meteorologists are actively working in the media. The National Weather Service employs approximately 5,000 professionals. At least 6,000 meteorologists are working in the rapidly expanding private sector.

What is the **private sector**?

With the end of World War II, thousands of returning military meteorologists were

Who hires meteorologists?

Meteorologists are employed by a wide variety of organizations besides the National Weather Service and television stations. They can be found serving in many government agencies including the Federal Aviation Administration (FAA), National Oceanic and Atmospheric Administration (NOAA), National Aeronautics and Space Administration (NASA), the Environmental Protection Agency (EPA), Department of Energy (DOE), the U.S. Air Force and Navy, U.S. Geological Survey, and the several National Laboratories. Many meteorologists work for state and county air pollution control departments, airlines, private forecasting services, meteorological equipment manufacturers, university research groups, litigation support companies, basic and applied research contractors, electric nuclear and gas utilities, flood control districts, fire control districts, climate research laboratories, weather modification companies, and marine services. Many environmental companies involved in helping industry comply with the Clean Air Act have large numbers of meteorologists on board.

looking for work. At that time the only significant employers were the (then) Weather Bureau and universities. Many introduced to the science during the war wanted to stay active in the field. Thus was born the private sector. Many small companies providing specialized forecasts were started. After a period of initial antipathy from the government establishment, the relationship between the two groups has generally been harmonious and mutually beneficial. A recent survey of private sector meteorologists listed their job responsibilities in general descending order of practitioners:

Weather forecasting
Broadcast meteorology
General consulting
Air quality
Computer programming
Research & development
Environmental impact studies
Systems integration
Climatology

Today in Europe there are many countries in which the laws are heavily skewed against the private practice of meteorology. Government weather services, facing declining budgets, try to enhance their funding by performing services normally reserved for the private sector in the United States.

Growth chambers used in studying the impact of air pollutants on crops at the University of Minnesota.

Where does the **Weather Channel** get its forecasts?

The 24-hour, all-weather cable channel, like most broadcast meteorology operations, receives the bulk of its weather data and forecast guidance from the National Weather Service. The real service provided by such an organization is to have skilled professional meteorologists and electronic graphic artists sort through the vast stream of incoming information and present it in a form that can be easily understood by the general public. Their ability to continuously update the weather reports as conditions change is especially valuable during severe weather.

What are the **salary scales** in the private sector?

According to a recent survey conducted of private sector meteorologists by the American Meteorological Society, the distribution of salaries (for all positions) as a function of experience is as follows:

Annual salary	Entry Level	1–2 years	3–5 years
< $15,000	7.1 percent	0.0 percent	0.0 percent
$15,000–20,000	27.5 percent	11.5 percent	2.1 percent
$20,000–25,000	28.1 percent	22.0 percent	6.4 percent
$25,000–30,000	24.4 percent	31.0 percent	13.8 percent

$30,000–40,000	9.9 percent	29.1 percent	39.8 percent
$40,000–50,000	1.5 percent	4.0 percent	27.8 percent
$50,000–75,000	1.2 percent	1.5 percent	6.4 percent
>$75,000	0.3 percent	0.9 percent	3.7 percent

Salary rates for government employees are generally set according to the standard civil service pay scales. Salaries in the television weather business are a strong function of the size of the media market and the weathercaster's longevity in that market (and the negotiating skill of his or her agent if one is used).

How do you get on **radio and television**?

Many aspiring meteorologists yearn to "get into television." It is possible, but it is not always easy. First, you need the basic training in meteorology, particularly in the practical aspects of forecasting. You should be reasonably telegenic, or at least have an engaging on-air personality. Computer skills are most valuable. College courses in communications, if available, are often very valuable. Starting out working at campus radio or television stations gets a foot in the door, as does volunteering for summer intern work (often without pay) at the local commercial television station. Having a mentor who already works in the business can be a big help. At some point you have to have a demo videotape of yourself to be able to send to news directors at various stations so they can see how you come across on the air. Getting that first demo tape is often a challenge, but sometimes it can be part of the deal for working as a summer intern or at the campus station. And getting on the air usually doesn't mean big city lights and lots of glamour at first. It usually means starting with the 6 A.M. weather cut-ins in a small town, working your way up to weekend weather in slightly larger markets, and then maybe the morning show in an even larger market. It usually takes ten years or more of service in the smaller markets before you can expect to become established in a major market. Plan to move a lot. And also plan for the fact that your job security may be minimal. But media work can be very professionally rewarding, some meteorologists do stay at the same station for decades, and even if you don't stay in the business forever, it is a good springboard to other activities.

What is the AMS **Radio and Television Seal of Approval**?

Not everyone presenting forecasts on radio or television is a trained meteorologist. Some are simply broadcasters who "rip and read," hopefully without their own embellishments, forecasts prepared by the National Weather Service or a private forecasting company. But there are many professional meteorologists working in the media. In order to help the viewers identify those weathercasters who indeed have the training, experience, and judgement to communicate the complex weather information in a professional and reliable manner, the American Meteorological Society has established its Seal of Approval. There are separate seals for radio and television. Each weathercaster presents their credentials and sample programs to a review board of his or her peers that accepts or declines their petition. Once granted, the Seal must be renewed periodically, encouraging the broadcasters to be involved with continuing education courses and upgrading the quality of their presentations.

What is the **American Meteorological Society**?

The primary scientific and professional society for atmospheric sciences in the United States is the American Meteorological Society. It has over 10,000 members who work in the various disciplines of meteorology, oceanography, and hydrology. The objectives of the Society are the development and dissemination of knowledge of the atmosphere and related oceanic and hydrological sciences, and the advancement of their professional applications. Membership is open to all. There is a grade of Associate Member for those who are interested in the goals of the Society but not educationally qualified for full membership. Student membership is available for those enrolled at least half-time at an accredited institution of higher learning. The AMS publishes a number of major technical journals including *Weather and Forecasting, Journal of Climate, Monthly Weather Review*, *Journal of the Atmospheric Sciences*, and the *Journal of Applied Meteorology*. The official publication of the AMS is the *Bulletin of the American Meteorological Society*. Their headquarters are located at 45 Beacon Street, Boston, Massachusetts, 02108-3693.

The AMS has a number of local charters scattered throughout the United States, many associated with universities and large National Weather Service installations. The Society publishes a number of technical journals, holds numerous conferences and technical meetings, promotes continuing education conferences and pre-college educational activities, and operates certification programs for consulting meteorologists and radio and television broadcasters.

Are there other **professional organizations** serving the atmospheric sciences?

Many atmospheric scientists are members of the American Geophysical Union (2000 Florida Avenue NW, Washington, D.C., 20009). The AGU conducts numerous scientific meetings that span the breadth of geophysics and publishes a number of well-regarded technical journals, including *Geophysical Research Letters* and the *Journal of Geophysical Research*. Those interested in forecasting and the operational aspects of meteorology often join the National Weather Association (6704 Wolke Court, Montgomery, Alabama, 36116-2134). The NWA publishes the *National Weather Digest*.

The National Center for Atmospheric Research (NCAR).

Is **financial aid** available to students interested in the atmospheric sciences?

Aside from the numerous financial aid programs offered by colleges and universities, there are several scholarship programs administered by the American Meteorological Society. Contact the AMS for information. Summer intern programs with a stipend are provided to undergraduate students through the National Council of Industrial Meteorologists. Check with the Meteorology Department office at the college and university for details.

The military weather services in the Air Force and Navy are a career option. If you choose to enter as a commissioned officer, you'll normally be awarded a ROTC scholarship to complete your bachelor's degree. After graduation, you attend Officer Training School.

Where do you find out about **jobs in meteorology**?

Networking and keeping your ear to the ground is probably the number one way to find out about positions. Asking people already in the business can be a big source of job leads. Starting out as an intern can be invaluable. The AMS publishes monthly listings of employment opportunities. Many college departments keep job announcements posted on the bulletin boards near the department office. Weathercasting positions are sometimes advertised in *Broadcasting*. Organizations such as the National Weather Service and the National Center for Atmospheric Research (NCAR) routinely publish their open positions (now often in electronic format). Many private sector jobs are arranged through personal contacts. If you have the training, good references, and healthy work ethic, chances are the work will find you. No matter what the position, you must be able to provide any employer with professional competence and good work habits.

What is aviation forecasting?

One of the earliest users of weather forecasting was the aviation industry. Today aviation weather forecasters work in the National Weather Service, the military, and private weather companies. Pilots need forecasts before they even take off to be sure that they will be able to land at their chosen destination or alternates. En route, avoidance of icing, severe turbulence, and thunderstorms is essential. Weather is still the leading cause of flight delays and cancellations (about half) and of aviation accidents, which result in 800 lives lost in the United States annually. Improvements are needed not only in the forecasts themselves but in the communication of the information to the pilots and their proper utilization of the forecasts. Accurate winds aloft forecasts are vital to economizing on fuel for long-haul jet traffic. Improved wind forecasting could save an additional quarter-billion dollars in aviation fuel each year.

What is **marine forecasting**?

The sea is a dangerous and unforgiving place. Mariners need the best weather information available. Forecasters prepare many products to aid both commercial and recreational maritime activities. By using winds to forecast waves and sea conditions, more economical routing of passenger and cargo vessels can save huge sums in fuel.

Avoiding areas of very rough seas is vital to safety. How high can waves get on the open ocean? One buoy in the North Atlantic reported one "swell" at 101 feet above mean water level. Rogue waves, some over 75 feet high, have long terrified mariners. In 1984, a 117-foot three-masted tall ship sunk in less than one minute north of Bermuda with a heavy loss of life after being struck by such a monster. Marine forecasters become involved in oil spill cleanup operations by assisting with predictions

A wind energy generator mines the air for electrical energy in the Great Plains.

about the spread of the slick. Warnings of tsunamis from earthquakes and landslides have saved countless lives.

Distributing weather forecasts and encouraging people to heed their message still remains a challenge. During the March 1993 "Storm of the Century," over 200 people perished. Many of these lives were lost needlessly. The storm was well fore-casted several days in advance. One ship's captain ignored a warning not to leave his Gulf Coast port. He paid for this imprudence with the loss of his ship, his crew of 33, and his own life.

What is **transportation forecasting**?

Railroads and especially highway traffic are seriously impacted by the weather. Each year weather causes or contributes to 6,000 fatalities on American highways. Keeping roads free of snow is critical. Over a half-billion dollars a year are spent preparing for snows that are predicted but never materialize. Clearly there is economic incentive for improvement. Many states are now establishing networks of weather stations specifi-cally to monitor highway weather conditions. Temperature forecasts are also impor-tant for trucking firms shipping perishable items. Sending an unheated truck through territory where the temperature can drop below zero can be a problem when there is a load of cut flowers on board. Railroads must be cognizant of severe weather (winds **371** and tornadoes) as well as deep and drifting snows during the winter. Private forecasting firms work closely with many transportation companies.

What is **fire weather forecasting**?

The condition of the atmosphere, especially wind, temperature, and relative humidity, greatly influence the spread of wildfires. Meteorologists work closely with forestry specialists in predicting when fire hazards are rising in order to mandate prevention measures (banning open burning in campgrounds, for example). Once a fire has started, forecasts become critical in determining where to make fire breaks, how to position (and evacuate) crews, and planning sorties by fire retardant bombing aircraft.

What is **agricultural forecasting**?

Crops are clearly sensitive to weather. And sometimes something can be done about it. Knowing that the next few days will be rain-free can prompt a farmer to make hay while the sun is shining, and thus maximize the yield. Forecasts of winds can assist in aerial application of chemicals while avoiding herbicide drift onto a neighbor's tomato field or organically grown fruit. Frost and freeze forecasts can be a call to action to fire up smudge pots, turn on wind machines or water sprinkling systems in the citrus groves. Cold, low lying spots in Wisconsin can have frost even during summer. Cranberries growing in the shallow bogs can be protected by flooding the field. The application of agricultural chemicals as well as irrigation water can be timed much more precisely and economically by monitoring past weather and matching it to predicted conditions.

Do **environmental firms** hire meteorologists?

Meteorologists have played a central role in much of the air quality research and control efforts in the United States over the past several decades. Atmospheric conditions play a key role in predicting the diffusion and transport (collectively called dispersion) of pollutants. If a new power plant is to be built, it is necessary to know the impact of the pollutants that it will release. Since one can't measure pollution concentrations before the plant is built, numerical models of pollution dispersion simulate the atmosphere's influence upon the plume once it leaves the proposed smoke stack. In an effort to control regional ozone, meteorologists work with chemists to create numerical "photochemical grid models" in which the known pollutant emissions are used to predict the ozone levels. Once these models are verified, then one can predict the consequences of planned emission controls, such as cutting automobile hydrocarbon emissions by 10 percent or power plant oxides of nitrogen releases by 30 percent. The complex models are necessary because the actual results of such controls can often be quite different from what might be expected.

What is **industrial forecasting**?

How much does a "bad winter" cost the U.S. economy? The harsh winter of 1976–77 was estimated to cause $37 billion in direct economic losses due to lost retail sales, increased energy consumption, difficulties in transportation and industrial production, and crop losses. Advanced warning can help reduce some of these losses. Cold weather, for instance, affects heating bills, thermal underwear sales, shipping—and video rentals. In many cities, video rentals have been known to double on weekends when the weather is exceptionally cold. Cold weather also means hot pizza. One Twin Cities pizza delivery establishment found that when it was bitterly cold, even normally hardy Minnesotans would rather someone else get the frostbite—his sales increased $400–500 on very cold evenings. Knowing that in advance means bringing in more help to meet demand. Baseball teams hire private forecasters to predict the beginning and end of rain to help the ground crews decide when to put on the tarps. On a larger scale, knowing that the temperature will jump 10°F in New York City on a summer day allows an electrical utility to purchase the needed extra power before the demand soars and the prices of power go up with it. The correct forecast of a few degree temperature rise or fall can save an electric company millions of dollars. Precipitation predictions for mountain reservoirs and drainage basins assist utility managers in planning hydroelectric power generation. Forecasts of temperature for snow making at a ski resort, rainfall on an outdoor movie set, relative humidity for a proscribed agricultural burn, and winds for a hot-air balloon rally are just a few of the many forecasts made for industry by the private weather forecasting firms in the United States.

What is **commodities forecasting**?

When weather forecasts can be used in conjunction with commodities trading, some significant profits can be generated (losses can be whopping, too). Three days' notice of a killing freeze in the Florida citrus belt can be the signal to buy orange juice futures. An overnight rain in the parched soybean belt can drive prices down quickly. Advance knowledge of the rainfall can prompt a profitable sale before prices collapse. In many ways this type of weather prediction deals with the complex connectedness of things in nature. How did a cool spell in Wisconsin in July 1994 affect the price of your Thanksgiving cranberries? Turns out the chilly mid-summer weather in the Wisconsin cranberry bogs hampered the pollination of the berries by the local bees. And while the final berry yield was 1.6 million tons, it was below a good year's output. Supply and demand takes over, and your Thanksgiving condiment goes up in price.

It was wet, then it was warm in Iowa in February 1992. The result? Mud. So much mud that school buses got bogged down, some schools even closed, and hogs couldn't be brought to market. Cattle had trouble putting on weight as they were working so hard carrying around all that mud. But commodities traders who know about potential problems like this before the rest of the market catches on can use it to their economic advantage.

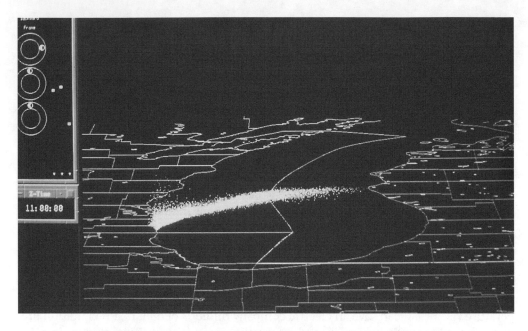

This computer simulation of the dispersion by winds of hypothetical radioactive pollutants released from a nuclear power plant was created for emergency response planning and evacuation studies.

What is **wind engineering**?

On 7 November 1940, brisk winds of 30–35 mph caused the structure of the Narrows Bridge in Tacoma, Washington, to vibrate so excessively, the entire structure collapsed into the water. And "Galloping Gertie," as the span was dubbed, entered the engineering Hall of Shame. But it also gave impetus to a field that could be called wind engineering. Meteorologists must work with structural and civil engineers when high-rise buildings, sports stadiums, bridges, and other large structures are designed and built. The failure to properly account for the characteristics of the wind in a given area can be a costly mistake. Also, the "wind proofing" of even ordinary homes can potentially save billions of dollars when the next major hurricane slams ashore.

What is **forensic meteorology**?

The forensic meteorologist, who may act as either a background consultant or an actual testifying expert, will collect, interpret, and analyze atmospheric data in support of insurance fraud claim investigations, civil and criminal trials, and environmental regulatory actions. The forensic meteorologist may be employed directly by an insurance company, the attorneys for either the plaintiff or defendant in a case, or, with increasing frequency, may be appointed by the court itself. Regardless of the employing party, it is not the role of the meteorologist to be an advocate for either side

Meteorologists can apply their skills to the operation of nuclear generating stations such as this one on the Lake Michigan shoreline.

in a dispute, but to assist the judge and/or jury in understanding the often complex facts in a case so that they may reach an appropriate verdict.

Some typical problems dealt with in forensic meteorology include the following: An automobile accident was caused by poor visibility—was that from a natural fog or pollutants from a nearby industrial plant? Was the building damaged by a tornado or a straight line thunderstorm wind? A person was found dead near a downed power line—was it a fault in the utilities' line or a lightning strike? How can we demonstrate that rain fell at a site that is located many miles distant from any National Weather Service reporting station?

The forensic meteorologist may collect standard weather observations, assemble weather radar and satellite imagery, process weather data taken by a party in the case, or locate non-standard sources of data such as lightning ground stroke reports or atmospheric data taken by air pollution monitoring networks. These data are then used in a comprehensive analysis of the meteorological facts pertinent to a case. There is increasing use of sophisticated computer graphics and video animation of weather information in trials and administrative hearings. Most forensic meteorologists have had long and varied careers in the atmospheric sciences, and it is their hard-earned expertise that is in demand. Few recent graduates can expect to be heavily engaged in such activities until they have significantly enhanced their resumes. Most successful forensic meteorologists have met the qualifications of Certified Consulting Meteorologist (CCM).

What is **emergency response planning**?

Weather is responsible for 85 percent of the U.S. areas declared a natural disaster area by the President. The average annual direct damage toll due to natural disasters in the United States has been running at about $6 billion. In recent years the toll has been much higher, such as 1989's $15+ billion and 1992's $30+ billion. While natural disasters usually cannot be prevented, adequate planning and timely warnings can greatly reduce the loss in human lives and property.

Federal, state, and local governments coordinate the distribution and response to weather warnings. The planning of evacuation routes and shelter locations in hurricane zones is one example. Conducting community awareness programs in flood and tornado-prone regions and upgrading building codes to make buildings less prone to wind failure are current activities. Dispersion meteorologists routinely simulate in computers the consequences of hypothetical accidents at nuclear and chemical plants so that the impact of any real accident can be greatly minimized.

What is **atmospheric chemistry**?

Atmospheric chemistry is a scientific discipline that deals with the chemical constituents in our air. The problems addressed, often at a very highly theoretical level, include understanding and predicting stratospheric ozone levels, which are now known to be strongly influenced by chemicals injected into the atmosphere by humans. Closer to the ground, unraveling the problem of regional smog has remained a major challenge. The fate of many chemicals released into the atmosphere and their interactions with ecosystems is under close study. The emissions of natural pollutants from trees, soil micro-organism, and geological processes is vital to understanding global chemical balances. There are now more than 10 million manufactured chemicals that have been identified. Many of them are released into the atmosphere with as yet unknown consequences.

What is **atmospheric physics**?

Many scientists with training in physics work on atmospheric-related problems. These can include issues related to radar and radio wave propagation, optical propagation, acoustics and spectroscopy, to name just a few. The field can be highly theoretical and mathematical. There are also many observationally oriented programs in which new sensing systems are developed such as Doppler lidar, radar, and acoustic profiler and satellite sounding systems.

What are **cooperative weather observers**?

While the federal government operates several hundred full- and part-time weather

What is the Skywarn Program?

Many professional and amateur meteorologists spend a lot of time watching the skies when severe weather threatens. Storm chasing has become an obsession with some. Since you're spending all that time cloud watching, you can also reap a very positive side benefit. The Skywarn Program is organized by the National Weather Service in cooperation with many local emergency management offices to establish a network of trained severe storm and tornado spotters. Even with the sophistication of the NEXRAD Doppler radar, nothing surpasses a reliable eyewitness report from a trained spotter of a tornado touchdown, large hail, or damaging winds. If you have an amateur radio license, so much the better. If you are interested in contributing your weather enthusiasm to the effort, contact your local NWS office or emergency management official, or check with the local amateur radio organizations. Training is provided and you will get to meet many fellow weather enthusiasts—and perhaps help save someone's life.

stations in the United States, given the size of the nation, many more stations are needed to fully define our weather and climate. The number of points is greatly expanded by the nation's team of cooperative weather observers. These volunteers take vital temperature, rainfall, and snow fall measurements on a daily basis at almost 10,000 sites across the nation. Their instruments are checked by the National Weather Service. Over five million cooperative reports are collected and assembled each year at the National Climatic Data Center.

Many of these cooperative stations remain within the same families for generations. Colorado has eleven cooperative weather observer stations that have been in operation for 100 years or more. Such continuous strings of weather data taken in the same placed are invaluable in assessing whether or not global and regional climate is in fact changing with time. The earliest known such weather records in America were kept by a clergyman at Swedes' Fort, near Wilmington, Delaware, in 1644 and 1645.

Weather observing is a major hobby in the United States. Two publications that would be of interest to hobbyists are the *American Weather Observer* (401 Whitney Blvd, Belvidere, Illinois, 61008-3772; 815/544-9811; awowx@aol.com) and *Weatherwise* magazine (published by Heldreff Publishers, Washington, D.C.; call 1-800-365-9753 to place a subscription).

What is a **State Climatologist**?

Once upon a time, each state had a federally appointed State Climatologist. Budget cutting ended that. But most states have found ways to re-establish this vitally important post, which helps relate weather and climate patterns to the state's economic activities. Only Pennsylvania and New Mexico do not have an official climatologist. They usually are associated with universities and play a key role in collecting weather data from a state's cooperative observers, assembling reports of severe weather, and performing studies relevant to a state's economic or safety interests.

What is a **Certified Consulting Meteorologist**?

The title of Certified Consulting Meteorologist (CCM) is generated by the American Meteorological Society. It is a formal recognition on the part of colleagues, acting through their Society, that an applicant is considered well qualified to carry on the work as a consulting meteorologist. The qualifications for certification are centered around the fundamental characteristics of knowledge, experience, and character. The CCM program is a service for the general public provided by the AMS, in order to certify that certain individuals have been tested and found to meet or exceed its high standards. The CCM designation provides a basis on which a client seeking assistance on problems of a meteorological nature may be assured a mature, competent, and ethical professional counsel. Only about 5 percent (500) of the Society's members have earned CCM certification.

How do you find a **Certified Consulting Meteorologist**?

Have a weather-related problem in your business and don't know how to locate a competent meteorological consultant or an expert witness for a trial? The American Meteorological Society will be pleased to provide a listing of Certified Consulting Meteorologists (CCMs) serving your area. Contact the AMS at 45 Beacon Street, Boston, Massachusetts, 02108 (617/227-2425) for assistance. A listing of meteorological consultants is also published monthly in the back of the *Bulletin of the American Meteorological Society*. Many private-sector firms also advertise in the yellow pages and professional and trade journal directories.

Further Reading

Books

Abbott, Patrick L. *Natural Disasters*. Wm. C. Brown Publishers, 1996.

Academic American Encyclopedia. Grolier, 1991. 21 vols.

Ahrens, C. Donald. *Meteorology Today: An Introduction to Weather, Climate and the Environment*. West Publishing, 1988.

Allen, Everett S. *A Wind to Shake the World*. Little, Brown and Company, 1976.

Allen, Oliver E. *Atmosphere*. (Planet Earth, Vol. 8) Time-Life Books, Inc., 1983.

Andrews, David G., James R. Holton, and Conway B. Leovy, Eds. *Middle Atmosphere Dynamics*. Academic Press, 1987.

Andrews, J. E. et al. *An Introduction to Environmental Chemistry*. Blackwell Science, 1996.

Asnani, G. C. *Tropical Meteorology*, Volumes 1 and 2. G. C. Asnani, 1994.

Atkinson, B. W. *Meso-Scale Atmospheric Circulations*. Academic Press, 1981.

Azad, Ram S. *Atmospheric Boundary Layer for Engineers*. Kluwer Academic Publishers, 1993.

Bach, Wilfrid. *Atmospheric Pollution*. McGraw-Hill, 1972.

Barnes, Jay. *North Carolina's Hurricane History*. The University of North Carolina Press, 1995.

Barry, Roger G., and R. J. Chorley. *Atmosphere, Weather, and Climate*. Methuen, 1987.

Battan, Louis J. *Cloud Physics and Cloud Seeding*. Anchor Books, Doubleday, 1962.

———. *Harvesting the Clouds*. Anchor Books, Doubleday, 1969.

———. *Radar Observations of the Atmosphere*. University of Chicago Press, 1981.

———. *Fundamentals of Meteorology*. Prentice-Hall, 1984.

———. *Weather*. Prentice-Hall, 1985.

Bentley, W. A., and W. J. Humphreys. *Snow Crystals*. Dover Publications, 1962.

Bernard, Harold W., Jr. *Global Warming Unchecked—Signs to Watch For*. Indiana University Press, 1994.

Birks, John W., Jack G. Calvert, and Robert E. Sievers, Eds. *The Chemistry of the Atmosphere: Its Impact on Global Change*. American Chemical Society, 1993.

Blanchard, Duncan C. *From Raindrops to Volcanoes*. Anchor Books, Doubleday, 1967.

Bluestein, Howard B. *Synoptic-Dynamic Meteorology in Midlatitudes. Volume I: Principles of Kinematics and Dynamics*. Oxford University Press, 1992.

379

———. *Synoptic-Dynamic Meteorology in Midlatitudes. Volume II. Observations and Theory of Weather Systems.* Oxford University Press, 1993.

Bohren, Craig F., *Clouds in a Glass of Beer.* Wiley, 1987.

Boubel, Richard W., Donald L. Fox, D. Bruce Turner, and Arthur C. Stern. *Fundamentals of Air Pollution.* Academic Press, 1994.

Bradley, R. S., and P. D. Jones, Eds. *Climate Since A.D. 1500.* Chapman & Hall, 1992.

Brasseur, Guy and Susan Salomon. *Aeronomy of the Middle Atmosphere.* Kluwer Academic, 1986.

Brindze, Ruth. *Hurricanes: Monster Storms from the Sea.* New York: Athenaeum, 1973.

Brodine, Virginia, Ed. *Air Pollution.* Harcourt Brace Jovanovich, 1973.

Brown, Ewing Franklin. *The Weathermen Let Them Fly.* Eight Air Force Historical Society, 1994.

Brutsaert, Wilfried. *Evaporation into the Atmosphere: Theory, History and Applications.* Kluwer Academic, 1982.

Budyko, M. I., A. B. Ronov, and A. L. Yanshin. *History of the Earth's Atmosphere.* Springer-Verlag, 1987.

Burgess, E., and D. Torr. *Into the Thermosphere.* U.S. Government Printing Office, 1987.

Burridge, D. M., and E. Kallen, Eds. *Problems and Prospects in Long and Medium Range Weather Forecasting.* Springer-Verlag, 1983.

Burroughs, W. J. *Weather Cycles: Real or Imaginary?* Cambridge University Press, 1992.

Caracena, F., R. L. Holle, and C. A. Doswell III. *Microbursts: A Handbook for Visual Identification.* U.S. Government Printing Office, 1989.

Carnegie Library of Pittsburgh. *The Handy Science Answer Book,* 2nd ed. Visible Ink Press, 1997.

Carpenter, C. *The Changing World of Weather.* Facts on File, 1991.

Cermak, Jack E., Alan G. Davenport, Erich J. Plate, and Domingos X. Viegas, Eds. *Wind Climate in Cities.* Springer-Verlag, 1994.

Chang, Jen-hu. *Atmospheric Circulation Systems and Climates.* Oriental Publishing Company, 1972.

Changnon, Stanley A., Ed. *The Great Flood of 1993.* Westview Press, 1996.

Chelius, Carl R., and Henry J. Frentz. *Basic Meteorology Exercise Manual.* Kendall-Hunt, 1984.

Church, C., D. Burgess, C. Doswell, and R. Davies-Jones, Eds. *The Tornado: Its Structure, Dynamics, Prediction, and Hazards.* American Geophysical Union, 1993.

Cole, Franklyn W. *Introduction to Meteorology.* Wiley, 1980.

Comfort, L. K., Ed. *Managing Disaster: Strategy and Policy Perspectives.* Duke University, 1988.

Committee on Earth and Environmental Science. *Predicting the Weather,* 1992.

Compton, Grant. *What Does A Meteorologist Do?* Mead, 1981.

Corliss, William R., Ed. *Handbook of Unusual Natural Phenomena.* Sourcebook Project, 1977.

———. *Tornadoes, Dark Days, Anomalous Precipitation, with Related Weather Phenomena.* Sourcebook Project, 1983.

———. *Rare Halos, Mirages, Anomalous Rainbows and Related Weather Phenomena.* Sourcebook Project, 1984.

———. *Science Frontiers: Some Anomalies and Curiosities of Nature.* Sourcebook Project, 1994.

Cothern, C. Richard, and N. Phillip Ross. *Environmental Statistics, Assessment, and Forecasting.* CRC Press, 1994.

Cotton, William R., and Richard A. Anthes. *Storm and Cloud Dynamics.* Academic Press, 1989.

Cotton, William R., and Roger A. Pielke. *Human Impacts on Weather and Climate.* Cambridge University Press, 1995.

Craig, Richard A. *The Edge of Space.* Anchor Books, Doubleday, 1968.

Crowder, Bob. *The Wonders of the Weather.* Australian Government Publishing Service, 1995.

Daley, Roger. *Atmospheric Data Analysis.* Cambridge University Press, 1993.

Darden, Lloyd. *The Earth in the Looking Glass.* Anchor Press, Doubleday, 1974.

De Blij, H. J. *Nature on the Rampage.* Smithsonian Books, 1994.

Diaz, Henry F., and Vera Markgraf, Eds. *El Niño: Historical and Paleoclimatic Aspects of the Southern Oscillation.* Cambridge University Press, 1992.

Donn, William L. *Meteorology.* McGraw-Hill, 1965.

Douglas, P. *Prairie Skies: The Minnesota Weather Book.* Voyageur Press, 1990.

Drewry, D. J., R. M. Laws, and P. A. Pyle, Eds. *Antarctica and Environmental Change.* Oxford University Press, 1994.

Dunlop, Storm, and Francis Wilson. *Weather and Forecasting.* Collier, 1987.

Dutton, John A. *The Ceaseless Wind: An Introduction to the Theory of Atmospheric Motion.* Dover, 1986.

———. *Dynamics of Atmospheric Motion.* Dover, 1995.

Duxbury, Alison, and Alyn C. Duxbury. *An Introduction to the World's Oceans.* Wm. C. Brown, 1988.

Eagleman, Joe R. *Severe and Unusual Weather.* Trimedia Publishing, 1988.

———. *Air Pollution Meteorology.* Trimedia Publishing, 1991.

———. *Meteorology: The Atmosphere in Action.* Trimedia Publishing, 1995.

Earth Works Group. *50 Simple Things You Can Do to Save the Earth.* Earthworks Press, 1989.

Edinger, James G. *Watching for the Wind.* Educational Services, 1967.

Eichenlaub, Val L., Jay R. Harman, Fred V. Nurberger, and Hans J. Stolle. *The Climatic Atlas of Michigan.* University of Notre Dame Press, 1990.

England, Gary. *United States Weather.* Oklahoma Edition. United States Weather Corporation, 1976.

———. *Weathering the Storm: Tornadoes, Television and Turmoil.* University of Oklahoma Press, 1996.

Epley, A. P. *The New Richmond Tornado of 1899: A Modern Herculaneum.* Michael G. Corenthal, 1988.

Fifield, R. *International Research in the Antarctic.* Oxford University Press, 1987.

Firor, J. *The Changing Atmosphere: A Global Challenge.* Yale University Press, 1990.

Fishman, Jack, and Robert Kalish. *The Weather Revolution: Innovations and Imminent Breakthroughs in Accurate Forecasting.* Plenum Publishing, 1994.

Flatow, Ira. *Rainbows, Curve Balls, and Other Wonders of the Natural World Explained.* Times Books, Random House, 1991.

Flavin, Christopher, and Nicholas Lenssen. *Power Surge.* W. W. Norton, 1994.

Fleagle, Robert G., and Joost A. Businger. *An Introduction to Atmospheric Physics.* Academic Press, 1963.

Fleagle, Robert G. *Global Environmental Change: Interactions of Science, Policy, and Politics in the United States.* Greenwood Publishing Group, 1994.

Fleming, J. R. *Meteorology in America, 1800–1870.* Johns Hopkins University Press, 1990.

Flohn, Hermann. *Climate and Weather.* McGraw-Hill, 1969.

Forrester, Frank H. *1001 Questions Answered about the Weather.* Dover, 1981.

Freier, George D. *Weather Proverbs.* Fisher Books, 1992.

Frenzel, Burkhard, Christian Pfister, and Birgit Gläser, Eds. *Climatic Trends and Anomalies in Europe 1675–1715.* Gustav Fisher Verlag, 1994.

Friday, L., and R. Laskey. *The Fragile Environment.* Cambridge University Press, 1989.

Fritts, Harold C. *Tree Rings and Climate.* Academic Press, 1976.

Gadsden, M., and W. Schröder. *Noctilucent Clouds.* Springer-Verlag, 1989.

Garratt, J. R. *The Atmospheric Boundary Layer.* Cambridge University Press, 1992.

Gates, David M. *Climate Change and Its Biological Consequences.* Sinauer Associates, 1993.

Gay, Kathlyn. *Ozone.* Franklin Watts, 1986.

Geigner, Rudolf, Robert H. Aron, and Paul Todhunter. *The Climate Near the Ground.* Vieweg Publishing, 1995.

Gipe, Paul. *Wind Power for Home & Business.* Chelsea Green Publishing, 1993.

Gill, Adrian. *Atmosphere-Ocean Dynamics.* Academic Press, 1982.

Gilmer, Maureen. *Living on Flood Plains and Wetlands: A Homeowner's Handbook.* Taylor Publishing, 1995.

Glantz, Michael H., Ed. *Drought Follows the Plow.* Cambridge University Press, 1994.

Glantz, M. H. *Societal Response to Regional Climatic Change: Forecasting by Analogy.* Westview Press, 1988.

Goals for Predicting Seasonal-to-Interannual Climate. National Research Council, National Academy Press, 1994.

Gore, Al. *Earth in Balance.* Houghton Mifflin, 1992.

Godish, Thad. *Indoor Air Pollution Control.* Lewis Publishers, 1989.

Graedel, Thomas E., and Paul J. Crutzen. *Atmospheric Change: An Earth System Perspective.* W. H. Freeman, 1993.

———. *Atmosphere, Climate and Change*. W. H. Freeman, 1995.

Greenler, Robert. *Rainbows, Halos and Glories*. Cambridge University Press, 1980.

Grefen, K., and J. Loebel. *Environmental Meteorology*. Springer-Verlag, 1989.

Gross, M. Grant. *Oceanography*. Merrill Publishing, 1971.

Groves, Don. *The Ocean Book*. Wiley, 1989.

Haltiner, George J., and Roger Terry Williams. *Numerical Prediction and Dynamic Meteorology*. Wiley, 1980.

Hanish, Ted, Ed. *Climate Change and the Agenda for Research*. Westview Press, 1994.

Hartmann, Dennis L. *Global Physical Climatology*. Academic Press, 1994.

Hasse, L., and F. Dobson. *Introductory Physics of the Ocean and Atmosphere*. Kluwer Academic, 1985.

Haynes, R. *Sentinels in the Sky: Weather Satellites*. U.S. Government Printing Office, 1988.

Henderson-Sellers, Ann, and Ann-Maree Hanson. *Climate Change Atlas*. Kluwer Academic Publishers, 1995.

Henry, James A., Kenneth M. Portier, and Jan Coyne. *The Climate and Weather of Florida*. Pineapple Press, 1994.

Hidore, J. J., and J. E. Oliver. *Climatology: An Atmospheric Science*. Macmillan, 1993.

Hidy, George M. *Atmospheric Sulfur and Nitrogen Oxides: Eastern North American Source-Receptor Relationships*. Academic Press, 1994.

Hobbs, Peter V. *Basic Physical Chemistry for the Atmospheric Sciences*. Cambridge University Press, 1995.

———, and A. Deepak, Eds. *Clouds, Their Formation, Optical Properties and Effects*. Academic Press, 1981.

———. *Aerosols and Climate*. A. Deepak Publishing, 1988.

Hoff, Mary, and Mary M. Rodgers. *Our Endangered Planet: Groundwater*. Lerner Publications, 1991.

Holton, James R. *The Dynamic Meteorology of the Stratosphere and Mesosphere*. American Meteorological Society, 1975.

———. *Introduction to Dynamic Meteorology*. New York, Academic Press, 1980.

Houghton, David D., Ed. *Handbook of Applied Meteorology*. John Wiley, 1985.

Houghton, Henry G. *Physical Meteorology*. MIT Press, 1985.

Houghton, John. *Global Warming: The Complete Briefing*. Lion Publishing, 1994.

Houze, Robert A., Jr. *Cloud Dynamics*. Academic Press, 1993.

Hughes, Patrick. *American Weather Stories*. U.S. Department of Commerce, 1976.

Inglish, Howard, Ed. *Year of the Storms: The Destructive Kansas Weather of 1990*. Hearth Publishing, 1990.

International Cloud Atlas: Volume II. World Meteorological Organization, 1987.

Isaksen, I. S. A., Ed. *Tropospheric Ozone: Regional and Global Scale Interactions*. D. Reidel Publishing, 1988.

Iver, David F. *Dictionary of Astronomy, Space, and Atmospheric Phenomena*. Van Nostrand Reinhold, 1979.

Johnson, Howard L., and Claude F. Duchon. *The Atlas of Oklahoma Climate*. University of Oklahoma Press, 1995.

Jones, Ian, and Joyce Jones. *Oceanography in the Days of Sail*. Hale & Iremonger, 1992.

Kahl, Jonathan D. *Thunderbolt: Learning about Lightning*. Lerner Publications, 1993.

———. *Storm Warning: Tornadoes and Hurricanes*. Lerner Publications, 1993.

———. *Weatherwise: Learning about the Weather*. Lerner Publications, 1993.

———. *Wet Weather: Rain Showers and Snowfall*. Lerner Publications, 1993.

Kaimal, J. C., and J. J. Finnigan. *Atmospheric Boundary Layer Flows: Their Structure and Measurement*. Oxford University Press, 1994.

Kessler, Edwin, Ed. *Thunderstorm Morphology and Dynamics*. University of Oklahoma Press, 1986.

Kidder, Stanley Q., and Thomas H. Vonder Haar. *Satellite Meteorology: An Introduction*. Academic Press, 1995.

Kirchheimer, Sid. *The Doctor's Book of Home Remedies*. Rodale Press, 1993.

Knowles, Middleton, W. E., and Athelstan F. Spilhaus. *Meteorological Instruments*. University of Toronto Press, 1953.

Komen et al. *Dynamics and Modelling of Ocean Waves*. Cambridge University Press, 1994.

Kotlyakov, V. M., A. Ushakov, and A. Glazovsky, Eds. *Glaciers-Ocean-Atmosphere Interactions.* International Association of Hydrological Science Press, 1991.

Kramer, Stephen. *Lightning.* Carolrhoda Books, 1992.

———. *Tornado.* Carolrhoda Books, 1992.

Kraus, Eric B., and Joost A. Businger. *Atmosphere-Ocean Interaction.* Oxford University Press, 1994.

Kristovich, and Robert W. Grumbine. *Discover Weather.* Publications International, 1992.

Lambert, David, and Ralph Hardy. *Weather and Its Works.* Facts on File, 1985.

Landsberg, Helmut E. *The Urban Climate.* Academic Press, 1981.

Laskin, David. *Braving the Elements: The Stormy History of American Weather.* Doubleday, 1996.

Lee, Sally. *Predicting Violent Storms.* New York: Franklin Watts, 1989.

Lefohn, A. S., and S. F. Krupa, Eds. *Acidic Precipitation: A Technical Amplification of NAPAP's Findings.* Air and Waste Management Assoc., 1988.

Leggett, Jeremy. *Global Warming.* Oxford University Press, 1990.

Lester, Peter F. *Aviation Weather.* Jeppesen Sanderson, 1995.

Levenson, T. *Ice Time: Climate, Science and Life on Earth.* Harper and Row, 1989.

Lewis, John S., and Ronald G. Prinn. *Planets and Their Atmospheres.* Academic Press, 1983.

Linke, Sigfried. *Know about Weather.* Academy Chicago Publications, 1979.

Liou, Kuo-Nan, Ed. *An Introduction to Atmospheric Radiation.* Academic Press, 1980.

Lockhart, G. *The Weather Companion: An Album of Meteorological History, Science, Legend and Folklore.* Wiley, 1988.

Lodge, James P., Jr., Ed. *The Smoke of London.* Maxwell Reprint Company, 1969.

Loffman, T., and R. Mann. *International Traveler's Weather Guide.* Weather Press, 1990.

Lorenz, Edward N. *The Essence of Chaos.* University of Washington Press, 1993.

Ludlum, David M. *The American Weather Book.* Houghton Mifflin Company, 1982.

Lutgens, Frederick, and Edward Tarbuck. *The Atmosphere: An Introduction to Meteorology.* Prentice-Hall, 1989.

Lynch, David. *Atmospheric Phenomena Readings from Scientific American.* W. H. Freeman, 1980.

———. *Color and Light in Nature.* Cambridge University Press, 1995.

Lyons, T., and W. Scott. *Principles of Air Pollution Meteorology.* Belhaven Press, 1990.

Macchio, William J. *Deadly Waters.* Historic Publications, 1994.

Magono, Choji. *Thunderstorms.* Elsevier, 1980.

Makower. *The Air and Space Catalog: The Complete Sourcebook to Everything in the Universe.* Vintage Books, 1990.

Mason, B. J. *Acid Rain: Its Causes and Its Effects on Inland Waters.* Oxford University Press, 1992.

Mather, John R., and Marie Sanderson. *The Genius of C. Warren Thornthwaite, Climatologist-Geographer.* University of Oklahoma Press, 1996.

Matsuno, T. *Short and Medium Range Numerical Weather Prediction.* Universal Academy Press, 1987.

Maunder, W. John, Ed. *Dictionary of Global Climate Change.* Chapman & Hall, 1992.

McCormac, B. M., Ed. *Introduction to the Scientific Study of Atmospheric Pollution.* D. Reidel Publishing, 1971.

McIlveen, R. *Basic Meteorology.* Van Nostrand Reinhold, 1986.

Meinel, Aden, and Marjorie Meinel. *Sunsets, Twilights and Evening Skies.* Cambridge University Press, 1983.

Miller, Albert, and Richard Anthes, Eds. *Meteorology.* Merrill Publishing, 1985.

Miller, Albert, Jack C. Thompson, and Richard E. Peterson, revised by Donald R. Haragen. *Elements of Meteorology.* Merrill Publishing, 1983.

Minsinger, William Elliott, and Charles Talcott Orloff, Eds. *The North Atlantic Hurricane.* Blue Hill Meteorological Observatory, 1994.

Mitsch, W. J., and S. E. Jorgensen, Eds. *Ecological Engineering: An Introduction to Ecotechnology.* Wiley, 1989.

Mogil, H. Michael, and Barbara G. Levine. *The Amateur Meteorologist: Explorations and Investigations.* Franklin Watts, 1994.

Monteith, J. L., and M. Unsworth. *Principles of Environmental Physics.* Routledge, 1989.

Moran, Joseph M., and Michael D. Morgan. *Meteorology: The Atmosphere and the Science of Weather.* Fourth Edition. Macmillan, 1994.

Moses, L., and John Tomikel. *Basic Meteorology, an Introduction to the Science.* Allegheny, 1981.

Munn, R. E. *Boundary Layer Studies and Applications.* Kluwer Academic Publishers, 1989.

Murphy, Allan H., and Richard Katz, Eds. *Probability, Statistics and Decision Making in Meteorology.* Westwiew, 1985.

National Research Council. *The Earth's Electrical Environment.* National Academy Press, 1986.

———. *Goals for Predicting Seasonal-to-Interannal Climate.* National Academy Press, 1994.

National Tuberculosis and Respiratory Disease Association. *Air Pollution Primer,* 1971.

Nebecker, Frederik. *Calculating the Weather: Meteorology in the 20th Century.* Academic Press, 1995.

Nese Jon M. et al. *A World of Weather: Fundamentals of Meteorology.* Kendal/Hunt Publishing, 1996.

Newman, Leonard, Ed. *Measurement Challenges in Atmospheric Chemistry.* American Chemical Society, 1993.

Nilsson, Annika. *Greenhouse Earth.* Wiley, 1992.

Nix, Stephan J. *Urban Stormwater Modeling and Simulation.* Lewis Publishers/CRC Press, 1994.

Officer, Charles, and Jake Page. *Tales of the Earth.* Oxford University Press, 1993.

Oke, T. R. *Boundary Layer Climates.* Methuen, 1987.

Ozone Depletion, Greenhouse Gases and Climate Change. National Academy of Sciences Press, 1989.

Padial, Antonio, Winifred H. Roderman, Ellen Beier, and Rawn McCloud, illustrators. *Weather.* Janus Books, 1982.

Panchev, S. *Dynamic Meteorology.* Kluwer Academic, 1984.

Panofsky, Hans, and Glemm W. Brier. *Some Applications of Statistics to Meteorology.* Pennsylvania State University Press, 1965.

Pearce, E. A., and Gordon Smith. *The Times Books World Weather Guide.* Times Books, 1984.

Pielke, Roger A. *Mesoscale Meteorological Modeling: An Introductory Survey.* Academic Press, 1984.

———. *The Hurricane.* Routledge, 1990.

Pomeroy, L. R., and J. J. Alberts, Eds. *Concepts of Ecosystem Ecology, Ecological Studies Series.* Vol 67. Springer-Verlag, 1988.

Predicting Our Weather. Report, Subcommittee on Atmospheric Research of the Earth and Environmental Sciences, 1992.

Putnam, William Lowell. *The Worst Weather on Earth: A History of the Mount Washington Observatory.* Mount Washington Observatory and the American Alpine Club, 1991.

Rao, P. K., S. J. Holmes, R. K. Anderson, J. S. Winston, and O. E. Lehr, Eds. *Weather Satellites.* American Meteorological Society, 1990.

Ray, Peter S. *Mesoscale Meteorology and Forecasting.* American Meteorological Society, 1986.

Reducing the Impacts of Natural Hazards. Report, Committee on Earth and Environmental Sciences.

Reifsnyder, William E. *Weathering the Wilderness: The Sierra Club Guide to Practical Meteorology.* Sierra, 1980.

Reiter, Reinhold. *Phenomena in Atmospheric and Environmental Electricity.* Elsevier Science Publishers, 1992.

Reiter, Elmar R. *Jet Streams.* Anchor Books, Doubleday, 1967.

Rethinking the Ozone Problem in Urban and Regional Air Pollution. National Academy Press, 1991.

Riegel, C. *Fundamentals of Atmospheric Dynamics and Thermodynamics.* World Scientific Publications, 1990.

Riehl, Herbert. *Introduction to the Atmosphere.* McGraw-Hill, 1965.

Roberts, Palmer W. *Winds of the World and the Many Cloud Forms.* Vantage Press, 1995.

Rogers, John H. *The Giant Planet Jupiter.* Cambridge University Press, 1993.

Rogers, R. R. *A Short Course in Cloud Physics.* Pergamon, 1989.

Rosen, L., and R. Glasser. *Climate Change and Energy Policy.* American Institute of Physics, 1992.

Rosen, Stephen. *Weathering.* M. Evans and Company, 1979.

Rossi, Giuseppe, Milgun Harmancioglu, and Vujica Yevevich, Eds. *Coping with Floods.* Kluwer Academic Publishers, 1994.

Rowland, F. A., and S. A. Isaksen, Eds. *The Changing Atmosphere.* Wiley, 1988.

Rudman, Jack. *Climatology-Meteorology*. National Learning, 1988.

Sagan, Carl. *Cosmos*. Random House, 1979.

Sanders, Ti. *Weather*. Icarus Press, 1985.

Sanderson, Marie, Ed. *Prevailing Trade Winds: Weather and Climate in Hawaii*. University of Hawaii Press, 1993.

Schackleton, N. J., F.R.S., R. G. West, F.R.S., and D. W. Bowen. *The Past Three Million Years*. The Royal Society, 1988.

Schaefer, Vincent J., John A. Day, and Christy E. Day. *A Field Guide to the Atmosphere*. Houghton Mifflin, 1983.

Schanda, E. *Physical Fundamentals of Remote Sensing*. Springer-Verlag, 1986.

Schneider, Stephen H. *Encyclopedia of Climate and Weather*. Oxford University Press, 1996.

Scorer, Richard. *Air Pollution*. Pergamon Press, 1968.

Scorer, R. and A. Verkaik. *Spacious Skies*. Sterling Publishing, 1989.

Seinfeld, John H. *Atmospheric Chemistry and Physics of Air Pollution*. Wiley, 1986.

Sellers, William D. *Physical Climatology*. University of Chicago Press, 1965.

Simpson, John E. *Sea Breeze and Local Wind*. Cambridge University Press, 1994.

Simpson, Robert H., and Herbert Riehl. *The Hurricane and Its Impact*. Louisiana State University Press, 1981.

Sloane, Eric. *Weather Book*. Hawthorn Books, 1952.

———. *Almanac and Weather Forecaster*. Hawthorn Books, 1955.

Spar, Jerome. *The Way of the Weather*. Creative Educational Society, 1967.

Spiegel, Herbert, and Arnold Gruber. *From Weather Vanes to Satellites: an Introduction to Meteorology*. Wiley, 1983.

Spinrad, Richard W., Kendall L. Carder, and Mary Jane Perry. *Ocean Optics*. Oxford University Press, 1994.

Stephens, Graeme L. *Remote Sensing of the Lower Atmosphere: An Introduction*. Oxford University Press, 1994.

Steurer, Pete, Ed. *World Weather Records: 1981–90 for Europe*. National Climatic Data Center 1996.

Stowe, Keith. *Exploring Ocean Science*. Wiley, 1995.

Strauss, Stephen. *The Sizesaurus*. Kodansha International, 1995.

Stull, Roland B. *Meteorology Today for Scientists and Engineers*. West Publishing Company, 1995.

———. *An Introduction to Boundary Layer Meteorology*. Kluwer Academic Publishers, 1988.

Tannehill, Ivan Ray. *The Hurricane Hunters*. Dodd, Mead, 1955.

Tape, Walter. *Atmospheric Halos*. American Geophysical Union, 1994.

Taylor, Barbara. *Mountains and Volcanoes*. Kingfisher Books, 1993.

———. *Rivers and Oceans*. Kingfisher Books, 1993.

———. *Weather and Climate*. Kingfisher Books, 1993.

Thiebaux, H. Jean. *Statistical Data Analysis for Ocean and Atmospheric Sciences*. Academic Press, 1994.

Tobin, W. E. *The Mariner's Pocket Companion 1989*. Naval Institute Press, 1989.

Tolbert, N. E., and Jack Preiss. *Regulation of Atmospheric CO2 and O2 by Photosynthetic Carbon Metabolism*. Oxford University Press, 1994.

Tomczak, Matthias, and J. Stuart Godfrey. *Regional Oceanography: An Introduction*. Pergamon Press, 1994.

Tufty, Barbara. *1001 Questions Answered about Hurricanes, Tornadoes and Other Natural Air Disasters*. Dover, 1987.

Turner, D. Bruce. *Workbook of Atmospheric Dispersion Estimates: An Introduction to Dispersion Modeling*. CRC Press, 1994.

Uman, Martin A. *Lightning*. Dover, 1969.

van Dop. H., and D. G. Steyn, Eds. *Air Pollution and Its Application VIII*. Plenum Press, 1991.

van Dorn, William C. *Oceanography and Seamanship*. Cornell Maritime Press, 1993.

Wagner, Ronald L., and Bill Adler, Jr. *The Weather Sourcebook*. Globe Pequot Press, 1994.

Walker, Sally M. *Water Up, Water Down: The Hydrologic Cycle*. Carolrhoda Books, 1992.

Wallace, John M., and Peter Hobbs. *Atmospheric Science: An Introductory Survey*. Academic Press, 1977.

Wang, B. *Sea Fog.* Springer-Verlag, 1985.

Warrick, R. A., E. M. Barrow, and T. M. L. Wigley, Eds. *Climate and Sea Level Change: Observations, Projections, and Implications.* Cambridge University Press, 1993.

Watts, Alan. *Instant Weather Forecasting.* Dodd, Mead, 1968.

———. *The Weather Handbook.* Sheridan House, 1994.

Weather for Those Who Fly. National Research Council, 1994.

Wells, N. *The Atmosphere and Ocean: A Physical Introduction.* Taylor and Francis, 1986.

Whipple, A. B. C. *Storm.* Planer Earth, Time-Life Books, 1982.

White, Donald A., Ed. *Drought Assessment, Management, and Planning: Theory and Case Studies.* Kluwer Academic Publishers, 1993.

Williams, Jack. *The Weather Book.* Vintage Books, 1992.

Williams, Richard S., Jr., and Jane G. Ferrigno, Eds. *Glaciers of Europe.* U.S. Geological Survey, 1993.

Williamson, Samuel J. *Fundamentals of Air Pollution.* Addison-Wesley, 1973.

Wolanski, Eric. *Physical Oceanographic Processes of the Great Barrier Reef.* CRC Press.

Wood, Richard A., Ed. *The Weather Almanac,* 7th ed., Gale Research, 1996.

Wright Jr., H. E. et al, Eds. *Global Climates Since the Last Glacial Maximum.* University of Minnesota Press, 1994.

Yihui, Ding. *Monsoons over China.* Kluwer Academic Publishers, 1994.

Yolen, Jane, Ed. *Weather Report.* Boyds Mills Press, 1993.

Young, K. C. *Microphysical Processes in Clouds.* Oxford University Press, 1993.

Zannetti, P. *Air Pollution Modeling.* Computational Mechanics, 1990.

———, et al., Eds. *Air Pollution.* Computational Mechanics, 1993.

Zepp, Richard G., Ed. *Climate-Biosphere Interactions: Biogenic Emissions and Environmental Effects of Climate Change.* Wiley, 1994.

Journals and Periodicals

American Weather Observer. Published monthly by AWO, 401 Whitney Blvd., Belvidere, IL, 61008-3772; 815/544-9811; awowx@aol.com.

Bulletin of the American Meteorological Society. Provided with membership in the American Meteorological Society, 45 Beacon St., Boston, MA 02108.

Discover. Published monthly by Walt Disney Magazine, Publishing Group, 500 S. Buena Vista, Burbank, CA 91521-6012.

Environment. Published ten times a year by Heldref Publications, 1319 Eighteenth St. NW, Washington, DC 20036.

National Geographic. Published monthly by the National Geographic Society, 17th & M Streets, NW, Washington, DC 20036.

Nature Magazine (now incorporated into *Natural History*). Published monthly by American Museum of Natural History, Central Park West at 79th St., New York, NY 10024-5192.

Popular Science. Published by Times Mirror Magazines, Inc., 2 Park Ave., New York, NY, 10016.

Science. Published weekly by American Association for the Advancement of Science, 1333 H St. NW, Washington, DC 20005.

Science News. Published weekly by Science Service, Inc. 1719 N St. NW, Washington, DC, 20036.

Scientific American. Published monthly by Scientific American, Inc. 415 Madison Ave., New York, NY 10017.

Smithsonian. Published monthly by Smithsonian Institution, Arts & Industries Bldg., 900 Jefferson Dr., Washington, DC 20560.

Weatherwise. Published bi-monthly by Heldref Publications, 1319 Eighteenth St. NW, Washington, DC 20036-1802.

Index

396

Date_____   Humidity _____
High/Low Temperature _____   Precipitation_____
Average Temperature _____   Sky_____
Barometric Pressure_____   Wind _____
Comments _____
_____

Date_____   Humidity _____
High/Low Temperature _____   Precipitation_____
Average Temperature _____   Sky_____
Barometric Pressure_____   Wind _____
Comments _____
_____

Date_____   Humidity _____
High/Low Temperature _____   Precipitation_____
Average Temperature _____   Sky_____
Barometric Pressure_____   Wind _____
Comments _____
_____

Date_____   Humidity _____
High/Low Temperature _____   Precipitation_____
Average Temperature _____   Sky_____
Barometric Pressure_____   Wind _____
Comments _____
_____

Date_____   Humidity _____
High/Low Temperature _____   Precipitation_____
Average Temperature _____   Sky_____
Barometric Pressure_____   Wind _____
Comments _____
_____

Date_____   Humidity _____
High/Low Temperature _____   Precipitation_____
Average Temperature _____   Sky_____
Barometric Pressure_____   Wind _____
Comments _____
_____

Date_____   Humidity _____
High/Low Temperature _____   Precipitation_____
Average Temperature _____   Sky_____
Barometric Pressure_____   Wind _____
Comments _____
_____

Date_____   Humidity _____

High/Low Temperature _____   Precipitation_____

Average Temperature _____   Sky_____

Barometric Pressure _____   Wind _____

Comments _____

_____

Date_____   Humidity _____

High/Low Temperature _____   Precipitation_____

Average Temperature _____   Sky_____

Barometric Pressure _____   Wind _____

Comments _____

_____

Date_____   Humidity _____

High/Low Temperature _____   Precipitation_____

Average Temperature _____   Sky_____

Barometric Pressure _____   Wind _____

Comments _____

_____

Date_____   Humidity _____

High/Low Temperature _____   Precipitation_____

Average Temperature _____   Sky_____

Barometric Pressure _____   Wind _____

Comments _____

_____

Date_____   Humidity _____

High/Low Temperature _____   Precipitation_____

Average Temperature _____   Sky_____

Barometric Pressure _____   Wind _____

Comments _____

_____

Date_____   Humidity _____

High/Low Temperature _____   Precipitation_____

Average Temperature _____   Sky_____

Barometric Pressure _____   Wind _____

Comments _____

_____

Date_____   Humidity _____

High/Low Temperature _____   Precipitation_____

Average Temperature _____   Sky_____

Barometric Pressure _____   Wind _____

Comments _____

_____

Date_____ Humidity _____
High/Low Temperature _____ Precipitation_____
Average Temperature _____ Sky_____
Barometric Pressure _____ Wind _____
Comments _____
_____

Date_____ Humidity _____
High/Low Temperature _____ Precipitation_____
Average Temperature _____ Sky_____
Barometric Pressure _____ Wind _____
Comments _____
_____

Date_____ Humidity _____
High/Low Temperature _____ Precipitation_____
Average Temperature _____ Sky_____
Barometric Pressure _____ Wind _____
Comments _____
_____

Date_____ Humidity _____
High/Low Temperature _____ Precipitation_____
Average Temperature _____ Sky_____
Barometric Pressure _____ Wind _____
Comments _____
_____

Date_____ Humidity _____
High/Low Temperature _____ Precipitation_____
Average Temperature _____ Sky_____
Barometric Pressure _____ Wind _____
Comments _____
_____

Date_____ Humidity _____
High/Low Temperature _____ Precipitation_____
Average Temperature _____ Sky_____
Barometric Pressure _____ Wind _____
Comments _____
_____

Date_____ Humidity _____
High/Low Temperature _____ Precipitation_____
Average Temperature _____ Sky_____
Barometric Pressure _____ Wind _____
Comments _____
_____

Date_____   Humidity _____

High/Low Temperature _____   Precipitation_____

Average Temperature _____   Sky_____

Barometric Pressure _____   Wind _____

Comments _____

_____

Date_____   Humidity _____

High/Low Temperature _____   Precipitation_____

Average Temperature _____   Sky_____

Barometric Pressure _____   Wind _____

Comments _____

_____

Date_____   Humidity _____

High/Low Temperature _____   Precipitation_____

Average Temperature _____   Sky_____

Barometric Pressure _____   Wind _____

Comments _____

_____

Date_____   Humidity _____

High/Low Temperature _____   Precipitation_____

Average Temperature _____   Sky_____

Barometric Pressure _____   Wind _____

Comments _____

_____

Date_____   Humidity _____

High/Low Temperature _____   Precipitation_____

Average Temperature _____   Sky_____

Barometric Pressure _____   Wind _____

Comments _____

_____

Date_____   Humidity _____

High/Low Temperature _____   Precipitation_____

Average Temperature _____   Sky_____

Barometric Pressure _____   Wind _____

Comments _____

_____

Date_____   Humidity _____

High/Low Temperature _____   Precipitation_____

Average Temperature _____   Sky_____

Barometric Pressure _____   Wind _____

Comments _____

_____

Date_____ Humidity _____
High/Low Temperature _____ Precipitation_____
Average Temperature _____ Sky_____
Barometric Pressure _____ Wind _____
Comments _____
_____

Date_____ Humidity _____
High/Low Temperature _____ Precipitation_____
Average Temperature _____ Sky_____
Barometric Pressure _____ Wind _____
Comments _____
_____

Date_____ Humidity _____
High/Low Temperature _____ Precipitation_____
Average Temperature _____ Sky_____
Barometric Pressure _____ Wind _____
Comments _____
_____

Date_____ Humidity _____
High/Low Temperature _____ Precipitation_____
Average Temperature _____ Sky_____
Barometric Pressure _____ Wind _____
Comments _____
_____

Date_____ Humidity _____
High/Low Temperature _____ Precipitation_____
Average Temperature _____ Sky_____
Barometric Pressure _____ Wind _____
Comments _____
_____

Date_____ Humidity _____
High/Low Temperature _____ Precipitation_____
Average Temperature _____ Sky_____
Barometric Pressure _____ Wind _____
Comments _____
_____

Date_____ Humidity _____
High/Low Temperature _____ Precipitation_____
Average Temperature _____ Sky_____
Barometric Pressure _____ Wind _____
Comments _____
_____

Date_____   Humidity _____
High/Low Temperature _____   Precipitation_____
Average Temperature _____   Sky_____
Barometric Pressure _____   Wind _____
Comments _____

_____

Date_____   Humidity _____
High/Low Temperature _____   Precipitation_____
Average Temperature _____   Sky_____
Barometric Pressure _____   Wind _____
Comments _____

_____

Date_____   Humidity _____
High/Low Temperature _____   Precipitation_____
Average Temperature _____   Sky_____
Barometric Pressure _____   Wind _____
Comments _____

_____

Date_____   Humidity _____
High/Low Temperature _____   Precipitation_____
Average Temperature _____   Sky_____
Barometric Pressure _____   Wind _____
Comments _____

_____

Date_____   Humidity _____
High/Low Temperature _____   Precipitation_____
Average Temperature _____   Sky_____
Barometric Pressure _____   Wind _____
Comments _____

_____

Date_____   Humidity _____
High/Low Temperature _____   Precipitation_____
Average Temperature _____   Sky_____
Barometric Pressure _____   Wind _____
Comments _____

_____

Date_____   Humidity _____
High/Low Temperature _____   Precipitation_____
Average Temperature _____   Sky_____
Barometric Pressure _____   Wind _____
Comments _____

_____

Date_____   Humidity _____
High/Low Temperature _____   Precipitation_____
Average Temperature _____   Sky_____
Barometric Pressure _____   Wind _____
Comments _____
_____

Date_____   Humidity _____
High/Low Temperature _____   Precipitation_____
Average Temperature _____   Sky_____
Barometric Pressure _____   Wind _____
Comments _____
_____

Date_____   Humidity _____
High/Low Temperature _____   Precipitation_____
Average Temperature _____   Sky_____
Barometric Pressure _____   Wind _____
Comments _____
_____

Date_____   Humidity _____
High/Low Temperature _____   Precipitation_____
Average Temperature _____   Sky_____
Barometric Pressure _____   Wind _____
Comments _____
_____

Date_____   Humidity _____
High/Low Temperature _____   Precipitation_____
Average Temperature _____   Sky_____
Barometric Pressure _____   Wind _____
Comments _____
_____

Date_____   Humidity _____
High/Low Temperature _____   Precipitation_____
Average Temperature _____   Sky_____
Barometric Pressure _____   Wind _____
Comments _____
_____

Date_____   Humidity _____
High/Low Temperature _____   Precipitation_____
Average Temperature _____   Sky_____
Barometric Pressure _____   Wind _____
Comments _____
_____

Date_____ Humidity _____
High/Low Temperature _____ Precipitation_____
Average Temperature _____ Sky_____
Barometric Pressure_____ Wind _____
Comments _____
_____

Date_____ Humidity _____
High/Low Temperature _____ Precipitation_____
Average Temperature _____ Sky_____
Barometric Pressure_____ Wind _____
Comments _____
_____

Date_____ Humidity _____
High/Low Temperature _____ Precipitation_____
Average Temperature _____ Sky_____
Barometric Pressure_____ Wind _____
Comments _____
_____

Date_____ Humidity _____
High/Low Temperature _____ Precipitation_____
Average Temperature _____ Sky_____
Barometric Pressure_____ Wind _____
Comments _____
_____

Date_____ Humidity _____
High/Low Temperature _____ Precipitation_____
Average Temperature _____ Sky_____
Barometric Pressure_____ Wind _____
Comments _____
_____

Date_____ Humidity _____
High/Low Temperature _____ Precipitation_____
Average Temperature _____ Sky_____
Barometric Pressure_____ Wind _____
Comments _____
_____

Date_____ Humidity _____
High/Low Temperature _____ Precipitation_____
Average Temperature _____ Sky_____
Barometric Pressure_____ Wind _____
Comments _____
_____

Date_____   Humidity _____
High/Low Temperature _____   Precipitation_____
Average Temperature _____   Sky_____
Barometric Pressure _____   Wind _____
Comments _____

_____

Date_____   Humidity _____
High/Low Temperature _____   Precipitation_____
Average Temperature _____   Sky_____
Barometric Pressure _____   Wind _____
Comments _____

_____

Date_____   Humidity _____
High/Low Temperature _____   Precipitation_____
Average Temperature _____   Sky_____
Barometric Pressure _____   Wind _____
Comments _____

_____

Date_____   Humidity _____
High/Low Temperature _____   Precipitation_____
Average Temperature _____   Sky_____
Barometric Pressure _____   Wind _____
Comments _____

_____

Date_____   Humidity _____
High/Low Temperature _____   Precipitation_____
Average Temperature _____   Sky_____
Barometric Pressure _____   Wind _____
Comments _____

_____

Date_____   Humidity _____
High/Low Temperature _____   Precipitation_____
Average Temperature _____   Sky_____
Barometric Pressure _____   Wind _____
Comments _____

_____

Date_____   Humidity _____
High/Low Temperature _____   Precipitation_____
Average Temperature _____   Sky_____
Barometric Pressure _____   Wind _____
Comments _____

_____

Date_____   Humidity _____
High/Low Temperature _____   Precipitation_____
Average Temperature _____   Sky_____
Barometric Pressure _____   Wind _____
Comments _____
_____

Date_____   Humidity _____
High/Low Temperature _____   Precipitation_____
Average Temperature _____   Sky_____
Barometric Pressure _____   Wind _____
Comments _____
_____

Date_____   Humidity _____
High/Low Temperature _____   Precipitation_____
Average Temperature _____   Sky_____
Barometric Pressure _____   Wind _____
Comments _____
_____

Date_____   Humidity _____
High/Low Temperature _____   Precipitation_____
Average Temperature _____   Sky_____
Barometric Pressure _____   Wind _____
Comments _____
_____

Date_____   Humidity _____
High/Low Temperature _____   Precipitation_____
Average Temperature _____   Sky_____
Barometric Pressure _____   Wind _____
Comments _____
_____

Date_____   Humidity _____
High/Low Temperature _____   Precipitation_____
Average Temperature _____   Sky_____
Barometric Pressure _____   Wind _____
Comments _____
_____

Date_____   Humidity _____
High/Low Temperature _____   Precipitation_____
Average Temperature _____   Sky_____
Barometric Pressure _____   Wind _____
Comments _____
_____

Date_____ Humidity _____
High/Low Temperature _____ Precipitation_____
Average Temperature _____ Sky_____
Barometric Pressure_____ Wind _____
Comments _____
_____

Date_____ Humidity _____
High/Low Temperature _____ Precipitation_____
Average Temperature _____ Sky_____
Barometric Pressure_____ Wind _____
Comments _____
_____

Date_____ Humidity _____
High/Low Temperature _____ Precipitation_____
Average Temperature _____ Sky_____
Barometric Pressure_____ Wind _____
Comments _____
_____

Date_____ Humidity _____
High/Low Temperature _____ Precipitation_____
Average Temperature _____ Sky_____
Barometric Pressure_____ Wind _____
Comments _____
_____

Date_____ Humidity _____
High/Low Temperature _____ Precipitation_____
Average Temperature _____ Sky_____
Barometric Pressure_____ Wind _____
Comments _____
_____

Date_____ Humidity _____
High/Low Temperature _____ Precipitation_____
Average Temperature _____ Sky_____
Barometric Pressure_____ Wind _____
Comments _____
_____

Date_____ Humidity _____
High/Low Temperature _____ Precipitation_____
Average Temperature _____ Sky_____
Barometric Pressure_____ Wind _____
Comments _____
_____

Date_____ Humidity _____
High/Low Temperature _____ Precipitation_____
Average Temperature _____ Sky_____
Barometric Pressure _____ Wind _____
Comments _____
_____

Date_____ Humidity _____
High/Low Temperature _____ Precipitation_____
Average Temperature _____ Sky_____
Barometric Pressure _____ Wind _____
Comments _____
_____

Date_____ Humidity _____
High/Low Temperature _____ Precipitation_____
Average Temperature _____ Sky_____
Barometric Pressure _____ Wind _____
Comments _____
_____

Date_____ Humidity _____
High/Low Temperature _____ Precipitation_____
Average Temperature _____ Sky_____
Barometric Pressure _____ Wind _____
Comments _____
_____

Date_____ Humidity _____
High/Low Temperature _____ Precipitation_____
Average Temperature _____ Sky_____
Barometric Pressure _____ Wind _____
Comments _____
_____

Date_____ Humidity _____
High/Low Temperature _____ Precipitation_____
Average Temperature _____ Sky_____
Barometric Pressure _____ Wind _____
Comments _____
_____

Date_____ Humidity _____
High/Low Temperature _____ Precipitation_____
Average Temperature _____ Sky_____
Barometric Pressure _____ Wind _____
Comments _____
_____